Knowledge Infrastructure and Knowledge Economy

Edited by

Karel Davids (*VU University, Amsterdam*)
Larry Stewart (*University of Saskatchewan, Saskatoon*)

VOLUME 7

The titles published in this series are listed at *brill.com/knie*

Knowledge, Patents, Power

Knowledge, Patents, Power

The Making of a Patent System in the Dutch Republic

By

Marius Buning

BRILL

LEIDEN | BOSTON

European Research Council
Established by the European Commission

Part of this research was supported by the ERC grant 818523, CREATIVE IPR- The History of Intellectual Property Rights in the Creative Industries.

The data for this project is available on Figshare at 10.6084/m9.figshare.16685788.

Cover illustration: Drawing from the patent request by Cornelis Cornelisz for the invention of a saw mechanism, 1601. NL-HaNA, States General, 1.01.02, inv. no. 7474

The Library of Congress Cataloging-in-Publication Data is available online at http://catalog.loc.gov

Typeface for the Latin, Greek, and Cyrillic scripts: "Brill". See and download: brill.com/brill-typeface.

ISSN 1877-2323
ISBN 978-90-04-32039-0 (hardback)
ISBN 978-90-04-32042-0 (e-book)

This book is printed on acid-free paper and produced in a sustainable manner.

For my wife Kalliopi and my son Fotios,
the light in my life.

Contents

Acknowledgements

This book came about through many conversations with friends, family, and colleagues. I am eternally grateful to them for that but fear that the acknowledgements would become too extensive if I were to thank them one by one. A few people, however, deserve a special mention: first, of course, I am grateful to my mother for all she has done for me over the years. Furthermore, I would like to thank Henk Looijesteijn for his support throughout this project, for his careful reading of the final manuscript, and for his tireless help in putting together the database. I am also infinitely grateful to Edouard Buning for his patient help in putting together the tables and graphs. I am highly indebted to the anonymous reviewers at Brill for their valuable suggestions, and to Sandra McElroy of The Language Gap for the wonderful job she did in editing my terrible English. Finally, I would like to thank Mario Biagioli, who has supported me intellectually over the years.

This book began as a PhD thesis, defended at the European University Institute in 2013. I have worked on it intermittently during my time as a postdoctoral fellow at the Max Planck Institute for the History of Science and as a DRS fellow at Dahlem Humanities Centre of the Freie Universität Berlin. The book finally came to fruition as I am moving again to a new country to take up a position in the ERC project CREATIVE IPR at the University of Oslo. I thank the staff at these institutions for their hospitality and for the wonderful support in putting together this manuscript.

In particular, I should like to express my gratitude to Véronique Pouillard, University of Oslo, for her patience and for making it possible to publish the first chapter of this book in Open Access. Finally, I also express my thanks to the Intellectual History Review for permission to reproduce parts of the following articles:

Buning, Marius (2014). "Inventing scientific method: the privilege system as a model for scientific knowledge-production." *Intellectual History Review* 24 (1): 59–70.

Buning, Marius (2014). "Between Imitation and Invention. Inventor Privileges and Technological Progress in the Early Dutch Republic (*c.* 1585–1625)." *Intellectual History Review* 24 (3): 415–427.

Figures and Tables

Figures

Tables

Introduction

Preliminary

Novelty is the motor of our modern economy. Day after day, we are bombarded with new technologies, new products, and new fashions, so that it sometimes seems as if the world would fade away in an instant if we did not keep up with all the novelties surrounding us. The desire to be at the forefront of things has become an integral part of our economic behavior. It is a typical characteristic of our modern mentality and to imagine a world without novelties is almost impossible.

Closely related to the omnipresent veneration of novelty is the possibility to own, protect, and exploit novel ideas. This is what legal scholars call patent law. This patent law is part of a larger family of intellectual property rights, which also includes copyrights, trademarks, and geographical indications. Patents are, as it were, a specific branch of this family. They give inventors the temporary and exclusive right to exploit their technological inventions on a commercial basis. Silently present in the background, patents are nearly everywhere. The microwave you use, the telephone you hold, the computer screen you might be looking at; all of them contain patented technology. Patents, in a word, are at the heart of our consumer society.

Despite their all-pervasiveness, we know surprisingly little about the history of patents. Just ask around. Most people will tell you, after a moment's silence in which they wonder why anyone would be interested in such an odd topic, that patents are probably somehow related to the Industrial Revolution. In fact, that is not such a bad guess. The steam engine, the light bulb, the phonograph, and many other inventions that poured on to the market during the long nineteenth century, did indeed herald the golden era of patent law. Yet its roots go back much further in time. At some point in the fifteenth and sixteenth centuries, in an era when new continents were being discovered and new ideas were mushrooming left and right, European authorities rather suddenly arrived at the conclusion that granting individual rights for the exploitation of novel inventions would positively affect the common good. It was a revolutionary moment in terms of moral thought. But, remarkably enough, this revolutionary moment has received little attention from historians.

This book is an attempt to change that. It analyses the making of a patent system in the Dutch Republic (*c.* 1581–1621), elucidating how the idea of protecting new inventions came about and how it functioned in practice. The choice of timepoint and location was made on the basis of two factors.

Firstly, the then Dutch Republic was also known as a pioneer in technological development within early modern Europe.[1] Even though, as we shall see in the following chapters, technological progress can in no way be equated with or measured on the basis of the number of patents that are awarded in a particular period, it is nevertheless interesting to take a closer look at the nature of technological development in the Republic on the basis of the patent system. The second reason for devoting attention to the Dutch context is because the Republic is also recognized as the "first modern economy", which is supposed to have been a defining factor in the evolution towards capitalism.[2] Within this form of thinking about economic interest, not only is the market important for determining the price of certain commodities, but there is also a firm emphasis on private property and capital accumulation. The obvious question is, therefore, how the institution of a Dutch patent system fits within the development of capitalist thinking.

Before we can get a grip on some of the particularities of the patent system in the early Dutch Republic, we must first acquire a better understanding of patents in our present day and age. Patents are exclusive rights with a limited timeframe for the commercial exploitation of inventions. They are granted after a rigorous examination by patent examiners, who take into account a number of conditions. The first of these conditions is that the subject matter of the invention satisfies the standards of patentability. This means, for instance, that mathematical models or discoveries without any direct practical purpose, cannot be patented. The second condition for patentability is that the invention is 'useful'. This means that it must be theoretically possible to 'reduce the invention to practice'. One cannot, for example, patent inventions that are contrary to the laws of nature, such as perpetual motion machines (there are, however, plenty of examples where 'impossible inventions' have slipped through the hands of the patent examiners). The third condition for patentability is 'novelty'. This clause is fairly absolute: if the subject matter of the invention has been revealed in any way, either earlier in time or anywhere in the world, whether by print or in actuality, the invention cannot be considered eligible for patent protection. Moreover, the invention must be sufficiently different from existing forms of knowledge, which is also called 'prior art' in this context. On the basis of prior art examples, patent examiners determine whether the invention satisfies the fourth condition for patentability, which is 'non-obviousness.' One cannot, for instance, make a mobile phone available in any number of colors and then claim rights to a new invention. To conclude, a formal

1 Davids, *The Rise and Decline*.
2 De Vries and Woude, *The First Modern Economy*.

condition for patentability is that an invention must be sufficiently revealed to the public. This is done by means of so-called patent specifications, which are written descriptions of the invention "in such full, clear, concise, and exact terms as to enable any person skilled in the art" to reproduce the invention.[3]

The definition of an 'invention' is a delicate matter, if only because of the lack of a universally valid system of patent law. Patent laws in the United States and continental Europe, in particular, show some significant differences in this respect. Whereas an invention in the European system is defined as a "an idea that permits the practical solution of a specific problem in a field of technology," the American system is much more permissive in this regard and allows for inventions to be made outside of the field of technology as well.[4] A detailed discussion of this issue, as well as of the other dissimilarities among patent systems around the world, would turn into an unfeasible exercise within the limited scope of this book. I have therefore chosen to follow the definitions used in the US system.

A patent, then, is an intellectual property right granted by the Government of the United States of America to an inventor "to exclude others from making, using, offering for sale, or selling the invention throughout the United States or importing the invention into the United States" for a limited time in exchange for public disclosure of the invention when the patent is granted.[5] In other words, patents are a kind of temporary monopoly right.[6] Whereas monopolies

---

3 35 U.S.C. 112 Specification.

4 World Intellectual Property Organization, *Introduction to Intellectual Property*, 7. Even if there are hardly any rules that determine what should be considered as eligible subject matter for patent protection, there is some agreement about what issues should under all circumstances be excluded from patent protection. Article 52 (2) of the European Patent Convention provides a concise overview of these exceptions. It reads:

(1) European patents shall be granted for any inventions, in all fields of technology, provided that they are new, involve an inventive step and are susceptible of industrial application.
(2) The following in particular shall not be regarded as inventions within the meaning of paragraph 1.
  (a) discoveries, scientific theories and mathematical methods.
  (b) aesthetic creations.
  (c) schemes, rules and methods for performing mental acts, playing games or doing business, and programs for computers
  (d) presentations of information.

The situation in US patent law is slightly different. Although scientific discoveries are excluded from patent protection, one can patent business models and computer software under a particular set of conditions. See Con Díaz, *Software Rights*; Sherman, "Intangible Machines."

5 Definition by the US Patent and Trademark Office (USPTO), "Glossary".

6 The exact relation between patents and monopolies is the subject of ongoing debate. From a historical point of view, there is however no doubt that 'patents' were originally interpreted as a type of temporary monopolies. For further discussion of this issue, see Chapter 4.

are condemned in other circumstances as unlawful interventions in the natural course of the economy, they are justified in the case of patent law on the basis that the introduction of new inventions is by definition to the advantage of the general public. The idea is that patents encourage disclosure. Inventors, it is believed, would keep their inventions secret if they were not guaranteed the possibility to gain something by putting their ideas at the disposal of the public. At the same time, it is believed that patents provide a general incentive for innovation.

The ideological underpinnings of modern patent law can largely be traced back to ideas that circulated in early modern Europe. In early modern Europe as well, people believed that the common good would be best served by granting inventors temporary monopoly rights for the commercial exploitation of new and useful inventions. Yet the use of the word 'patent' in this context can lead to misunderstandings. The word 'patent' was used as an adjective in the expression 'letters patent', which was a literal translation of the Latin *litterae patentae,* meaning 'open letters' (from the verb *pateo*).[7] These open letters were used to disclose legal decisions to the general public and they were identified by a seal hanging from the bottom of a letter to confirm its authenticity. In addition, the authorities made use of so-called 'letters close', which were sealed closed letters used to convey more sensitive information. Patent letters, in short, alluded to the form of legal decision, not to their content.

The actual right to establish a temporary monopoly was instead called a 'privilege'.[8] Privileges were exclusive rights granted when the authorities wanted to make an exception to the *ius gentium*, or the 'law of the land'. In the words of a seventeenth-century legal scholar, privileges were:

> Acts of grace or benefice granted to someone by the Highest Authority or the Prince of the Land, because of provided services: slightly at disposition to the common laws and ordinances of the country, without doing considerable harm to third parties.[9]

---

7 Hill, "Origin and Development," 406.

8 See also pp. 156-167. For a general overview on the notion of privilege, see Dölemeyer and Mohnhaupt, *Das Privileg*.

9 "Gratien of Beneficien, die yemand by de Hooge Overigheyd, of Prince van den Lande, om desselfs verdiensten verleend werden: eenigsints tegens de Dispositie vande Gemeene Rechten ende Ordinantien des Lands, sonder merckelik achterdeel eens derden." Merula, *Manier van procederen*, 40–41. A more detailed analysis of the work by Merula will follow in Chapter 4.

Because of their nature, these acts of grace were used for a wide variety of purposes that went well beyond our present understanding of a patent.[10] Thus, privileges were not only used to give inventors specific rights, but also for instance to guarantee the rights of Jews, creditors, and artificial persons such as the guilds.

In addition, other words were used in Dutch as well. Privilege rights were granted by means of *opene brieven van octroij*, which can be translated as 'letters patent of permission'. Soon, however, the words 'privilege' and '*octrooij*' were used interchangeably. As a result, authorities often granted inventors "permission and privilege" (*octroy ende privilege*) to exploit their technological inventions to the exclusion of others. The tautology need not concern us further. Instead, it is important to note that from this point onwards I will use the expression 'invention privileges' rather than 'patents' when speaking about the early modern period to avoid misunderstandings and anachronistic interpretations.

History and Historiography

The first privileges that explicitly obliged individuals to put new methods or machines to use were granted by the Republic of Venice around the beginning of the fifteenth century.[11] Other Italian city-states soon followed the Venetian example. The city of Florence, for example, decided in 1421 to grant a three-year privilege to the artist Filippo Brunelleschi (1377–1446) for the development of a new type of boat that would haul loads of marble along the Arno to Pisa (an enterprise that failed miserably, by the way).[12] Still, invention privileges remained a fairly sporadic phenomenon until the Venetian authorities promulgated something that resembled a real 'patent law' in 1474. They argued that:

---

10 See Mohnhaupt, "Unendlichheit des Privilegienbegriffs," 10–11.

11 As far as I know, the first privilege that mentioned the exclusive use of a new machine was granted to Franciscus Petri from Rhodes in 1416. See Mandich, "Primi Riconoscimenti Veneziani," 115–16, 149–50; Mandich, "Venetian Origins of Inventors' Rights," 379; Long, "Invention, Authorship, 'Intellectual Property,'" 877.

12 See Prager, "Brunelleschi's Patent." The three-year privilege (1421) specified that no one except Brunelleschi could "have or use any new ship or machine designed for transporting goods" on Florentine waters. Only "familiar and usual" ships could be used, whereas any "new or newly shaped machine would be burned." Moreover, all goods transported by the newly invented ship would be free of levies and taxes for the duration of the privilege.

> There are men in this city, and also there come other persons every day from different places [...who...] should it be legislated that the works and contrivances invented by them could not be copied and made by others so that they are deprived of their honor [...] would exert their minds, invent and make things that would be of no small utility and benefit to our State.[13]

As the decree indicated, the primary goal of invention privileges was to incite innovation and to domesticate profitable technologies from abroad. In this context, it was of lesser importance whether a privilege holder had first conceived of the idea behind his invention. Instead, what mattered was that the inventor was the first to reveal the technical details of an invention to the local authorities and thus render the invention useful to the local economy.

The ambiguous attitude towards 'novelty' explains, among other things, why in 1469 a certain Johann von Speyer (1420–1470) obtained a six-year privilege from the Venetian authorities for his 'invention' of the printing press. In the eyes of the Venetian authorities, it did not matter that Johannes Gutenberg (1395–1468) had invented such a press more than a decade earlier (*c.* 1440), since the technical details of Gutenberg's invention were still unknown in Venice at the time of Speyer's application. Today, one might be inclined to interpret the actions of the Venetians as a form of industrial espionage; but the Venetians quite openly concluded that Speyer deserved the credit (and hence the privilege) for having brought the knowledge of the printing press to Venice.

The Italian experiences, it is presumed, served as an example for other intellectual property regimes throughout Europe.[14] This presumption is supported by the relatively large number of Italians among the first inventors to obtain privileges in other European sovereignties. This suggests that, as they wandered around Europe, inventors transmitted not only technological knowledge, but also a particular type of legal knowledge.[15] Nonetheless, it took

---

13 "Venetian Statute on Industrial Brevets, Venice (1474)," *Primary Sources on Copyright (1450–1900)*, eds L. Bently & M. Kretschmer, www.copyrighthistory.org (last accessed 3 March 2018).

14 For Venetian patents, see Mandich, "Le privative industriali"; Mandich, "Primi riconoscimenti"; Berveglieri, *Inventori stranieri*; May, "The Venetian Moment"; Molà, *The Silk Industry*, 186–214. Both articles by Mandich have been translated in abridgement by Frank Prager: Mandich, "Venetian Patents"; Mandich, "Venetian Origins." For Florentine privileges, see: De Mullenheim, "Les privilèges"; Lamberini, "Inventori di macchine."

15 Doorman, *Octrooien*, 12–13. The theory that Italian craftsmen spread the idea of invention privileges across Europe along their journeys and migrations does not explain how invention privileges came to be so widely accepted and implemented in a number of states *at the same time*. This fact strongly suggests the diffusion of the idea passed through learned canals, and in particular legal treatises. On Italian craftsmen who emigrated abroad:

almost a century before invention privileges were employed on a grand scale, remarkably enough in a concentrated timeframe during the years 1535–1550. Examples from that period of privileges for the exploitation of technological inventions can be found in the Holy Roman Empire,[16] Habsburg Spain,[17] the French Kingdom,[18] and even the Papal States.[19] Most attention, however, has been focused on the privilege practices in the Kingdom of England, which ultimately resulted in the 1624 Statute of Monopolies.[20] The intention of the Statute of Monopolies was actually to put a halt to the interminable and honorific privileges granted by King James I (1566–1621).[21] But it also provided a clause that offered the 'true and first inventor' the possibility to obtain a privilege for fourteen years, if he was to work or make "any manner of new manufactures within this realm." In the wake of the grandness of the British Empire, this clause has in the past been wrongfully interpreted as the 'true precursor' of our modern-day patent law system.

Besides some sporadic publications, interest in the history of patent law only really got going in the twentieth century.[22] The most productive scholars

---

Long, "Invention" 879–81; Mandich, "Le privative industriali," 536; Heller, *Labour*, 251–52; Belfanti, "Between Mercantilism and Market," 324. See also Jeremy Philips, who argued that the Italian engineer Jacob Acontius (1492–1566) should be accredited with the introduction of privilege practices in England: Phillips, "The English Patent."

16 Gleitsmann, "'Wir Wissen"; Pohlmann, "The Inventor's Right"; Meldau, "Hauptwurzeln"; Müller, "Patentschutz"; Pohlmann, "Neue materialien"; Popplow, *Neu, nützlich und erfindungsreich*; Silberstein, *Erfindungsschutz*, 41–90; Creutz, "Herausbildung des Erfindungsschutzes"; Flechsig, *Frühneuzeitlicher Erfindungsschutz*; Hartig, "Zur Statistik".

17 Barrera-Osorio, *Experiencing Nature*, 68–72, 140–46; García Tapia, *Patentes de invención españolas*, 39–52; García Tapia, *Tecnica y poder en Castilla*; Goodman, *Power and Penury*, 129–41; Moreno, *Historia de la tecnología en España*, 89–96; Pérez, "Los inventos llevados de España"; Royo, "State Policy". On the Habsburg Netherlands, see Doorman, *Octrooien*, 15–17.

18 Braunstein, "A l'origine"; Beltran and Galvez-Behar, *Des brevets*; Frumkin, "Les anciens brevets"; Isoré, "De l'existence"; Plasseraud and Savignon, *L'État et l'invention*.

19 Lincoln, "The Jew and the Worms."

20 Federico, "Origin and Early History"; Duncan, "Monopolies under Elizabeth I"; Frumkin, "Early History"; Gomme, *Patents of Invention*; Hill, "Origin and Development"; Hulme, "The History of the Patent System"; Hulme, "The History of the Patent System: A Sequel"; MacLeod, *Inventing the Industrial Revolution*, 10–39; Palliser, *The Age of Elizabeth*, 374–79.

21 The (scanned) source, as well as a transcription, is available online: "Statute of Monopolies, Westminster (1624)," *Primary Sources on Copyright (1450–1900)*, eds L. Bently & M. Kretschmer, www.copyrighthistory.org (last accessed 3 March 2018).

22 Interest in the history of patent law first surfaced when, by the end of the nineteenth century, a number of legal scholars were trying to develop a more systematic and universally coherent form of intellectual property law. For a Dutch example from the period, see Star Busmann, *Octrooien van uitvinding*. References remained however somewhat sporadic, and the first historians to seriously engage with the history of patent law were active

in the field were Frank Prager and Gulio Mandich. Whereas Prager argued that there had always been an idea of intellectual property, but that it had been perfected in the United States, Mandich provided the final proof for the existence of patent-privileges in Venice.[23] Other scholars, in turn, tried to put Germany on the map as an important area for the development of modern patent law by drawing attention to the importance of early mining privileges.[24] In these somewhat nationalistic debates, the role of the English Statute of Monopolies as a precursor to modern patent law was slated in the awareness that similar laws had existed elsewhere at an earlier date.[25]

As regards the history of patent law in the Dutch Republic, the work of Gerard Doorman deserves pride of place. In 1940, Doorman decided to publish a list of 'patents' that he had traced in the Dutch archives. His intention was thus to "cast a great light [...] on the great many things done in this country in the technological sense."[26] As the vice-president of the Netherlands Patent Board, Doorman was mainly looking for the historical roots of his own discipline and he therefore emphasized the notion of 'technological progress'. Accordingly, his monograph contained a number of highly interesting chapters on the gutting of herrings, the gilt leather technique, drainage mills, and so forth. But it left little room for the analysis of the socio-economic circumstances in which privileges could prevail. Moreover, it soon became clear during my research that Doorman had overlooked a substantial number of documents that could have provided better insight into the nature of the Dutch privilege system. Given that all later histories of that patent law were based on Doorman, revising this aspect of Doorman's work turned into an important feature of my work.

---

during the interbellum period. It is striking that the increased historical interest for patent law paralleled efforts by the League of Nations to internationalize patent regulations.

23 Mandich's pre-war article (1936) remained unnoticed until a translation in 1948 by Prager. Mandich, "Le privative industriali"; Mandich, "Venetian Patents (1450--1550)." Then, Mandich, "Primi riconoscimenti"; Mandich, "Venetian Origins."

24 This idea is clearly formulated in Silberstein, *Erfindungsschutz*, 41–61. Silberstein based himself predominantly on the works by Adolf Zycha (1871–1948).

25 The debate on the origins of patent law shifted its focus once more in the aftermath of the Second World War, when the unmasking 'the first real modern patent' came to be equaled with its reformulation from a privilege to a right. Consequently, it was claimed that 'the real breakthrough' had only come with the American Constitution (1790) which defined all intellectual property as an inalienable right. Cf. Khan, *The Democratization of Invention*. The first U.S. patent law was enacted by Congress in 1790 as part of the Constitution. It stated: "The Congress shall have power [...] to promote the Progress of Science and useful Arts, by securing for limited Times to Authors and Inventors the exclusive Right to their respective Writings and Discoveries." U.S. Constitution (art. 1, sec. 8, cl.8.)

26 Doorman, *Patents*, 8.

Sources

The main reason why Doorman had a restricted view on the total number of documents relating to the privilege business is that his research was predominantly based on the so-called Books of Acts (*Akteboeken*), which contained copies of all the acts issued by the Assembly of the States General; the highest authority responsible for granting privileges in the Dutch Republic (more detailed information on the handling of privileges will follow in the next chapter). This meant that Doorman only recorded the privileges that had actually been granted, not the applications that had been denied.[27] To get an idea of the total number of privilege applications filed with the States General, one has to turn to the daily records of the Assembly, which have been recorded in the resolutions (*resolutiën*).[28] When doing so, it appears there were many more applications than has hitherto been assumed.

27 In defense of Doorman, it should be mentioned that he did not claim to have looked further than the *akteboeken*. Doorman, *Octrooien*, 5–6. Based on the work of Japikse, Doorman sometimes actually did note down a number of applications as they appeared in the resolutions (up until 1608). Generally speaking, he rightly did not count these applications as a privilege. But sometimes an application is suddenly counted as an actual privilege, as in the case of Francois Desplanquez's invention of a dredging machine. Doorman, *Octrooien*, 90 (G10, 14 December 1590). The reasons for this are unclear.

28 As the upcoming chapters will show, invention privileges were mainly granted by the States of Holland and the States General of the Dutch Republic. The archives of these institutions are kept at the Dutch National Archives, hereafter abbreviated as: NL-HaNA. For the reader's convenience, I've also kept a reference to the "Doorman number" in the notes, wherever that was possible, for example like this: [G122, 5 April 1613]. For the States of Holland, I have made use of the (published) resolutions, both as a number of additional (financial) registers. There are two printed series of the resolutions: I have mainly used the second series. Until 1621, the resolutions of the Delegated States were published in the same series. For the States General, I have made use of the financial registers, petitions, the Books of Acts, as well as the resolutions. Fortunately, these resolutions have been indexed in the twentieth century though a collective effort by a number of historians. This resulted in two series of publications in the so-called *Geschiedkundige Publicatiën*, which each provide a short summary of the original documents. The so-called *Oude Reeks* (*1576–1609*) of these indexes, hereafter abbreviated as RSG, was ordered thematically under topics such as "Foreign affairs", "the Prince of Orange", "Patents", etc. But for some reason not all content and not all resolutions were included. In the so-called *Nieuwe Reeks* (*1610–1625*), hereafter abbreviated as RSG NR, all the resolutions were included and ordered chronologically. For a useful guide on these issues, see Roelevink, "Navigating New Waters." In addition to these abbreviated entries, I have made extensive use of the work of Johannes Jacobus Dodt van Flensburg (1800–1847), once an assistant at the academic library of Utrecht, who published in the mid-nineteenth century a transcription of a number of resolutions "related to the development of civilization" (*betreffende de geschiedenis der beschaving*). Dodt van Flensburg, *Archief voor kerkelijke en wereldsche geschiedenissen, inzonderheid van Utrecht*. 7 vols. Utrecht: N. van der Monde, 1838–1848.

Moreover, Doorman was not meticulous in his numbering of inventions. His so-called G-numbers and H-numbers refer to the number of granted privileges, regardless of the number of inventions that were described in one privilege act. A good example is probably the case of the engineer Simon Stevin, who obtained protection for nine different inventions that were described in one privilege [G7, 28 November 1589].[29] On the other hand, for example, G184 and G175 in Doorman's numbering concern the same invention (the duration of the privilege was adjusted and hence a new privilege was issued). In short, counting the entries in Doorman's system gives little insight into the actual number of privileged inventions (compare Figure 1).

In order to tackle the inconveniences of Doorman's data set I made a new digital database for the years 1581–1621. This database has been set up to to compile and verify the tables in this book. Perhaps it needs saying at this juncture that not all of the information in the database could be discussed in the text. Rather than trying to be extensive, with the risk of becoming superficial, I have chosen to pick out some key cases and analyze them in more detail. The aim is thus to put a different complexion on the idea that "the central

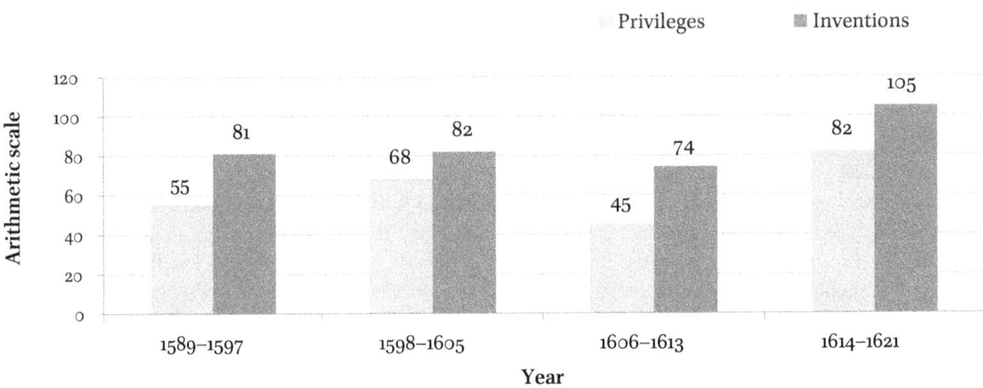

FIGURE 1 Number of privilege acts vs. the number of privileged inventions

Hereafter abbreviated as: Dodt, *Archief.* Especially these more extensive summaries of the resolutions (for the years 1601–1619) proved extremely useful for my research. Where the registers gave enough indications to follow the path to the original text, I have consulted the original resolutions to follow up on this lead. Primary sources keep to the original spelling and grammar, with the exception of the long 's' which is replaced by its modern equivalent. All names have been standardised for spelling. Patronymics have been abbreviated; for example, Cornelissoon (meaning "the son of Cornelis") is rendered as Cornelisz. Translations are by the author unless otherwise stated.

29 NL-HaNA, States General, 1.01.02, inv. no. 12298, fol. 9 [G7, 28 November 1589]. For a transcript of the privilege, see Stevin, *Principal Works*, 5:30–32.

authorities in the Dutch Republic (or the provincial estates for that matter) did not pursue a deliberate policy of promoting the emergence and diffusion of inventions for the sake of increasing the wealth and power of the state."[30]

Scope and Objectives

Central to this book is the question of how invention privileges fitted within the Dutch way of thinking about state management. To answer that question, the following chapters examine who made the law and to what end they did so; who or what had to be protected; which theories were used to legitimise and institutionalise exclusive ownership of technological innovation; and what the effect of all this was on society. Privileges provide an unconventional historical source for understanding the diversity of ideas circulating in the early modern Republic. With this book I hope to take a new approach to these sources and contribute to debates concerning the distribution of political power, the relationship between technology and progress, and the emergence of sustainable knowledge infrastructures on the European continent.

Privileges for invention reflect an ideology that positively valued novelty because of its effect on the common good. This ideology was preserved under threat of a punishment imposed by the state. Yet the state could not rely entirely on the fear factor, and it is clear that a certain amount of collaboration on the part of the population was essential for the successful implementation of privileges as an institutional reality. If only for that reason, it was important to present the creation of property comprising the effective control over the use of technological innovation as self-evident and justifiable. This line of logic, by and large, defined the form of the privilege: there was a problem that could be better solved by a specific technology than hitherto had been the case, and it was only right that an individual should be rewarded for his efforts to provide this solution. A revealing silence reigned over the question whether it was indeed possible to solve a problem by means of individual commitment (it was all about the 'decisive step', or so it was presented). Negative features of the institution of exclusive property, or any negative aspects of an invention, were not readily revealed in a privilege. It is precisely these silences that reveal the ideology that has had a lasting effect on our understanding of the nature of technological knowledge. Individual property became linked to technological progress and the purpose of technology was reformulated in terms of general

30 Davids, "Patents and Patentees," 268.

utility. It is in this nexus that the materialization of a new relationship between state and technology can be discerned.

With this in mind, this book also questions what the study of privileges can contribute to our understanding of the Dutch Republic as "the first modern economy", i.e. as an economy "with features that assist in the process of institutional, organizational, and technological change that improve the efficiency of production and distribution."[31] It should be clear that this is not economic history in the traditional sense of the word. Rather, the work rubs up against new institutionalism, and has been inspired in particular by the work of Charles Tilly, Theda Skocpol, and Douglass C. North, among many others.[32] Although I do not start from the assumption that the state converges exclusively with its institutions,[33] it cannot be denied that state institutions play an important role in determining "the rules of the game."[34] I apply Douglass North's definition here of institutions as including "any form of constraint that human beings devise to shape human interaction."[35] Thanks to the work of James C. Scott, among others, I came to understand that the institutions of the state collectively can turn into a quasi-independent actor making decisions (whether consciously or otherwise) that have a lasting impact on the course of history.[36] But it is precisely because of the tendency of the abovementioned authors to analyse changes related to "big structures, large processes, and huge comparisons" that historical specificities are sometimes somewhat underestimated.[37]

For this reason, I also looked extensively at the literature on state formation in the early modern period. Over the last few decades, there has been a

31 De Vries and Woude, *The First Modern Economy*, 713.

32 A useful overview in Bevir and Rhodes, *The State as Cultural Practice*, 25–41.

33 As Steve Hindle rightly put it: "The state is not to be viewed exclusively as a set of institutions; rather it is a network of power relations which become institutionalized to a greater or lesser extent over time. It therefore follows that state formation should not be viewed simply as a matter of teleological development towards 'modern', 'sovereign', 'rational', 'bureaucratic' state forms, but as a continuous, though not necessarily linear, process in the development of institutionalized power relations." Hindle, *The State and Social Change*, 19.

34 North, *Institutions, Institutional Change*, 3.

35 North, 4.

36 Scott, *Seeing Like a State*. I say "quasi-independent" because I do not believe in the state as an abstract concept: the state consists of people. However, once rules and regulations are in place, they can influence the options and limitations within which a later decision-making process will take place. For the sake of being complete, I wish to add that there are of course many institutions outside of the state; I do not see the state as "the institution of institutions" in the way that for instance Carl Schmitt did so. Schmitt, *On the Three Types of Juristic Thought*, 79.

37 Tilly, *Big Structures, Large Processes, Huge Comparisons*. Cf. Tilly, *Coercion*, 35–36.

lot of interesting research in that field on the implementation of rulership in particular.[38] Moving away from a teleological definition of state, and to complement the idea of a fiscal-military state that is fostered by growing taxes and bureaucracies, there has been a tendency to place more and more emphasis on the importance of personal networks for the exercise of power, how power was (culturally) mediated to different parts of the population, and how the less powerful were involved and shaped the superstructure of the state, for example by means of petitions and opposition (some have also spoken in this context of "empowering interactions" or "state-building from below").[39] The advantage of these approaches is that they avoid the *a priori* acceptance of the category of 'state' (and of the 'law' for that matter too). Questioning the relationship between agents and structure, as well as how the political economy worked in practice, becomes more important than focusing unduly on abstract political thought. The making of any durable system is seldom a completely conscious phenomenon, and for that reason it is also more sensible to speak of state formation rather than state-building.

Still, the precise definition of 'state' and 'sovereignty' continues to be problematic.[40] It appears difficult to come up with an unequivocal definition that covers each and every variation of state formation. In particular, the Dutch state raises problems for historians as it is assumed that the Dutch Republic, because of the high degree of autonomy of the cities, has to be seen as an exception in the early modern European landscape of political power. I have decided not to deal with the particulars of that discussion here, but to save the specifics for the following chapters, where they will be more relevant. Nonetheless, this is the right place to mention that I entirely agree with Michael Braddick when he says that it makes no sense to define the early modern state "in terms of form – for example, bureaucracy or centrally located institutions – or in terms of particular functions – making war or keeping peace for example."[41] Instead, in Braddick's definition, the state is a "distinctive form of political association" that can be defined "by the kind of power that is distinctive to it. The forms assumed by this power, and the uses to which it was put, are the object of the

38 For an excellent overview, see Brakensiek, "New Perspectives."

39 Blockmans, Holenstein, and Mathieu, *Empowering Interactions*. See also Braddick and Walter, "Grids of Power."

40 For a genealogy of the concept of the State, see Skinner, *The Foundations*; Harding, "The Origins." However, see also Skinner's later work, where he maintained that "To investigate the genealogy of the state is to discover that there has never been any agreed concept to which the word state has answered." Skinner, "A Genealogy of the Modern State."

41 Braddick, *State Formation*, 18.

enquiry, rather than part of its terms."[42] In brief, the state is best understood as "a coordinated network of agencies exercising political power. The precise form assumed by these agencies was a product of their territorial and functional bounds, but also of the wider beliefs in terms of which their activities were justified."[43]

In the process of allowing state power to become active, the law played a crucial role. The major importance of the law in this regard obviously extends much further than just the awarding of a certain sort of privilege: for a full analysis of political power, state formation, and legal transformations, it would be necessary to look at the exact composition of commissions, the nature of the fiscal system, the state's defence system, the local courts, the city civil service, communication technologies, provincial infrastructures, and so on (the networks and translations, in ANT terms, are endless).[44] Often, existing studies select a few elements to make a certain point. The results are undeniably fascinating. Yet, I have opted for a different approach in this study. Aware of my limitations, I have chosen a limited number of sources (privileges for inventions in the mechanical arts in a specific timeframe) as moorings, upon which the network was temporarily halted, fixed in a certain form.[45] Invention privileges function as a lens through which to see the constellation of political power in a particular light.

For one thing, invention privileges enable us to look at the state's attempts to determine how the market should be shaped (whether the state always succeeded in doing so is another question). The law did not simply reflect the economic interests of the ruling class, yet it did shape possibilities. On this point, my understanding of the law is also influenced by legal anthropology, which

42 Braddick, 18.

43 Braddick, 90. It is worth noting that Braddick is following a different yet very interesting line of approach in which he emphasizes the embodiment of the state in (local) political offices. It would be challenging to examine whether this approach could be applied to the Dutch Republic as well; yet, my emphasis lies on the embodiment of the state in legal documents that we call privileges for invention.

44 On Actor-network theory, see Latour, *Reassembling the Social*. The subtitle of this book, "The Making of a Patent System in the Dutch Republic", is of course a clear nod to the work of Latour, who analyzed law as a 'mode of existence' in his *The Making of Law: An Ethnography of the Conseil d'Etat* (Cambridge: Polity, 2010). See also Latour, *An Inquiry into Modes of Existence*, 357–79. Even if I do not always agree with the outcome, I find Latours' attempts to come to terms with the 'social' inspiring. Being a field-based method, ethnography or ANT is however not particularly suited to study early modern history.

45 Compare the remarks in Strathern, "Cutting the Network."

has shown how the law can serve as a shifting mechanism to "sort things out."[46] Invention privileges offered the possibility to legitimize specific forms of knowledge, to place specific knowledge "into a canon of certified knowledge."[47] In that sense, the law can also be considered a technology that actively shaped industrial and societal developments, instead of being merely a response to these developments.[48] To borrow a term from linguistics, the law determined the felicity conditions under which specific utterances had to be interpreted.[49] It follows that discursive communities were deemed to react in a specific way; after all, their utterances had to be sensible in order to become part of a discourse. The lawmakers determined not only which knowledge was to be produced, but also the framework within which contestation could take place.

The idea of resocialization as well as a disciplining of knowledge, as I use it here, clearly resonates with Foucault's notion of 'governmentality' ('governmental rationality'). I entirely agree with Foucault that:

> We frequently speak of the technical inventions of the seventeenth century – chemical, metallurgical technology – yet we do not mention the technical invention of this new form of governing man, controlling his multiplicity, utilizing him to the maximum, and improving the products of his labour, of his activities thanks to a system of power which permits controlling them.[50]

Yet a very important methodological difference is that, in contrast to Foucault, I do not think in terms of a 'dispositif' or an 'apparatus', but rather confine myself to one specific kind of historical source, invention privileges, from

46 For a very useful bibliography, see Kelly, "Legal Anthropology." I am thinking here in particular of the work by Rosemary C. Coombe, Marilyn Strathern, Alain Pottage, and Annelise Riles (particularly her work on legal technicalities): Coombe, *The Cultural Life*; Pottage, Copeman, and Street, "Law after Anthropology"; Pottage and Mundy, *Law, Anthropology, and the Constitution of the Social*; Riles, "A New Agenda for the Cultural Study of Law"; Riles, "Afterword." The expression "sorting things out" does not come from legal anthropologists, however, but from Bowker and Star, *Sorting Things Out.*

47 Burk and Reyman, "Patents as Genre," 167.

48 For a more extensive exposition of intellectual property law as a technology, see Biagioli and Buning, "Introduction."

49 Austin, *How to Do Things with Words.* More specifically on the use of speech-act theory in legal settings is Constable, *Our Word Is Our Bond.*

50 Foucault, "The Incorporation of the Hospital," 146. On Foucault's understanding of the law, see for instance Baxter "Bringing Foucault into Law"; Hunt and Wickham, *Foucault and Law.* For a contrasting opinion, see Golder and Peter Fitzpatrick, *Foucault's Law*, but also see the excellent review by Pottage, "Foucault's Law".

whence to expose an existing power constellation.[51] As much as anything else, this book can therefore be read as an exercise in how to do things with sources.

In all of this, it should be borne in mind that invention privileges are textual constructs in a specific material form. Privileges are the precipitation of certain ideas and habits, but at the same time give shape to those ideas and habits, if only because the ownership rights were mediated on paper and thus, for instance, lend a different rhythm to the way in which knowledge was marketed. By means of the patent system, inventions became part of a capitalistic market that was regulated largely anonymously via obscure instruments of the law instead of in one concrete place (such as a market in the traditional sense of the word).[52] Still, the privileges themselves are only one element within the collection of text documents that give rise to the law, and they can only be studied meaningfully in relation to other documents and data. To give an illustration: a privilege applicant tried to convince the authorities to award a privilege, while another privilege holder tried to prevent the award by means of an appeal; in the meantime, reports on the efficacy of a certain invention were written, and behind the scenes a higher official tried to persuade his colleague that the privilege should be granted in the interests of the state; a testimonial was left with a notary, while at the same time books and pamphlets were written in an attempt to convince the public that it was entirely legitimate (or entirely wrong) to grant privileges to begin with. Would one only look at the invention privileges themselves, one would undoubtedly miss parts of the rich ecosystem in which invention privileges claim their place as a constituent element.

The following chapters will each highlight a different aspect of privileges issued for the exclusive use of new technologies. Where necessary, there will also be more emphasis on the relevant historiography. Nonetheless, this book obviously does not cover all the possible dimensions of invention privileges. So, there are a number of things this book explicitly does not address. First of all, it does not discuss the technology of the privileged inventions in much detail. In that sense, this is not a contribution to a conventional history of technology. I am not concerned with the emergence of new technologies, nor

---

51 If I was to think in terms of a dispositif, probably it would be 'monopolisation'. Foucault argued that we should "break free … of the theoretical privilege of law and sovereignty" (Foucault 1976:118–19). However, following Agamben, one might argue "daß die Produktion eines biopolitischen Körpers die ursprüngliche Leistung der souveränen Macht ist." Agamben, *Homo sacer*, 16. Biopolitics is as old as the sovereign exception. The privileges for monopolies described in this book fit within Agamben's notion of exception.

52 Braudel, *Civilization and Capitalism*, 456. For a discussion, see Howell, *Commerce before Capitalism*, 29–34; Prak, "Early Modern Capitalism."

with their use and afterlife, or with their maintenance.[53] Although one might argue that invention privileges are an integral part of technological systems and sociotechnical ensembles, I do not wish to talk about "how technology is made"[54] but about the manifestation of political power in the making of the law. One of the central concerns in this exercise is to analyse the (administrative) decision-making process leading up to a privilege. But also on that point, this work has its limitations. In particular, despite the necessary attention to biographical detail, it was often difficult to find much information on the life histories of the approximately 335 inventors who appear in the study; perhaps, over time, more particulars will come to light, allowing me to give more depth to the lives of these actors. One possibility could be to add genealogical data and employ the method of prosopography. For example, it would certainly be worth studying the family relations between the inventors and the authorities in more detail, or to look at the religious networks that connected the different inventors.[55]

Finally, it should be emphasized that this book is not a study of innovation or inventions in the Dutch Republic in general. It is well known, after all, that many successful inventions were never privileged at all and that many innovative activities actually took place outside of the privilege system.[56] In this context, one might highlight the importance of innovation prizes and rewards, both on an urban and interprovincial level, or the innovative character of guild apprenticeships.[57] Although particular topics relating to these broader issues will be discussed when necessary, they are not the central focus here. This book has a more modest objective: it attempts to reconstruct and analyse the nature of privileged knowledge during the early days of the Dutch Republic.

The structure of the book is as follows. Chapter 1 explores how the Dutch managed to give their own twist to the privilege system they inherited from the Burgundian overlord. It examines the degree of coherence within the Dutch system as well as the relationship between cities in the Republic and central organs such as the States General. The Dutch are often regarded as an absolute anomaly in early modern Europe, where rivalry among cities and provinces frustrated every claim to central authority. This chapter posits that the argument must be more subtle. It shows that the management of

53 Russell and Vinsel, "After Innovation."

54 Bijker, "How Is Technology Made?"

55 Cf. Adams, "The Familial State," 1994; Adams, *The Familial State*, 2005.

56 De Vries and Woude, *The First Modern Economy*, 345.

57 On prizes and innovation awards, although mostly with a focus on later periods, see Burton and Nicholas, "Prizes, Patents and the Search for Longitude"; Jonkers, "Rewards and Prizes"; Khan, "Inventing Prizes"; Moser, "Innovation without Patents."

invention privileges was not about decentralization or about the concentration of knowledge resources by the central government; it was about the strong links between local politics and the central state.

In mapping these ties, Chapter 2 scrutinizes the administrative procedures entailed in the procurement of invention privileges. Following the route of a standard application, attention will be given to the financial settlement of invention privileges, the areas of usage, specifications, and the remuneration practices targeted more generally at the production of innovative knowledge. The principal conclusion is that 'warring' and 'colonialism' were certainly important instigators for the production of innovative machinery and methods; however, concrete local issues were equally important. In the Dutch scenario, this meant finding a response to the continuous threat of water that had to be controlled in some way. It also meant finding methods to make more efficient use of wind and finding solutions to make life in the cities more bearable.

Chapter 3 then analyses the economic and social dynamic of the Dutch privilege regime. It discusses the nature of trade agreements between inventors and investors, paying special attention to the notion of projects. It then moves on to examine the social composition of the actors, demonstrating how the privilege system functioned as a place of encounter where different ways of understanding technological inventions could merge. Inventors, merchants, visionaries, and administrators: they all had to find a common language to reach an agreement concerning privilege rights. It was here that a new notion of utility was crystallised that overlapped with a new way of thinking about the relationship between state, society and the economy.

Chapter 4 looks at how the exclusive use of 'useful knowledge' was legitimised in the legal literature of the time. It collects some of the scattered references that gradually formalized the practice of granting privileges to inventors and provides an overview of the different means of furnishing legal proof in the early Republic. It then examines how this theory worked in practice. Comparing the examination methods prevalent in the system of invention privileges with procedures used to prove the understanding of natural phenomena, I put forward the argument that specific elements of the so-called Scientific Revolution can be related to the introduction of legal procedures in the process of knowledge evaluation.

Chapter 5 again takes up a different perspective and frames the Dutch system of invention privileges against a background of geopolitical competition. The chapter unravels the unstable notion of 'invention' at the turn of the seventeenth century, before discussing the multifarious relationship between novelty and imitation. The chapter then turns to the economic utility of invention privileges as seen from the perspective of the authorities. It shows how the system of invention privileges in the Dutch Republic revolved around the organization of resources formulated upon a complex notion of effectiveness.

By way of conclusion, the final chapter puts the Dutch experiences in a broader framework and takes up the challenge to rethink the position of the early Dutch Republic in the wider history of European processes of state-formation, knowledge management, and economic development. The main argument is that the very existence of a system of invention privileges illustrates that the Dutch federal authorities managed to implement a long-lasting institution tailored to the production of knowledge. The shape of this knowledge infrastructure was an integral part of the institutional configuration of the Republic. After tackling the question whether one can speak of a 'republican patent' in this regard, I conclude by sketching out the after-effects of the Dutch system of invention privileges on the long-term relationship between patents, power, and knowledge.

CHAPTER 1

Inventing a System

This chapter situates the Dutch system of invention privileges in its political context. It explores some of the continuities and discontinuities between the Dutch system and the system previously in place under Burgundian-Habsburg rule, before analyzing the relationship between the local authorities in the United Provinces and the central government seated in The Hague. Despite the prominent role of the cities in the Dutch state structure, on an overarching level there was a centralized policy dealing with the stimulation of technological innovation in the Dutch Republic: invention privileges were a distinctive part of the ongoing processes of state formation whereby local authorities relinquished some of their power in the interests of a greater good. To set up this broader argument it will be useful to first clearly situate our study object in time and space. I have decided to do so by means of a short prelude that offers a narrative of the Dutch Revolt.

Setting the Stage: The Dutch Revolt

Up until the end of the fifteenth century, the region that would later become the Dutch Republic had never been more than a somewhat loose collection of territories. It was only in 1543, when Charles V finally managed to conquer the Duchy of Gueldres, that the territory later to become the Dutch Republic was, for the first time in history, brought under the control of one ruler. The ruling prince of the Habsburg Netherlands was also the King of Spain (Castile and Aragon), the Holy Roman Emperor, the Archduke of Austria, and in command of virtually all of the New World that had recently been discovered.

For the purposes of ruling more conveniently over his enormous empire, Charles V implemented a number of institutional changes that were aimed at centralizing his power. The most important of these changes for the Burgundian domains was the appointment of a permanent governor, assisted by three permanent councils.[1] Formerly, the tasks of the governor had included taking

1 The reorganization of church lands in 1559, as well as the tightening of the Inquisitional power, formed an integral part of the centralization process. I focus here mainly on the political aspects in terms of institution building; for an in-depth study of the religious unification strategies under Charles V, see Fühner, *Die Religionspolitik Kaiser Karls V.*

 | DOI:10.1163/9789004320420_003

on the responsibilities of the highest authority, when the King was temporarily absent for whatever reason. But Charles V turned governorship into a permanent office in 1531, when he appointed his sister Mary of Hungary (1505–1558) as the governor of Habsburg Netherlands. In addition, he institutionalized three permanent councils, the Collateral Councils (*Collaterale Raden*), that were supposed to provide the governor with advice. They consisted of a Council of State (*Raad van State, Conseil d'Etat*) providing advice on matters of general policy; a Privy Council (*Geheime raad, Conseil Secret*) that administered affairs of justice and police; and a Council of Finances (*Rekenkamer, Conseil des Finances*) that looked over fiscal matters.[2]

The provisional tailpiece of Charles V's political strategies to unify the Low Countries was the so-called Pragmatic Sanction (*Pragmatieke Sanctie*, 1549). The goal of this Sanction was to prevent the fragmentation of the former (but now expanded) territory of the Duchy of Burgundy, renamed as the Seventeen Provinces, at the hands of the heirs of its respective ruler.[3] Thus, the Seventeen Provinces had been successfully united when Charles stepped down in favor of his son Philip II in 1555. At least, so it seemed.

In line with the initiatives developed by his father, Philip II continued to centralize political and economic power in the Seventeen Provinces. His enterprise was ill-fated from the start. In accordance with general practice, Philip II had assembled the representatives of the different provinces in the course of his rise to power with the goal of negotiating a *bede* (a 'demand for money') to cover some of his enormous debts. But the negotiation process proved unsatisfactory for both sides. When it finally came to an agreement in 1559, the different provinces had managed to obtain extensive freedoms in return for their money. On the other hand, they had been weakened in their power by Philip's decision to prevent any general assemblies in the near future. Philip II concluded the negotiations by installing his sister Margaret of Parma (1522–1586) as the new governess of the Burgundian Lands.[4] He then left the Seventeen Provinces, once and for all, to take up residence in his Spanish palaces.

2 Both governorship and the councils were to a considerable degree the continuation of preexisting power structures, but they were now given more important and specific responsibilities. For more information about the genesis of the Collateral Councils, see Aerts and others, *De Centrale Overheidsinstellingen van de Habsburgse Nederlanden (1482–1795)*, 1:265–82, 295–324, 497–552.

3 Fruin, *Staatsinstellingen*, 27.

4 Mary of Hungary had stepped down as governor in favor of Emanuel Filibert of Savoy (1528–1580) in 1555, the year that Charles V abdicated.

The Provinces were left in turmoil. In 1566, a league of about 400 members of the high nobility lodged a petition (*Smeekschrift*) to the governor in which they demanded the suspension of the persecution of heretics and the revocation of the States General. The petition was forwarded to Philip II, who refused to adapt his policies. Within a matter of months, the so-called Iconoclastic Fury (1566) broke out. Protestants from the Southern Provinces descended upon Catholic churches in a burst of anger to destroy every form of religious imagery, which in their eyes represented Papal foolishness.

Philip II reacted by replacing his sister Margaret of Parma with the Army General Fernando Álvarez Duke of Alva (1507–1582, in power 1567–73). The Duke reigned with an iron fist, setting up a Council of Troubles (*Bloedraad*) to prosecute the King's opponents. At first, the measures seemed to have the desired effect, and by 1569 the rebellion had been more or less suppressed.[5] Then, however, Alva proposed the introduction of new taxes, including a ten per cent levy on all sales other than landed property (the so-called 'Tenth Penny'). Even though the proposal was heavily contested by the Provinces, the Duke was eager to push the tax measures through. This only strengthened resistance, which was relentlessly fanned by a large group of refugees who had fled the country during Alva's rule.[6]

1572 was the major turning point. A group of pirates, who later became known as the Sea-Beggars (*Zeegeuzen*), captured the town of Brill in Southern Holland.[7] The event was carefully exploited by the rebels and became a sign for Protestants all over the Seventeen Provinces to instigate revolt. Among these rebels was William the Silent, Prince of Orange (1533–84). As the former right-hand man of Charles V, he had retreated to his estate in Dillenburg (Nassau, present-day Germany) upon the rise of Alva.[8] During a meeting in Dordrecht in July 1572, he was confirmed as the Stadtholder of the provinces of Holland, Zeeland, Friesland and Utrecht.[9] Thus, effectively, he became the leader of

5 In 1568, the Catholic nobles Egmont and Horne, were beheaded in the square of Brussels as a precautionary measure. This act met with a lot of resistance and led to several uprisings. Alva kept tight control, however, sentencing over 9000 people by default and executing over a 1000 opponents. Van Gelderen, *The Political Thought*, 49.

6 Israel, *The Dutch Republic*, 161–73. Between 1566 and 1572 there was a sort of guerrilla war in the Northern provinces, but Alva's army proved much stronger

7 In fact, the Sea-baggers captured the town Brill without any political objects; they were forced to go ashore because of heavy winds. Van Gelderen, *The Political Thought*, 42.

8 Van Gelderen, 34.

9 The meeting at Dordrecht also settled the division of power between the House of Orange and the States; a division that would remain precarious during the remaining days of the Republic. See Price, *Holland and the Dutch Republic*, 154–71.

the revolt. Shortly afterward, the Seventeen Provinces negotiated an internal peace treaty (the Pacification of Ghent, 1576) in which they agreed to stand up for religious tolerance and to fight against the mutinous Spanish forces who had sacked the city of Antwerp.[10]

The Revolt now gained a more permanent form. Its focus shifted to the Northern Provinces when Alessandro Farnese Duke of Parma (1545–1592) came to power as the new Governor of the Seventeen Provinces in 1579.[11] The Duke of Parma had returned a number of privileges to the Walloon nobles at his accession to office, which had resulted in several Southern Provinces expressing their loyalty to Philip II by signing the Union of Arras (1579).[12] As an immediate reaction, the Northern provinces united themselves in the Union of Utrecht (1579), which became something of a 'founding charter' of the Dutch Republic.[13]

This is, more or less, where our story begins. Events during the years 1579–1621, such as the rise and fall of Oldenbarnevelt, the expanding powers of Stadtholder Prince Maurice, or the Truce Troubles, shall be dealt with as and when required (see Figure 2). As an impermanent end station for my research, I have chosen the year 1621, because it signified the end of a Twelve-Year Truce between Habsburg Spain and the Republic. This marked a new phase in the Dutch Revolt, as it now became intermingled with the wider conflict of the Thirty Years War (1618–1648). Initially, the regenerated war worked out unfavorably for the rebels; Spanish forces recaptured large parts of the country for most of the 1620s under the guidance of their skilled army general Ambrogio Spinola (1569–1630). By 1628, however, the cards turned again in favor of the Republic. The Republicans recaptured some important areas of the Southern provinces and somehow managed to confirm their borders.[14] The Dutch Revolt would eventually be settled with the Peace of Münster, when the Spanish Habsburgs recognized the sovereignty of the Dutch Republic.

---

10 Following financial troubles of the Spanish Crown in 1576, Spanish mercenaries decided to disburse themselves by plundering the city of Antwerp at the cost of some 8.000 lives. The event became known as the 'Spanish Fury'.

11 The Duke of Parma succeeded to John of Austria (1547–1578), an illegitimate son of Charles V. John of Austria had succeeded Luis de Zúñiga y Requesens (1528–1576), who had been the governor of the Netherlands since the departure of the Duke of Alva.

12 Van Gelderen, *The Political Thought*, 50.

13 Price, *Holland and the Dutch Republic*, 221. Southern cities like Bruges, Ghent, Brussels and Antwerp later joined the Union of Utrecht.

14 Prak, *The Dutch Republic*, 40–41.

Key events of the Dutch Revolt, with a focus on 1581–1621

1566 – Iconoclastic fury
1567 – Duke of Alva replaces Margaret of Austria Duchess of Parma as governor
1572 – Sea-Beggars (Zeegeuzen) take control of the town of Den Briel (Brill)
1576 – 'Spanish Fury'; Pacification of Ghent
1579 – Union of Arras (Atrecht). Southern provinces return to the Habsburg Netherlands
– Union of Utrecht: 'founding charter' of The Dutch Republic
1581 – 'Act of Abjuration': United Provinces depose Philip II as sovereign
1582 – Francis, Duke of Anjou becomes the new leader of the Provinces. He leaves in 1583.
1584 – William of Orange assassinated
– William Louis of Nassau-Dillenburg Stadtholder of Frisia
1585 – Treaty of Nonsuch; English aid for the Dutch Rebels
– Robert Dudley Earl of Leicester Governor-General of the United Provinces
– Maurice of Nassau new Stadtholder of Holland, West-Frisia & Zeeland
1586 – Johan van Oldenbarnevelt becomes Land's Advocate of Holland
1587 – Leicester leaves the United Provinces
1588 – New Instructions Council of State
1596 – Triple Alliance; France, England and the Republic unite against Philip II
– Philip II cedes sovereignty over the Habsburg Netherlands to Albert of Austria and Isabella
1598 – Death of Philip II; long live Philip III
1601 – Spanish Army General Ambrogio Spinola (1596-1630) replaces General Mendoza (1546–1623), who had led the field from 1598
1609 – Twelve-year truce
– Death of Arminius; beginning of the Truce Troubles
1618 – Maurice of Nassau becomes Prince of Orange; arrest of Oldenbarnevelt
1619 – Final session Synod of Dort
– Execution of Oldenbarnevelt
1620 – Death William Louis of Nassau-Dillenburg; long live Ernst Casimir I of Nassau-Dietz
1621 – Death Philip III; long live Philip IV
– War re-emerges between the United Provinces and the Spanish Habsburg Empire
– Don Ambrogio returns as Army General to the Republic until 1628
1625 – Death of Prince Maurice of Orange; long live Frederick Henry

1648 – The Revolt ends with the Peace of Münster (Peace of Westphalia)

FIGURE 2 Timeline of key events during the Dutch Revolt

Privileges in the Burgundian Netherlands

The increased use of invention privileges in the Low Countries was part of the consolidation of power over large areas of land controlled by a single ruler. Indeed, the first privileges securing exclusive exploitation rights date from the year 1515, the year of Charles V's rise to power as Duke of Burgundy and

Count of Flanders.[15] These early privileges related mainly to the exploitation of newfound land, and in particular mines, for defined periods of time.[16] Privileges were part of a well-advised settlement policy; they were used as a means to incite investments and to attract foreign entrepreneurs by offering them monopoly rights in exchange for their knowledge. The merchant Jehan de Lame from Cremona, for example, managed to obtain a privilege from the Burgundian-Habsburg authorities on 17 September 1549 that gave him exclusive rights to produce Venetian style glassworks in any of the Burgundian Lands.[17] The inventor promised to import the necessary workers and materials from Italy and to introduce his art in the city of Lier.[18]

Although invention privileges were officially granted in the name of the sovereign overlord, they were not granted at his spontaneous will. Prior to awarding exclusive rights, the economic and social impact of the temporary monopoly was carefully considered. First of all, local authorities such as burgomasters or guild representatives gave their advice on the matter. When, for instance, Philip II decided to grant Willem Aemissen a privilege for the exploitation of an economical boiler, it was adduced that the device worked well "as proven by certification of the burgomaster and the Board of Aldermen of Leeuwarden, because the brewers in that city have considered the named art good by experience."[19] Then, at a more federal level, advice was usually sought from the Financial Council and its *commis de finances*, who sometimes followed the advice of the Privy Council.[20] The careful examination process shows that privileges were subject to an elaborate system of checks and balances aimed at steering the economy in a particular direction.

---

15 This section of my research is almost exclusively based on the data provided by Doorman, who admitted that his research on the Spanish side was far from complete. Doorman, *Octrooien*, 16, note 1. For that reason, it is quite possible that further research will shine a different light on the early days of invention privileges in the Low Countries.

16 For examples, see Doorman, *Octrooien*, 65. Also in later periods, mining remained an important reason pass privileges. See, for example, Doorman, 71 [K11, 7-1-1567].

17 Doorman, 65.

18 Goris, *Etude sur les colonies marchandes*, 435. Jehan de Lame probably did not succeed in working his privilege. In 1551, he received a letter from the governor Mary of Hungary that "jusques ores l'on n'a fait devoir mectre ledit privilège en exécucion. [...] A ceste cause ferez bien de nous faire entendre à la vérité à quoi il a tenu ou tient que ledit devoir n'a encoires esté fait, sans en vouloir faire faulte." Document taken from the *Collection des papiers d'État, aux Archives du royaume*, published by M. Pinchart, quoted in Edouard Garnier, *Histoire de la verrerie et de l'émaillerie*, 283.

19 "[...] gelyck tselve blyckende is by certificaten van burgemeester en scepenenen de raden onser voornoemde stad Leeuwarden aldaer de de brouwers de voorseyde conste by experiencie goed ende expresse bevonden hebben." Doorman, *Octrooien*, 74. [K13, 1568].

20 Doorman, *Octrooien*, 67, 70, 73 [K2, 1561; K6, 1563; K13, 1568].

Privileges for the commercial exploitation of inventions in the Burgundian Netherlands were usually part of a 'package deal' with which the authorities tried to lure inventors to their territory. In exchange for the implementation of a profitable machine or method that might serve to boost the local economy, the inventor would obtain such things as specific tax cuts, free housing, and a temporary monopoly. This sustained relationship between monopolies and invention privileges does not always come clearly to the fore in the secondary literature, where it has become commonplace to distinguish between invention privileges (i.e. proper 'patents') and trade monopolies (*privileges pro arte introducenda*).[21] This is not a very useful distinction and leads to serious misunderstandings regarding the period system. The aim of invention privileges was, and had always been, to legitimize trade monopolies for the commercial exploitation of inventions on the local market.

The Emergence of a Dutch Body Politic

In order to understand how the privilege policies in the Burgundian Lands affected later policies in the Dutch Republic, it is important to get a better understanding of the political structure of the newly formed state that emerged rather suddenly at the end of the sixteenth century. That Dutch state essentially consisted of seven provinces that were united on the basis of the Union of Utrecht. These provinces, which were also called the United Provinces, were the duchy of Guelders, the counties of Holland and Zeeland, the former bishopric of Utrecht, the lordship of Overijssel, the province of

21 The discrepancy between invention privileges and monopolies is in part related to the way in which the historiography of the field has developed. Until the 1950s, the history of invention privileges (i.e. 'patents') was namely predominantly studied by lawyers and professionals in the patent industry. These scholars were interested in the antecedents of their own discipline, where it was – and is – not by definition obvious to consider a patent as a monopoly. Cf. Kitch, "Elementary and Persistent Errors." When in the second half of the twentieth century, the study of patents entered the field of the history of technology and economic history, scholars came to stress the use of invention privileges (i.e. 'patents') as tools for technology transfer. In that process, the notion of invention privileges as economic monopolies for a local market has retreated somewhat to the background. An important exception was the work by Joan Thirsk, who embedded invention privileges in the wider context of state sponsored projects to spur the development of a consumer society: Thirsk, *Economic Policy and Projects*. For a useful discussion of early patents as monopolies, see Walterscheid, *The Nature of the Intellectual Property Clause*, 241–55. Walterscheid discusses the American situation; yet, the observations are useful as well for a broader understanding of the nature of exclusive rights.

Friesland and (from 1595 onwards) the province of Groningen (see Figure 3 and 4).[22] Although nominally the individual provinces were sovereign, they gave up these sovereign rights on a number of issues that were of common interest, such as defense and finances. These issues were decided upon in the assembly of the States General in consultation with the Council of State.

The individual provinces were governed by the Provincial States (*Provinciale Staten*).[23] In this assembly, which convened only a few times per year, the nobility, the cities, and the smaller towns had a representative vote.[24] For routine matters, however, the Provincial States delegated much of its political power to permanent commissions, which were called the Delegated States (*Gedelegeerde Staten* or *Gecommiteerde Raden*). The Delegated States was primarily responsible for the execution of the resolutions of the Provincial States; to govern the finances; to maintain the military system and, in some cases, to administer the law.[25]

The Provincial States was, in turn, spearheaded by a Stadtholder (*Stadthouder*). Although each province could in theory have its own Stadtholder, in reality Stadtholdership over several provinces was often combined. During the period 1581–1625, the power in the Republic fell effectively to two Stadtholders who happened to be cousins: Count Maurice of Nassau (1567–1625) and Count William Louis of Nassau-Dillenburg (1560–1620).[26] Maurice became the Stadtholder of the Province of Holland & West Frisia[27]

22 In point of fact, there was an eighth province, the lordship of Drenthe, but the area was considered so poor that it was exempt from paying taxes. Consequently, it was also denied access to the assembly of the States General. Israel, *The Dutch Republic*, 277. The Generality Lands, which consisted of the territory conquered on the Spanish Habsburgs, did not have any official representation either. Other than that, it should be noted that the contemporary name for Groningen was *Stad en Lande*. The domain *Stad en Ommelanden* was conquered by the Republic in 1594 who renamed it *Stad en Lande*. I continue instead to use the name Groningen, as it resonates better with current geographical understanding.

23 The institution of a Provincial States dated back to the time before Burgundian-Habsburg rule. For the long-term history, see Blockmans, "De representatieve instellingen."

24 The exact representation differed from province to province, and was liable to the necessary changes during the Revolt. It would lead too far afield to deal here with all the provincial particularities. For an excellent overview, see Fruin, *Staatsinstellingen*, 214–57.

25 Fruin, 219–77.

26 Jan VI of Nassau-Dillenburg (1535–1606) was the father of William Louis and the brother of William the Silent. He saw to the education of his nephew Count Maurice. For a biography of Maurice of Nassau, see Van Deursen, *Maurits van Nassau*. William Louis has received somewhat less scholarly attention, but see Bergsma, "Willem Lodewijk"; Waterbolk, "Met Willem Lodewijk aan tafel."

27 The States of Holland and West Friesland was a novelty that came into being because of the Revolt. Holland and West Friesland (roughly speaking the area north of Amsterdam) had been united since the days of Count Floris V but, because the cities of Amsterdam and Haarlem had remained loyal to the King in the early days of the Revolt, West Friesland had gained some independency. Throughout the years 1576 to 1585, West Friesland set

FIGURE 3 Map of the United Provinces, situation around *c.* 1600

and Zeeland, succeeding his father William the Silent (1533–1584) in 1585. Later (from 1589 onwards) he assumed the Stadtholdership of Utrecht, Guelders, and Overijssel as well. William Louis, on the other hand, became the Stadtholder of Friesland (1584), Groningen (1594), and Drenthe (1596).[28]

up a number of independent institutions, which later remained intact. It run its own finances, disposed over a separate admiralty, and even retained an independent Delegated States, but it was united with Holland under one vote in the Assembly of the States General. Fruin, *Staatsinstellingen*, 230–32.

28 The position of Stadtholder would only become hereditary by the end of the seventeenth century. Utrecht, Overijssel, and Guelders were shortly under the Stadtholdership of Adolf van Nieuwenaar (1584–1589) before he blew himself to pieces while testing some new artillery.

FIGURE 4 Major cities in and around the United Provinces

The role of the Stadtholders remained somewhat vague after the separation of the United Provinces from the Spanish Habsburg Empire. Originally, the Stadtholder had been the representative of the sovereign monarch; since sovereignty had fallen to the Provincial States and the cities, the office of Stadtholder was thus strictly speaking somewhat superfluous. The result was inevitable squabbling about the exact role the Stadtholder was supposed to play. The position of the Stadtholder is best summarized as being subordinate to the States, while retaining certain sovereign rights, such as naming the magistrates or the prerogative of mercy. The political circumstances of the time often determined how far the Stadtholder could go in strengthening his personal power.

The Stadtholder of Holland, Maurice of Nassau in this case, was moreover the Admiral General and Captain General of all the troops under the flag of the Republic. In that capacity, he participated as a full member of the Republican Council of State (*Raad van State*).[29] Besides the Stadtholder, the Council of State consisted of twelve members from the different provinces complemented by a registrar. The institution had emerged from the Council of State that had existed under Burgundian-Habsburg rule. But instead of having a purely advisory role, it took on executive responsibilities in the newly founded independent Republic. The Council was in charge of all international relationships of the Republic and of managing the finances, and it was in control of army command.[30]

The importance of the Council of State was however soon counterbalanced by the rising prominence of the States General of the United Provinces. The States General was again one of those remnants of the Burgundian-Habsburg period, whose function was completely reinterpreted after the abjuration of Philip II in 1581. Originally a space for negotiations between the Burgundian overlord and the provinces, it soon began to function as the central and coordinating authority in which each of the seven provinces constituting the Republic was represented with one vote.[31] Like the Provincial States, the States General traditionally represented the nobility and the burghers of the principal towns, but appointment procedures differed considerably from province to province. In the case of Holland, for instance, the *Ridderschap* ('knighthood') was represented by one member who was appointed for life, whereas the other members changed regularly. In the case of Zeeland, all members were appointed for life.

As we shall see in the next section, the States General assumed the authority to issue privileges that related to the introduction of new technologies soon after a short moment of 'independence' in 1587. In addition, it chiefly had the authority to decide over five issues: 1. foreign affairs 2. defense 3. the system of coinage 4. differences between provinces 5. the Generality Lands (about one fifth of the territory of the Republic, conquered on the Spanish Habsburgs and with no official representation). On those issues, decisions had to be taken

---

29 Fockema Andreae, *De Nederlandse Staat Onder de Republiek*, 19–22.

30 Whereas the States General was vested with the highest authority and held supervision over the actual conduct of war, the Council of State supervised the finances, pay, fortifications, oats, etc. For the general division of tasks between the States General and the Council of State, see 't Hart, *The Making of a Bourgeois State*, 32–39.

31 The States General had been a relatively new political body, coming into being around 1475 to facilitate consultations between the ruler and the provinces. Fruin, *Staatsinstellingen*, 105. In the beginning it had disposed of relatively little power, but as the power of the sovereign was waning, the power of the States General increased. Blockmans, "De representatieve instellingen," 36.

with unanimous consent.[32] In practice, however, the province of Holland often managed to impose its views, if only because it contributed much more financially than any of the other provinces.

The Assembly of the States General began to convene regularly from 1582 (when, incidentally, it changed the official language to Dutch). From 1588, it was hardly ever out of session.[33] Around the same time, it settled in the Inner Court (*Binnenhof*) in The Hague, which housed, besides the States General, the Council of State and the General Accounting Office, as well as the States of Holland, the Delegated states of the South Holland (*Zuiderkwartier*), the auditor offices of Holland, the Court of Holland, and the High Court. In short, it was not long after the abjuration of Philip II in 1581, that the large concentration of powerful institutions turned The Hague into the center of political power in the Dutch Republic.

Still, officialdom was never overwhelmingly big. Even on a busy day at the Inner Court there were no more than 250 public servants present, who sometimes worked in relatively poor conditions.[34] The States General was, for that matter, housed in several rooms on the first floor of a side building. The historian Paul Knevel has described such a vivid image of its main assembly room that it deserves to be included here *in extenso*:

> The room was dominated by a rectangular table with round edges and covered with a green cloth, which was standing in the middle of the assembly room. There were twenty-eight chairs around this table, thirteen on each side and one on each end. The most remarkable was perhaps the chair of the Stadtholder, with armrests and a pillow in green velvet, situated on the left end of the table, in front of the fireplace and close to the entrance of the room. The representatives had to satisfy themselves with simpler seats. Only the chairman of the assembly – which came from a difference province every week – had a somewhat wider chair with armrests. He was seated in the middle of the length of the table, facing the windows. On

32 In case a province did not agree with a decision taken by the States General, it could register a divergent opinion. One example from the privilege business was the case of Jeremias de Brunes and his company, who obtained an exclusive privilege for making silver and gold wire; the States of Zeeland indicated "not to be able to consent" (*niet te kunnen consenteren*). Dodt, *Archief*, 6:362 (21 Jan 1615).

33 From 1593, it convened permanently, even on Sundays. Fruin, *Staatsinstellingen*, 183. The registry settled in The Hague as well in 1588. It consisted of two tenured clerks, one agent, one messenger, several translators and freelance employees. These employees were mostly house clerks who were privately employed by either Van Aerssen or Oldenbarnevelt. Knevel, *Het Haagse bureau*, 21.

34 For a very vivid description of the working conditions, see Knevel, *Het Haagse bureau*, 11–19.

FIGURE 5 View on the Inner Court and the Hofvijver in The Hague, in three plates

> his right-hand side there were six chairs, of which the first was reserved for the representative of the presiding province, and then three for those of Friesland and (since 1595) two for Groningen. On the left side, there were the six representatives for Holland. Facing them were the chairs for the representatives of the remaining provinces: six for Guelders, three for Zeeland, two for Utrecht and two for Overijssel. At the far-right end, the registrar was seated, who was the most prominent official of the States.[35]

The general assembly room was flanked by two anterooms (the *groote* and the *kleine vertrekkamer*) that were used, among other things, for commission meetings, possibly in connection with the examination of inventions as well. Moreover, to test inventions, commissioners would in some cases move to the official residence and headquarters of the Stadtholder (*Stadhouderlijk Kwartier*), which was located on a corner of the Inner Court (see Figure 5).[36]

35 "[…] de kamer werd gedomineerd door een met groen kleed bedekte langwerpige tafel met ronde uiteinden, die in het midden van de vergaderzaal stond opgesteld. Rondom deze tafel stonden achtentwintig stoelen, dertien aan elke zijde en één aan elk uiteinde. Het meest opvallend was nog de stoel van de stadhouder, met armleuningen en een met groen velours gestoffeerd kussen, gesitueerd aan het linkeruiteinde van de tafel, vlak voor de schoorsteenmantel en nabij de ingang van de kamer. De gedeputeerden moesten het met simpeler zetel doen. Alleen de president van de vergadering – die ieder week uit een andere provincie afkomstig was – beschikte over een bredere stoel met armleuningen. Hij zat in het midden van de lange zijde, met zijn gezicht naar de ramen. Aan zijn rechterhand stonden zes stoelen, waarvan de eerste bestemd was voor een afgevaardigde van het presiderende gewest, vervolgens drie voor die van Friesland en (sinds 1995) twee voor Stad en Lande. Links van de president zaten de zes gedeputeerden van Holland. Tegenover hem stonden stoelen voor de afgevaadigden van de overige gewesten: zes voor Gelderland, drie voor Zeeland, twee voor Utrecht en twee voor Overijssel. Aan het rechteruiteinde zat de griffier, de voornaamste ambtenaar van de Staten." Knevel, 14.

36 For instance, when testing of the new invention of the telescope, see pp. 91–92.

Constitutional Confusion: The States General Takes Control

During the first decades of the Revolt, the Dutch still presented themselves in official documents as subjects of the Burgundian-Habsburg overlord, whom they recognized as their formal leader.[37] Philip II was said to have been misinformed by his entourage about the real issues at stake. This meant that, as far as privilege practices were concerned, there was little urge to change the usual routine: privileges continued to be issued in the name of the overlord, and the control over privileged monopolies remained under the umbrella of the Collateral Courts in Brussels.

By signing the Union of Utrecht, however, it became increasingly clear that the rift between the Spanish Habsburg crown and the Dutch authorities was going to be a lasting affair. The Provinces went on the prowl for a new leader. After rejections by the English and the French monarchs, the revolt's leaders eventually decided to offer the leadership of the United Provinces to Duke Francis of Anjou of the House of Valois (1555–1584). Against that background the so-called Act of Abjuration (*Placcaat van verlaetinge,* 1581) was proclaimed, in which the Provinces declared that Philip II had not upheld his responsibilities to the Dutch population and hence had forfeited his right to act as the legitimate overlord.[38] The Act of Abjuration turned out to be a radical move in terms of thinking about political representation. But it also permanently changed the nature of the privilege system in the United Provinces of the Netherlands.

Changes did not come to light immediately. The Duke of Anjou, as far as we know, did not grant any invention privileges in his name during his short time in power (1582–1583). This was probably because the Duke was not given the title of sovereign during his appointment as lord of the Netherlands.[39] But when Anjou left the United Provinces in June 1583, the States of Holland saw their chance and quickly granted an invention privilege without referring to any other authority.[40] This was a perfect reflection of the distribution of power in the United Provinces, where Holland was pulling the strings.

---

37 For illustration, the University of Leiden (1575) was still founded with reference to Philip II.

38 For a more in-depth analysis, see Gelderen, "Freedom Fighters."

39 "In the negotiations at Plessis-lès-Tours, which made Anjou the lord of the Netherlands, the States Deputies under the leadership of Marnix van St Aldegonde refused to include the word 'sovereign' in Anjou's title, arguing that the word did not exist in Dutch language (which was completely nonsense)." Van Gelderen, *The Political Thought,* 180.

40 The privilege was granted to the city-carpenter of Delft, Cornelis Dircksz Muys. See NL-HaNA, States of Holland, 3.01.04.01, inv. nr. 334, fol. 186 [H2, 9 June 1583]. Holland had earlier extended a privilege that had been granted to the mason Corstiaen Anthonisz of

Meanwhile, the search for a new leader was under way. Plans were made to entrust the Stadtholder of Holland with high and supreme powers, but to complicate matters Prince William of Orange was shot on 10 July 1584. Then, Queen Elizabeth I of England (1533–1603) decided to step in by sending one of her suitors to the United Provinces as the new Lieutenant-general: Robert Dudley, 1st Earl of Leicester (1533–1588).

The appointment of Leicester was part of a vigorous power struggle. According to the 1585 Treaty of Nonsuch, whereby Elizabeth I agreed to support the Dutch, the English Crown reserved the right to appoint two councilors in a new Council of State that was presided over by the Lieutenant-General, who was in turn empowered to name the Stadtholder.[41] But to complicate matters for Leicester, the States of Holland and Zeeland had frustrated his claims to power by quickly appointing the eighteen-year-old son of William the Silent, Maurice of Nassau, as their new Stadtholder on 14 November 1585, shortly before the official arrival of the Lieutenant-general in the United Provinces (see Figure 7). It was the beginning of a strained relationship between Leicester and the Provinces that would last until the end of English rule.

The documents concerning invention privileges clearly reflect the tensed relations. On 24 November 1586, Leicester returned to England for consultation after ratifying an act that his Council of State should observe governance during the absence. The very same day, the States General decided to issue an invention privilege to the renowned engineer Simon Stevin (1549–1620) for a newly devised water mill, without referring to Leicester or his Council of State.[42] The privilege was later officially re-issued by the Council of State in

---

Delft in 1582, without making reference to any other sovereign: it illustrates the precarious position of Anjou that had existed since the beginning. On the period of Anjou in the United Provinces, see for instance Koenigsberger, *Monarchies, States Generals*, 298–304. An important event that marked the fall of Anjou was the so-called 'French Fury' of January 1583, when Anjou wanted to enforce his *Joyeuse Entrée* in Antwerp. He was entrapped within the city walls and some 1500 of his troops perished.

41 The Council of State besides Leicester was appointed on 10 February 1586; it consisted, aside from the two members appointed by the English Queen, of six members appointed by Leicester. For the original documents, see Whatley, *A General Collection of Treatys*, 2:84–97. See especially article 16–20 of the 1585 Treaty of Nonsuch.

42 Both privileges were inscribed in the *Rekenkamer* (Audit Office) of Holland on 15 September 1588 (*Camer van Rekening in Hollant, Witte register met de Rode roos*, fol. 51). On 23 August 1588, Stevin entered into a contract with his friend Johan Cornets de Groot (1554–1640), the later burgomaster of Delft, for the exploitation of his inventions. De Groot would have an equal share in the proceeds of the two privileges, and together Stevin and De Groot implemented the invention at several locations. Stevin, *Principal Works*, 5:13–14. On the fate and fortune of the mills invented by Stevin, see ibid., 5:309–412, especially 324–332.

the name of Leicester on 22 February 1588, nota bene at a time when the Earl was no longer in the United Provinces but in England to sign a kind of Act of Abjuration.[43] These details about the conferment of Stevin's privilege provide a first impression of how the control of profitable technologies was part of a bigger power game; they illustrate the direct interference on the English side to gain control over technological innovation in the Republic and the importance of the issue of sovereignty in this respect. Nonetheless, in terms of the overall administrative settlement, little changed. As was the case under Spanish-Habsburg rule, the Council of State played a central role in settling the introduction of novel technologies on Dutch territory.[44]

The president of the Council of State, Leicester, was for that matter well acquainted with the privileges system that had developed in England earlier in the sixteenth century.[45] He had taken out several privileges under his own name, including one for the exploitation of a salt mine (though the privilege for a theatre group, *Leicester's men*, is probably more well-known).[46] In England, Leicester had been directly involved in giving advice on new projects and the accompanying patronage practices around the court. In the Republic too, Leicester tried to support his clients, among them the Italian engineer, Federigo Giambelli.[47] Giambelli had played an important role in the Fall of Antwerp

43 On 17 December 1587, Leicester left for England where he signed a sort of "Act of Abjuration." But the Dutch States General only took notice of his actions on 1 April 1588. Meanwhile, the Dutch Council of State (formed by Leicester) continued to meet: the last meeting was on 12 March 1588, although incoming letters were still communicated among the members.

44 To be sure, I am talking here about the Council of State in the Northern Netherlands; the Council of State in the Southern Netherlands remained in function in Brussels. Leicester was assisted by his own Council of State as well as by a new Chamber of Finances (1586) with Jaques Reingoud (1530–1595) as the treasurer-general. Van Riemsdijk, *De Griffie van Hare Hoog Mogenden*, 5–8. Knevel, *Het Haagse Bureau*, 39. Oosterhoff, *Leicester and the Netherlands*, 93–97. Formally, he "asked advice" from the various institutions. In the 1588 privilege issued to Simon Stevin, for example, one reads that the Leicester had the "advice of those from the Audit office in Holland" (*advys van de luyden van de Reekeninghe in Holland*) as well as the "advice and deliberation of the Council of State standing beside us" (*hebben bij advys ende deliberatie van de Raede van Staete neffen ons wesende*). Stevin, *Principal Works*, 4:28.

45 The literature on the early days of the English patent system is voluminous, but see especially Gomme, *Patents of Invention*; Duncan, "Monopolies under Elizabeth I, 1558–1585"; Heal and Holmes, "The Economic Patronage of William Cecil"; Thirsk, *Economic Policy and Projects*.

46 Duncan, "Monopolies under Elizabeth I, 1558–1585," 361.

47 Information about Federigo Giambelli (also spelled Genibelli; Genebelli) can be found in Van der Aa, *Biographisch Woordenboek*, 7:174–175.

(1585) and entered the service of Leicester as his Postmaster General (*Postmeester Generaal*) in 1586. On 10 April of that year, he obtained a privilege from his patron for the invention of a drainage mill that "could be of great benefit for the Land, and especially Waterland [= an area north of Amsterdam]."[48] When the relationship between the Lieutenant-general and the Provinces quickly deteriorated in the course of 1587, the presence of Giambelli in Amsterdam invoked the belief that the Lieutenant-general might want to restore his power by forging an attack on Holland's major trading capital. The burgomasters of Amsterdam even decided at some point to send out a spy to keep an eye on the engineer.[49] But nothing happened, and Leicester left the United Provinces though the backdoor on 17 December 1587, once and for all.[50]

The departure of Leicester gave rise to a power vacuum that, again, directly reflected in the privilege practices.[51] As we just saw, the Council of State

48 This can be deduced from a later document: Federigo Genebelli "vesoek om te hebben approbatie van 't octroy van syne Exc. daerop. [...] daar het werck [...] tot grooten weldaet vanden Lande, bysunder van Waterlandt soude cunnen strecken soo heeft de Supplianten tot vorderinge van de gemeende beste ende om te mogen genyeten de vruchten van zyne invente ons ootmoedelijk gebeden hem te gunnen privilege [!] voor de tyt van thien Jaeren [...]." 6 February 1587, 3.01.04.01, inv. no. 342, fol. 280. The privilege issued by Leicester in Utrecht on 10 April 1586 was approved by the States of Holland on 24 March 1587, on the strong recommendation of Count Maurice and lieutenant-general Count of Hohenlohe. RSG, 5:610. Later that year, however, the relation between Leicester and the Provinces quickly deteriorated.

49 Van der Aa, *Biographisch Woordenboek*, 7:174.

50 After the departure of Leicester, Giambelli returned to England, where he was involved in a "number of large-scale projects in 1591, to the benefit of the realm, including plans to increase London's water supply and decrease its risk of fire." Harkness, *The Jewel House*, 155. In 1597, Giambelli offered his services anew to Sir Robert Cecil, proposing him to "assign me any share you please [in the benefits arising from certain inventions]. During the term of the privilege, I will not show my said inventions to anyone without first advertising your Lordship and showing you the drawings and models." British History Online, *Cecil Papers*, December 1597, 1–15. In 1598, the inventor reappeared again in the Netherlands, this time to obtain approbation from the States General for his privilege granted earlier by Leicester. RSG, 10:371 (30 May 1598). Giambelli was passed over with the excuse that he should first produce a model of his invention, and the privilege was probably never granted. Giambelli kept working in England and, amongst others, drew a plan in 1602 of Plymouth Fort on the instruction of the Privy Council.

51 In 1589, a Dordrecht-based inventor consortium tried to obtain a new privilege because they had heard that some cities would not respect their rights based on the argument "that the privileges issued by the Count of Leicester (being only Governor-General and not a Sovereign Lord of the Land) should not be allowed to exist, as it was not in his powers to distribute any privilege" (*dat d'octroyen by den Graeff van Leycester (zijnde maer Gouverneur-Generael ende geen Souverain Heere van den Lande) niet en soude moegen bestaen, ende dattet in zynen vermogen niet en soude zijn geweest, om eenich octroy te mogen*

initially took on the responsibility of issuing privileges for technological innovation. When on 24 December 1588, Albert de Veer, a secretary for the city of Amsterdam, petitioned for a privilege to protect his newly invented mill to cut Brazilian wood, he initially addressed himself to the Council of State as well. This was, after all, the most logical continuation of the existing situation. The Council of State then forwarded the petition to the States of Holland for their advice on the matter, which was not unusual either. But in their response to the Council of State, the States of Holland pointed out that in order:

> to obviate any misunderstanding, the Suppliant's petition and similar [petitions] ought to be referred to the States General, which has reserved such matters for itself as well as everything that concerns the police [= 'management'] of the Lands in general. Therefore, the Council of State is requested kindly to refer all such affairs to the States General and to the States of the provinces all affairs concerning such a province in particular.[52]

This was the reflection of a new political reality, in which the States General tried to establish a form of political unity by assuming some of the supreme powers that the previous landlords had enjoyed under the combination of all of their titles.[53] Thus, from 1588 onwards, all privilege applications relating to more than one provincial jurisdiction were dealt with by the States General. It established an entirely new practice with regards to technological

---

geven). They indeed obtained a new privilege: NL-HaNA, States General, 1.01.02, inv. no. 12298, fol. 15.

52 "[…] alle confusie en de misverstanden te verhoeden des suplliants verzoek ende dierlijken behoort gerenvoyeert te worden aan the Staten Generaal, die hetzelfde aan hen hebben gereserveerd ende ook alle het gund de generaliteit in het stuk van de politie van den landen is concerneerende zulks, dat die van de Rade van Staten al zulke zake aan de Staten Generaal zullen gelieven te renvoyeeren ende aan den Staten van de provincien in 't particulier in diergelijke zaken die den provincien in particulier zijn concerneerende." Doorman, *Octrooien*, 82–83 (16 November 1588). On 24 December 1588, Amsterdam was asked for their advise on the matter. RSG, 6:287. For the six-year privilege, see NL-HaNA, States General, 1.01.02, inv. no. 12298, fol. 3 [G3, 30 January 1589].

53 Holland's reprimand was based on the *Instruction for the Council of State of the United Netherlands* (*Instructie Raad van State*), dated 12 April 1588. Article 25 stated that: "[The Council of State] shall not grant any privileges nor continue any of those that have been granted by the States General for a certain amount of years after they have expired […]." (*Sullen geene Octroyen verleenen, nochte oock die by de Staten Generael voor sekeren tijd van jaren verleent zijn, na de expiratie van deselve continueren, nochte eenige pensioenen of beneficien tot laste van den lande, noch resignatien van Officien mogen toestaen*). Cau and others, *Groot placaet-boeck*, 4:122.

policy: instead of the Council of State, the States General now came to play a central role regarding the introduction of new technologies in the United Provinces.

The central link in this clever political move by the States General was undoubtedly the eminent statesman Johan van Oldenbarnevelt (1547–1619, see Figure 6). Oldenbarnevelt had been a pensionary of Rotterdam before becoming the Land's Advocate for the States of Holland on 16 March 1586.[54] This meant that he automatically got a seat on the States General as well.[55] The freshly appointed Advocate then quickly took charge in the political landscape. No decision was made without him, even if he was nominally just a servant of the States.[56] Notorious for his cunningness, Oldenbarnevelt was known for frequently re-editing the resolutions of the States General with the help of the registrar of the meeting, his friend and ally Cornelis van Aerssen (1545–1627). Alas, he was also known for his secretive attitude, and it is perhaps for that reason that no detailed documentation on Albert van de Veer's privilege application has survived.[57]

Yet one can still conjecture as to why Oldenbarnevelt might have wanted to sideline the Council of State. First of all, the Council of State was an institution in its own right, whereas the States of Holland could more easily put pressure up in the States General.[58] Second of all, Maurice had a permanent seat on the Council of State, whereas he was not eligible to vote in the States General. One might therefore consider to what extent Oldenbarnevelt consciously tried to limit the power of the Stadtholder in favor of the States General, and of Holland in particular, when it came to the introduction of commercially interesting inventions. Most importantly, however, it should be taken into account that the English Crown had maintained the privilege of having two permanent

54 Oldenbarnevelt succeeded to Paulus Buys (1531–1594), who had been arrested by Leicester for treason. The son of Paulus Buys, Cornelis Buys, would play an important role as a member of the General Chamber of Auditors of the County of Holland. Detailed information on the struggles between Oldenbarnevelt and Leicester can be found in Den Tex, *Oldenbarnevelt*, 1:289–435.

55 Den Tex, *Oldenbarnevelt*, 2:52.

56 Janssen, *Het stokje van Oldenbarnevelt*, 38. Sir Thomas Bodley, founder of the Bodleian library in Oxford and Elizabeth's I representative to the Dutch Republic, noted that "[The States General] proceed very slowly, unless those of Holland and especially the Advocate are present." Den Tex, *Oldenbarnevelt*, 2:52. See also Van Deursen, *Maurits Van Nassau*, 42; Fruin, *Staatsinstellingen*, 63. One might compare Oldenbarnevelt with the later "information master" Jean-Baptiste Colbert (1619–1683). See Soll, *The Information Master*.

57 For Oldenbarnevelt's secretive attitude, see Janssen, *Het stokje van Oldenbarnevelt*, 37–38.

58 For the tensed relation between Oldenbarnevelt and the Council of State, see Den Tex, *Oldenbarnevelt*, 2:29–62.

FIGURE 6
Portret of Johan van Oldenbarnevelt

FIGURE 7
Equestrian portrait of Maurice, prince of Orange

members in the Council of State after the departure of Leicester.[59] From a competitive point of view, it would therefore not have been wise to keep privileges under the umbrella of the Council of State, thereby informing the English of every new technological invention introduced in the Dutch Republic.

Scope and Validity

An important question in the settlement of early modern invention privileges was their exact validity. Official international agreements for the protection of intellectual property did not exist and, in contrast to modern patent law, there was also no notion of "novelty in its modern sense of new and original."[60] What mattered was that an inventor was the first to practice a certain craft within the jurisdiction of the privilege-granting authority; that he would be the first to reveal the technical details of an invention and thus render the invention useful to the local economy. Novelty, in short, was at all times geographically determined.

This makes it relevant to examine the exact scope of privileges as well as the associated nomenclature. Under the Habsburgs, invention privileges had been issued in the name of the overlord, who ruled by means of his separate titles.[61] Even if most of the privileges were valid in all of the Seventeen Provinces, each of the jurisdictions, where the privilege was valid, had to be separately mentioned.[62] In addition, the individual provinces apparently still had to give their permission for the admissibility of a privilege granted by the overlord. This would at least explain why certain privileges valid in the entire Netherlands were later issued separately for individual Provinces as well (notably Brabant and Land van Overmaze).[63]

59 Fruin, *Staatsinstellingen*, 109–11. This privilege was rooted in the Treaty of Nonsuch (1585).

60 Walterscheid, "Novelty in Historical Perspective (Part 1)," 692.

61 Charles V had created the title "Lord of the Netherlands" in 1555, but the position as sovereign ruler was a precarious one in which the king constantly had to negotiate his power with the local landlords and vassals.

62 The expression used in the documents was "par tous noz pays de pardesa" (K4; K5), "en noz pays de par deçà" (K7; K10), which indicated that these were arrangements valid in all the provinces of the Burgundian Netherlands (*de landen van herwaertsover*). The expression was used in contrast to the "Pays de par delà" (*landen van hierwaertsover*), which meant only the duchy of Bourgogne and Franche-Comté. On the ambiguous use of the expression of "pays de par deçà", see Cockshaw, "A propos des Pays."

63 For illustration, in the case of Frederik van den Sande, the States of Brabant had to give its consent. Doorman, *Octrooien*, 76. Other examples are K15, K16, K18, K19, 31-5-1570 and 4-4 -1570.

In addition to the general privileges that were valid in all of the Seventeen Provinces, there were also privileges that were valid in a specific jurisdiction only: the so-called 'special' privileges.[64] The inventor Roelant de Bruste, for example, obtained a privilege for using mills and pumps in copper mines, valid exclusively for Brabant. Also, the Company of Guillaume van Dorsten obtained a privilege, this time for the manufacturing of sulfide, that was valid only for Brabant.[65] These examples suggest that 'special privileges' were mainly used in relation to mining. In other cases, particularly when new inventions could be implemented at various locations, the inventor tried to amass a number of different privileges that were valid in the individual provinces. The Venetian inventor Paulo Pergolato, for instance, obtained a privilege on 4 January 1566 for the production of a greasy oil from a specific type of trees: his privilege was only valid in Burgundy, but he obtained two more privileges on the same day for Brabant and the Land van Overmaze.[66]

The practice of granting both general and special privileges continued in the phase of transition towards the establishment of the Dutch Republic. General privileges were valid in each of the seven provinces, plus the Generality Lands, whereas special privileges were only valid in the jurisdiction of one particular province. As Figure 8 attests, however, only the States of Holland managed to grant a fairly significant number of particular privileges in the period under discussion. As far as we know, only two other provinces, Utrecht and Zeeland, issued a number of privileges on their own behalf: though only a mere five privileges in total (over a forty-year period, which was meagre for instance in comparison to the patenting activity under the rule of Leicester, who issued three patents in only two years). Also in the field of technological advancement, Holland clearly assumed its prominent position of power among the other provinces. The main difference between 'general privileges' issued under Dutch rule, and the earlier privileges granted under Burgundian-Habsburg rule, was that the individual jurisdictions did not have to be mentioned separately: privileges granted by the States General were valid throughout the entire United Provinces.

---

64 Examples are [K15, 1570] for Brabant and [K19, 1582] for Flanders, Brabant and Land van Overmaze. Doorman, *Octrooien*, 75, 77. More detailed information on the legal history of special privileges will follow in Chapter 4.

65 Doorman, *Octrooien*, 73 [K14, 22 October 1569].

66 Doorman, *Octrooien*, 70 [K8, 4 January 1566].

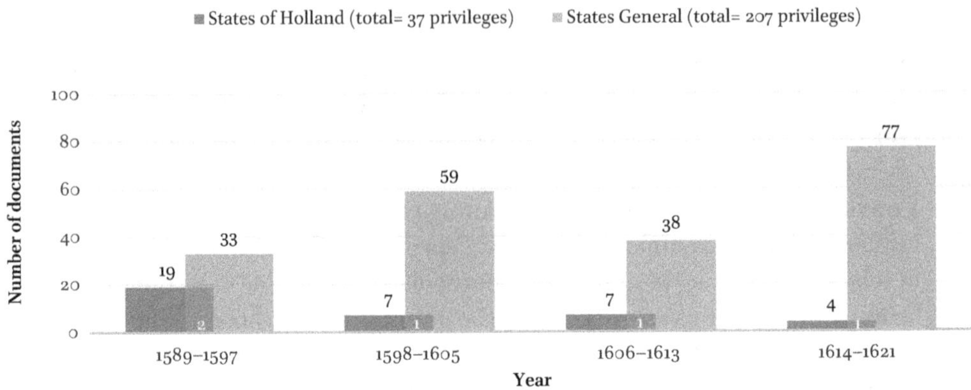

FIGURE 8 Total number of privileges issued by the States of Holland and the States General

The first privileges with a general validity in all of the United Provinces appear in the resolutions of the States General in the year 1584. This means that they were settled right in the period of the power vacuum between Anjou and Leicester. In fact, these first privileges in the resolutions of the States General were not original privileges. They were confirmations of privileges that had been granted earlier by the States of Holland. It seems as if, amidst the constitutional confusion, there was a need for general approval in order to confirm the validity of the privileges. For example, a privilege granted to Simon Stevin by the States of Holland on 17 February 1584 (for the exclusive exploitation of a new type of watermill) was confirmed by the States General on 22 February 1584. Simon Fransz van Merwen then obtained a privilege for a similar invention from the States of Holland on 10 April 1584, which was confirmed by the States General on 31 May 1584.

The practice of the 'confirmation' continued throughout the early years of the Dutch privilege system. A number of privileges granted by the States of Holland were thus later granted in general (see overlapping areas in Figure 8). Aside from being a form of reassurance regarding legal validity, there were other reasons why inventors preferred a general privilege to a special privilege. One inventor, for example, who had already obtained a privilege from the States of Holland, argued that he was

> concerned that this invention will be imitated and counterfeited outside of the Province of Holland and West Frisia [...].[Therefore] he felt obliged to come to us gentlemen. [He obtained a privilege from the States General] which conformed to the privilege granted by the gentlemen of

> the States of Holland etc. not to imitate, sell, or distribute [his invention on a penalty of 100 pounds of 40 groats].[67]

Considering the close proximity of the provinces, protecting an invention only within one province was somewhat futile. At any rate, the States General clearly took charge from 1588 onwards and the number of privileges issued by the individual provinces remained negligible until the 1630s.

Around that time, one starts to encounter a clause stating that a privilege was only valid provided the inventor obtained *attaché* (permission) from the province or city where he planned to install his business. Although the clause had also appeared on occasion earlier in the seventeenth century, it only became standard later on.[68] A new practice thus came about whereby the local authorities had to consent to the invention privileges granted by the States General. A major turning point was somewhere around the year 1633, when Hans LeMaire and Jacob Dircxz de Swart complained at the Court of Holland about infringement on a privilege granted to them on 5 August 1628.[69] De Swart was at that point the gilt-leather maker of the Stadtholder of Holland Frederick Henry of Orange (1584–1647), who had taken over after the death of his half-brother Maurice in 1625. His privilege for the "art to print flowers, honorable persons, animals and birds" on all sorts of leather had indicated that both His Excellency Frederick Henry and His Majesty of Bohemia had found such pleasure in seeing some examples of the art that they had desired to drape several of his rooms with the new material.[70] After complaining to the Court

67 "[...] grote costen [...] omme alles behoorlyck op zyn Italiaens te doen backen ende zy bedugt waeren dat deselven inventie buyten de Provincie van Hollant ende WestVrieslant nagemaeckt ende geconterfeyt souden wordden [...] genootsaeckt waeren hun tot ons te keren. [...] alles conform de voors bedingingen van octroy gegeven by de heeren Staten van Hollant [...] nae te maken, vercopen of te distribueren." NL-HaNA, States General, 1.01.02, inv. nr. 12299, fol. 117 [G79, 8 April 1603].

68 There had been four cases during the period 1581–1621.

69 The privilege had originally been made out on the name of De Swart. Hans LeMaire then tried to infringe upon the privilege of De Swart by bribing one of his workers in order to produce the machinery needed for printing figures on textile. Van Dilllen, *Bronnen*, 2:677 [no. 1215, 31 May 1629]. The two parties later decided to work together.

70 Jacob Dircxz de Swart was referred to as the 'gold leather maker of His Excellency' (*goude ledermaker van Syn Excelletie*) Prince Frederick Henry of Orange. On 5 August 1628 he obtained privilege for the "conste, van allerley bloemen, loffwercken personnagen, dieren gevogelte ende anders verheven ende gebootseert gewys sonder Jucht ofte swart te drucken, printen, ende formeeren, op vergulde silvere ende allerley gecoleurde Leeren,

of Holland, De Swart and LeMaire obtained a warrant "to cease the work of the citizens of Amsterdam" who had infringed upon the privilege.[71] But the Delegates of Amsterdam did not submit to this decision and raised the issue in the Provincial States. They demanded that "the privileges of the States General will have no effect in this province unless agreed upon by [the Provincial States]."[72] The States of Holland adopted the request, stating that they:

> do not agree because the privilege [of Jacob Dircxz de Swart] is granted without the knowledge of this province and also because other provinces [try to] attract this Maerten Swart [= the Amsterdam based copycat], promising to maintain him, in order to thus incite their business [...].[73]

The outcome of the story was "no longer to grant privileges at the Generality without prior communication with this Assembly."[74] Thus a new era began in

---

[...] waer van voor desen aen syne hoochgeduchte Excellentie eenige proefkens verthoont synde, hadde niet alleen syn Excellentie maer oock Syne Majesteit van Bohemen daer inne soodanich goet behagen genomen, dat de selve daer van behangsel tot eenige Camers hadde begeert, ende den Suppliant verder geanimeert ende vermaent de selve syne Inventie metten eersten int werck te stellen." NL-HaNA, States General, 1.01.02, inv. nr. 12303, fol. 287v [G286, 5 August 1628]. For the resolution, dated 4 August 1628, see Nijenhuis and others. *Resolutiën Staten-Generaal 1626–1630*. Incidentally, the 'Majesty of Bohemia' refers to Frederick V (1596–1632), the Winter King who was living in exile in the Republic during the years 1622–1632. De Swart had earlier obtained a privilege for printing "all kinds of persons, forests, hunting scenes, wild animals, birds, both with gold and silver alonse, and with color if needed" (*allerley personagien, historien, bossgaigen, Jachten, wilde dieren, gevogelte ende anders, soo met goudt ende selver alleen, als oock met aller coleuren na den Eysch van t werck seer cierlyck te drucken*). NL-HaNA, States General, 1.01.02, inv. no. 12300, fol. 435v [G127, 17 December 1613]. The privilege indicated that, after seeing several examples, "not only the Princely Excellency but also the Princess of Orange" had wanted to have several of their rooms draped with the material in question. Incidentally, in this case the 'Princess of Orange' referred to Louise de Coligny (1555–1620), fourth wife of William the Silent, stepmother of Count Maurice (not his wife), and a close friend of French Queen Mother Maria de' Medici (1575–1642).

71 "Mandement penael omme 't voorz werck vanden burgheren tot Amsterdam [...] te doen cesseren." Doorman, *Aanvullingen I*, 11. Doorman refers to a sentence by the Court of Holland on 10 November 1633 (Inv. No. 705 *Sententiën*).

72 "[...] de octroyen van de heeren Staten Generael geen effect sullen hebben omme in dese provincie gebruyckt te werden, tensy met consent van dese vergaederinge [...]." Ibid.

73 "Datte heeren Staten nyet verstaen dewijl dit octroy buyten kenisse van deese provintie gegeven is ende meede andere provintiën deesen Maerten Swart aenlocken, belovende hem te maincteneren, om alsso de neeringe te diverteren." Stellingwerf, *Notulen*, 6:468 (17 January 1634).

74 "ter Generaliteit geen octroyen te verleenen sonder voorghaende communicatie van deese vergaderinghe [...]." Stellingwerf, *Notulen*, 6:467 (17 January 1634).

the precious balance of power between particularism and central governance: privileges granted by the States General would first have to be confirmed by the Provincial States or else would be put aside.

The Role of the Cities

The Assembly of the States General in The Hague was filled with delegates from the provinces, who continuously held consultations with the local representatives and factions. Contrary to the situation in some other European sovereignties, the delegates to the States General were not entitled to make their own decisions without the consent of the individual States.[75] It resulted in a continuous back-and-forth between the central and local authorities.

The exact relationship between the representatives in the various local and interprovincial institutions was highly complex. Primarily during the formative years of the Dutch Republic, the theoretical discussion was thrashed out around the question of 'sovereignty': whether a new monarchy, an aristocratic oligarchy, or a *respublica mixta* should be introduced; whether the example of Venice or Switzerland should be followed; or whether an entirely new form of government should be conceived that combined various elements of representation.[76] Bookshelves are full of tomes written about this subject, and yet no satisfactory answer has ever been found to the sovereignty question in the early Republic. Historians of political theory often take a seminal text by François Vranck (1555–1617) as a point of departure; Vranck argued that from medieval times the sovereignty was with the nobles, magistrates and town councilors, and not with the (Provincial) States.[77] On the other hand, it was argued even back then that "The cities in Holland and Zeeland are not

75 See also Koenigsberger, "The Power of the Deputies."

76 The literature on this topic is too big to list exhaustively, but see De Bruin, "De soevereiniteit"; Van Deursen, "De Raad van State en de Generaliteit"; Van Deursen, "Tussen eenheid en zelfstandigheid"; Kossmann, "Volkssouvereiniteit"; Mout, "Ideales Muster"; Weststeijn, *Commercial Republicanism*; Van Gelderen, "Aristotelians, Monarchomachs and Republicans"; Van Gelderen, "Het gemenebest"; De Bom, Lesaffer, and Thomas, *Early Modern Sovereignties*.

77 Vranck, *Short Exposition* (1587), in *The Dutch Revolt*, 227–38, in particular 233–234. In his famous *Short Exposition*, Fançois Vranck argued against the English councilor Thomas Wilkens (one of the English members of the Council of State) and against the sovereignty of Leicester. For the context in which his remarks were made, see Ibid. xxvii–xxviii; Koenigsberger, "The Power of the Deputies," 204–7. I believe that the remarks by Vranck are sometimes all too readily taken as a true representation of things, instead of being part of a cunning plan to dispose of Leicester.

sovereign as some cities in Italy, etc. [...] Sovereignty has not been made common by the Union but has remained with the provinces without talk about the sovereignty of the towns."[78] Without going into all the aspects of the theoretical discussion, a glance at the privilege system will demonstrate that there is probably no straightforward answer to the question of who wielded the most power in the Republic. Rather, it was a question that changed on a case-by-case, subject-by-subject, and sector-by-sector basis. Because although the cities certainly granted individual privileges, for example for appointments or specific exemptions, it was apparently not the case that the cities themselves awarded invention privileges that imposed a ban on counterfeit on the basis of the condition of novelty. Very occasionally one comes across something that resembles a "patent", but in general the cities spent most of their time regulating the local market by means of pure trade monopolies. The Dutch policy of promoting new inventions was in this respect less unique than previously thought, and there is no question of any consistent independence of the cities when it came to privileges for inventions.

This does not mean, in any sense, that cities were any less important. Aside from the fact that the most of the political power games surrounding privileges played out within the bounds of city walls, inventions served the needs of city dwellers above all. The city, for instance, figured prominently in the case of Daniel Note, who had:

> found and invented a certain Invention [= a hose pipe] that will clean all the Streets in the Cities of dirt, and which will take away many causes of fire, and which is particularly useful to silence and extinguish the fire in some houses or cities in case they are on fire [...]. [Moreover, he had found another] Invention to take away all the evil smells of the jakes in the houses, Rooms, and other places, from which in populous cities and places great diseases arise.[79]

78 "De Steden van Holland, en Zeeland zyn niet Souverain als eenige Steden van Italië, enz. [...] In het I. Kap leerd de Groot dat de Souveraniteit by de Unie niet is gemeen gemaakt, maar gebleven bij de respective provincien; zonder dat 'er van een souveraniteit van de steden gerept wordt." Van der Schelling and Van Zurck, *Codex Batavus* [Rotterdam, 1758], 703.

79 "Oock daerenboven gevonden ende geinventeert seeckere Inventie daerdoor alle Straten in de Steden sullen werden gereynicht van alle vuylicheyt, waerdoor oock veele oorsaecken ende occasien van brant sullen werden wech genomen, ende Sonderlinge dienstich, zoo wanneer den brant in eenige huysen ofte Steden souden mogen syn omme den selven brandt te stillen ende uuyt te blusschen, ende dat alles door seecker wagen Sleed ende water werck daertoe ende meer andere dienstlickeden bequaem, ende lestelyck gevonden eene Inventie omme alle quade luchten van de secreten in de huysen, Camers, ende

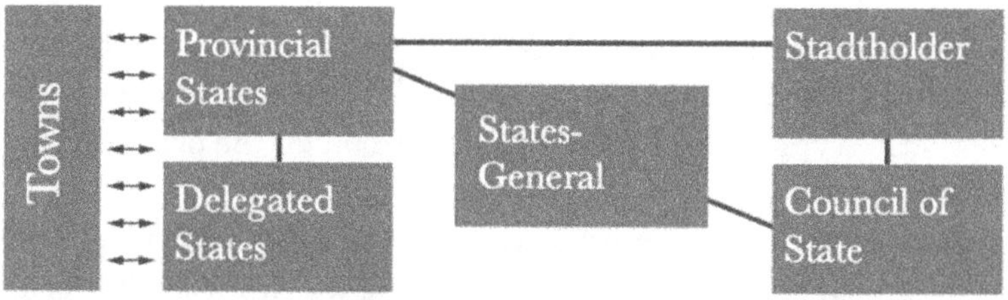

FIGURE 9 Schematic overview of the main political bodies

The city gave rise to new demands: fire extinguishers, for example, but also soot filters to clean the air produced by 'factories', or installations to supply sufficient drinking water. The stakes were often high for urban authorities supporting a particular invention.[80]

It was also easy for the cities to present their interests, because the entire Dutch state apparatus was grafted onto urban interests (see also Figure 9).[81] Where the Assembly of the States General consisted of representatives from the Provincial States, the Provincial States consisted of nobility (represented with one vote in the case of Holland) and the cities (represented with no less than eighteen votes in the case of Holland). Relationships in the Republic were therefore already determined by the cities by means of the voting ratios. The cities were, for that matter, governed by a council and a group of magistrates. Practices differed from place to place, but often the magistrate consisted of four burgomasters and a variable number of aldermen who assisted them in the day-to-day management of the city. Magistrates only served for one to two years and were elected by and from the council, whose members were elected for life. The council, which had 10 to 40 members, only convened to discuss important matters, usually on a weekly basis or even less.

One of the things the urban authorities were particularly interested in, was the implementation of new technologies that promoted employment. Authorities would often close special deals with inventors to ensure that this type of businesses would be set up on their premises, offering accommodation, laborers, tax exemptions, and suchlike. In return, the inventor promised to work

andere plaetsen wech te nemen, waer uuyt in populeuse Steden ende plaetsen groote sieckten causeren." NL-HaNA, States General, 1.01.02, inv. no. 12302, fol. 65 [G165, 16 February 1618]. On the carreer of Note, see also van den Heuvel, "Tot meerder bewijs".

80 For an example of a fresh water supply system, see NL-HaNA, States General, 1.01.02, inv. no. 12298, fol. 143 [G31, 28 May 1596]. For an example of building trade, see NL-HaNA, 1.01.02, inv. no. 12300, fol. 232 [G108, 12 February 1612]. For examples of inventions meant to reduce (the complaints about) coal pollution, see Dodt, *Archief*, 6:372, 5:262.

81 See also Prak, "Urbanization".

exclusively within the bounds of the city walls.[82] How this negotiation process worked in practice can be illustrated by examining the deal struck between the inventor Claes Jansz Wytmans and the city of Rotterdam for the production of a new type of window glass.

Wytmans obtained an invention privilege from the States General for the erection of a glass 'factory' (*Glashuis*) on 9 January 1614.[83] At this juncture, no deal had yet been struck for the large-scale production of his invention. On 29 January 1614, however, it appears that Wytmans had approached the city council of Rotterdam and petitioned for several parcels to erect his factory on-site as well as a "freedom of city excises, journeys, and mounting guard, for both the master and the workmen."[84] The city fathers then decided to send out two representatives "to deal with the suppliant [...] in order to obtain the most profit and benefit for the city."[85]

On 2 February 1614, a "concept of the agreement" between Claes Janz & Co. and Rotterdam was read aloud to the Aldermen. Wytmans had "obliged himself to execute the business of window glass for the duration of the privilege, namely 20 years, nowhere else than here in this city."[86] In return, he would obtain a building lot and 800 guilders, as well as "the freedom of journeys and

---

82 See also Davids, "Beginning Entrepeneurs," 173. "When a city government decided to give support to beginning entrepreneur, the parties usually entered into a formal contract specifying the rights and obligations on both sides." One should think of cheap loans; lower physical capital costs; lower costs of raw materials; and the free use of buildings. See also Klein, *De Trippen in de 17e eeuw*, 37–38. The custom of city authorities to attract inventors with special deals was nothing new or exceptional to the Dutch Republic, by the way. See, for instance, Molà, "States and Crafts," 136.

83 NL-HaNA, States General, 1.01.02, inv. no. 12301, fol. 1v [G129, 9 January 1614]. The privilege contained an exploration period of one year. Claes Jansz Wytmans was a skilled painter who, among others, had produced a stained glass window representing *The Adulterous Woman in Front of Christ*, which was presented as a suitable gift by the Aldermen of Rotterdam to the St. Jans Church in Gouda. Information about the case of Wytmans (also spelled Witmans or Wystmans) is mainly derived from Obreen and Scheffer, *Rotterdamsche Historiebladen*, 1:740–53.

84 "[...] vrijdom van Stadts exchijnsen, tochten en wachten voor den meester ende arbeijders [...]." Minutes of the city council, as cited in Obreen and Scheffer, 1:741.

85 "[...] om inspectie te nemen van de plaatsen ende metten suppliant te handelen opt welbehaegen van de Heeren vroetschappen so sij ten meesten dienste ende profite van de stadt sullen bevinden te behooren." Ibid.

86 "medebrengende dat denselven Claes Jansz ende syn Compagnie haer verbonden de voorsz neeringe van huijs ofte veijnsterglas den tyt van haer octroy, naementlick den tyt van xx jaeren geduyrende, nergens anders als hier ter stede te mogen exerceren ofte doen exerceren ende alle contraventien naer haer uiterste vermooghen affdoen." Ibid., 742. The deal was made after further consultation with a number of 'carpet-makers' (*Tapijtwerckers*).

mounting guard [...] but not of property tax."[87] The final contract followed on 5 February.[88] It clarified furthermore that Wytmans was under the obligation to "employ no less than eleven person for the duration of the privilege"[89] and that Wytmans had:

> committed himself by this [contract] not to make or have made any window glass, whether within or outside of the United Netherlands other than in the city of Rotterdam, but [that he] would try to preclude and prevent with all his powers in case something like that is done by someone else.[90]

This fragment of text is interesting because it shows how cities extorted a type of exclusiveness that was not, as such, part of the privilege grant. More often than not, inventors would first obtain a privilege from the States General and then use that to negotiate with individual urban authorities. When a certain Caspar Benoist, for instance, obtained a privilege for the implementation of a new type of silk weaving mill, the central authorities indicated that the privilege would only take effect "provided that Benoist erects his mill within five months to come in any one of the United Provinces, and gives proof thereof by means of reliable attestation."[91] A similar twist can be found in the privilege of Jacob van Aaken, who got permission to erect a new type of water mill for

87 "Meester en Arbeiders [...] genieten vrijheijt van tocht ende wacht, mitsgaders van stadts wijn - ende bierexcijns, maer niet van verpondinge." Ibid. It was furthermore stipulated that Wytmans would acquire the ownership of the parcel (*eijgendom van deselve erven sal hebben ende behouden*) in case he would fulfill his obligations. However, if Wytmans would not keep to the contract for any reason, or if he would not fulfil his full twenty years of service, the parcel would return to assets of the city.

88 Ibid., 746. The final contract was signed by all the members of Wytmans' Company: Wytmans himself, Lambrecht de Hoogh, and Henrick Carelsz van Cracouw ('from the city of Cracow'). The burgomasters, the municipal architects (*fabriekmeesters*), and two commissioners of the city council (*vroedschapsleden*) signed on behalf of the city of Rotterdam.

89 "[...] continueren gedurende de hele tijt van haer Octroy, houdende tot dieneijnde in gestaedigen arbeijt nijet minder als elff persoonen." Ibid., 745.

90 "[...] verbonden gelijck hij hem verbindt mits desen geen huijs ofte veijnsterglas te maecken, te doen ofte laten maecken in eenige plaetsen, 't sij binnen ofte buijten de Vereenichde Nederlanden anders als binnen de stat van Rotterdam, maar soo wanneer sulcx bij yemandt anders soude geschieden tselve bij alle middelen sooveel in sijn ende sijns compaignies macht is te verhinderen ende te beletten, [...]." Ibid., 744. About the later proceedings of the glass factory, see Ibid., 748–754.

91 "Mits dat hy Benoit gehouden wordt ten lancxsten binnen den tydt van vyff maenden naestcommende den voors. moeten te stellen in een van de vereenichde Provincien, ende daervan te doen blycken by behoorlycke attestatie [...]." NL-HaNA, States General, 1.01.02, inv. no. 12299, fol. 188 [G88, 23 August 1604].

melting metals "in all places in these united Lands."[92] This, then, was the general procedure: the inventor would first obtain his privilege 'in general' from the States General and then strike a deal with particular cities or provinces where he was to implement his invention.

Still, it was a two-way thing. In other situations, the inventor had already managed to sign an implementation contract before obtaining a privilege, or he had been employed by urban authorities to find a solution to a particular problem. When, for example, the inventor Cornelis Eeuwoutsz Proot obtained a privilege for a new method of ringing church bells, it was explicitly noted that "the Magistrate of Middleburg in Zeeland had requested [this] from him."[93] In addition, recommendation letters were often an implicit sign that the city authorities wanted to develop a certain technology. When Pieter Pietersz Enten, for example, obtained a privilege on 22 February 1602 for a new method to pump up water, both the City of Amsterdam and the City of Haarlem decided to recommend the use of the invention.[94] The dynamic of the Dutch privilege system was thus entirely based on a continuous interaction between local authorities and the central government seated in The Hague.

The close interaction between the cities and the central authorities led to a system in which knowledge was easily diffused. But, as one can imagine, and as we shall see repeatedly throughout this book, the practice of luring inventors could lead to fierce competition, and even conflict, between the different cities too. A pertinent example in this regard is that of Dordrecht merchant Jacques de Cagnioncle, who petitioned the States General for a privilege for the exclusive production of saltpeter on 2 March 1618.[95] After intensive consultation with

92 "[...] aleene in alle plaetsen deser vereenichde Landen [...] te mogen opmaken ende opstellen [...] onder sulcke voorwaerden ende condition als hy [...] sonder consent express van hem Suppliant." NL-HaNA, States General, 1.01.02, inv. no. 12298, fol. 368 [G60, 28 July 1601]. The application is available in NL-HaNA, States General, 1576–1796, 1.01.02, inv. no. 7474.

93 "[...] van Weghen den Magistaet der Stede van Middelbrugh in Zeelandt versocht wesende [...]." NL-HaNA, States General, 1.01.02, inv. no. 12299, fol. 40 [G69, 22 February 1602]. Proot was the ammunition master in the city of Geertruidenberg. The invention was ready three years after Middelburg had asked for a solution; the inventor had "labored with great expense and effort" and asked "to be compensated and reimbursed" (*gearbeit met grote kosten en moeite* [...] *ons daarvan vergoet ende gerembourssеert te worden*).

94 RSG 12:321; NL-HaNA, States General, 1.01.02, inv. no. 12298, 391v [G68, 22 February 1602]. A later request for prologation of the privilege was denied: RSG NR 2:586. See also Van Dillen, *Bronnen*, 2:33 [no. 64]. Enten's drilling method was a topic that reverberated in scholarly discussions for more than 50 years; it was discussed by Simon Stevin, in the correspondence between Marin Mersenne and Constantijn Huygens, and came up in a lecture that Huygens delivered for the the Royal Society in 1663. Van den Heuvel, "Mapping Knowledge Exchange", 98; Van den Heuvel, *De Huydbou*, 75–77, 475–478.

95 Dodt, *Archief*, 7:23. The States General decided that further research was needed prior to issuing the privilege.

Stadtholder Maurice of Nassau, the army commander-in-chief, the privilege was issued on 10 March 1618, for the duration of twenty years.[96] One year later, De Cagnioncle reappeared in the resolutions of the States, requesting 10,000 guilders to put his invention into practice. In return he would deliver 100,000 pounds of saltpeter free of charge.[97] The States turned down this request and, for a while, De Cagnioncle disappeared from the official documents.

Then, on 21 January 1627, a complaint was lodged by the Delft-based company of Willem Willemsz van Lijnschoten.[98] The company claimed that De Cagnioncle was hindering them in their business of saltpeter production based on the privilege that had been issued to him in 1618. Van Lijnschoten maintained that the Dordrecht merchant had not kept to the condition in his privilege to provide a sample of his work within a year, and thus had forfeited his claim to the privilege. In the absence of concrete results, the Lijnschoten company had felt free to petition for a privilege themselves, which indeed they had obtained from the States General on 9 September 1624 (for the duration of thirty-one years, see Figure 10).[99]

To resolve the conflict, the States General commissioned its representatives Nicolaas van den Bouchorst (for Holland) and Simon van Beaumont (for Zeeland) to hear both parties and so to find an agreement.[100] After a month or so, they reported that it had not been possible to find a satisfactory solution; yet the States insisted they try again. Shortly after, the States General received an official letter drawn up by the mighty city of Dordrecht, strongly urging them to uphold the De Cagnioncle's privilege.[101] After further interference by the commissioners, it was then decided that De Cagnioncle would be allowed to use his privilege next to, and for the same duration, as the company of

96 NL-HaNA, States General, 1.01.02, inv. no. 12302, fol. 58v [G 168, 10 March 1618].

97 Dodt, *Archief*, 7:77 (19 July 1619).

98 In a Rotterdam notary archive, mention is made of the company of Willem Willemsz van Linschoten, the younger, his brother Matheus Willemsz van Linschoten, their father Willem Willemsz van Linschoten, the older, and their brother-in-law, Adryaen Jacobsz Balbiaen, who together produced gunpowder during the years 1623–1626. All profits and losses were divided by four. Information from Jansssen, "Maria".

99 For the documents regarding the application of the Company of Matthias and Willem Willemsz van Lijnschoten, see NL-HaNA, States General, 1.01.02, inv. no. 12579.21.

100 Nicolaas van den Bouchorst was a representative to the States General for Holland. He was a member of the Knighthood of Holland, bailiff and dike reeve of Rijnland, lord of Noordwijk. Dr. Simon van Beaumont was a representative to the States General for province of Zeeland. He was also a poet and the pensioner of the city of Middelburg. NNWB, 1:265 (column).

101 Letter dated 15 March; as Noortwyck and Beaumont were on the case, they were summoned to investigate the letter further. For the resolution, dated 23 March 1627, see Nijenhuis and others. *Resolutiën Staten-Generaal 1626–1630*.

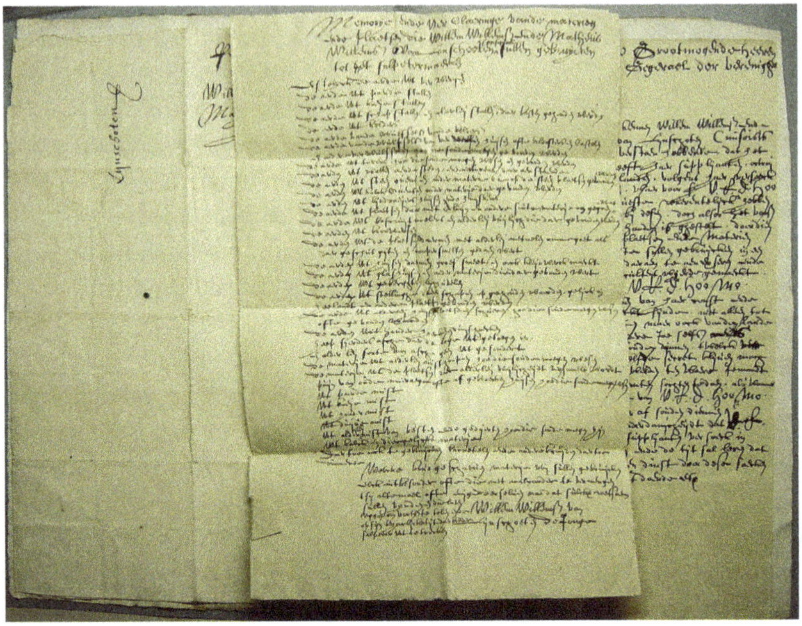

FIGURE 10 Van Lijnschoten's list of ingredients for the production of saltpeter, 1624

Lijnschoten.[102] Moreover, the Dordrecht merchant was given the opportunity to limit the validity of the Lijnschoten's privilege in the province of Holland to south of the rivers Maas and IJssel for a fee of 500 guilders. However, this would only be allowed in the event that Lijnschoten & Co decided not to let Cagnioncle partake of half of the profits resulting from their privilege.[103] The further outcome of the case remains unclear. Yet the case of *De Cagnioncle v Van Lijnschoten* is an excellent illustration of how various authorities directly interfered with conflicts over invention privileges and how urban authorities pushed their own interests at the expense of neighboring cities.

To conclude this section on the validity of invention privileges, a few words should be said on the scope of the Dutch legal regime. Generally, the prohibition on counterfeiting privileged inventions extended to all the "inhabitants of the Lands", in other words the seven provinces in Northwestern Europe. Yet there were a couple of occurrences where the validity of a privilege specifically

102 Resolution dated 23 March 1627. Ibid.

103 The company of Lijnschoten was free to decide whether it wanted to share part of their privilege; yet, the final decision regarding the settlement was up to the jurors. Those jurors would also decide whether either party was to obtain similar conditions in case the States General would decide to prolong the privilege upon expiry.

extended to the inhabitants of the Dutch colonial empire.[104] One of those instances concerned the case of Dirck Pietersz, a shipwright and resident of the City of Amsterdam, who had obtained a privilege for a device that would prevent ships from sinking or capsizing. The invention, which he had found with "great labor, effort, and research by the grace of God," could not be counterfeited "for the time of thirty years with the interdiction to all inhabitants, and others from these United Provinces, as well as those Subjects of these Provinces in the East and West Indies, and other Lands being under our territory."[105] The penalty for "making or imitating, either entirely or in parts, or importing [the technique] if imitated outside the United Netherlands" was fixed at 1000 guilders.[106] In addition, any ships that made use of this invention would be confiscated too, and it was forbidden for the technique to be "transplanted from here to other Lands."[107] A similar inclination towards secrecy can be found in the case of three Amsterdam-based inventors, who obtained a thirty-year privilege for the production of a new type of ship that served "to find and discover unknown and foreign Coasts, Rivers, Bays, and sand bars."[108] The exclusive exploitation rights extended to both the inhabitants of these Lands as well as to our "subjects in the East and West Indies as well as elsewhere."[109] Since the authorities considered it "important that strangers [...] did not find

104 As far as I am aware of, the first privilege protecting an instrument for use in the New World appeared in 1524, in Habsburg Spain. Barrera-Osorio, *Experiencing Nature*, 60.

105 "groote arbeyt, moeijte, ende ondersoek door Gods genade gepractiseert ende gevonden hadde de Conste, middel, ende gelegenheyt [...] voorden tyt van dertich eerstcommende Jahren, met Interdictie aen alle de Inwoonders, ende andere deser Vereenigde Provincien, als oock die Ondersaten van dese Provintien in Oost ende WestIndien, ende andere Landden synde ende frequenterende die onder ons gebiet syn staende [...]." NL-HaNA, States General, 1.01.02, inv. no. 12302, fols. 211v-212 [G194, 11 March 1621].

106 "int geheel ofte ten deelen na te maecken, ofte doen namaecken, ofte elders buyten de Vereenigde Nederlanden nagemaeckt, in deselve te brengen, om gebruyckt te werden van gelijcken oyck de Ondersaten van dese Landen in Oost ende Westtindien, ende andere Landen zijnde, ende frequenteerenden, die onder ons gebiedt syn staende, de voorz. Inventie, ofte stucx vandeselve verandert, in dese, ofte van hier in andere Landern te planten [...]." Ibid.

107 Ibid.

108 "omme te doene vliegende tochten, ende volck te Landen, omme alle vrembde ende uuytheemsche Custen, Rivieren, Bayen, ende droochten op te soecken, ende te ontdecken [...]." NL-HaNA, States General, 1.01.02, inv. no. 12302, fol. 206 [G192, 16 December 1620]. The invention is mentioned in Baudaert, *Memoryen ofte Cort Verhael*, 146–148. See also p. 107.

109 "aen alle de Inwoonderen ende andere dese Vereenichde Provincien hanterende/ als oock aende Ondersaten van dese Provincien in Oost ende WestIndien, ende andere Landen van dese synde en frequenterende die onder ons gebiedt syn staende [...]." Ibid.

out about the details of this invention and art," they authorized the magistrates of those places where the boats were made to administer an oath of secrecy from the craftsmen involved in the construction. [110]

Conclusions

The Dutch system of invention privileges emerged in the mid-1580s, that is to say, at the height of the Dutch Revolt. For the most part, it was a continuation of the system that had existed under Burgundian-Habsburg rule, except that privileges were no longer granted by the overlord but by the States General of the United Provinces. The States General thereby assumed a form of sovereignty that previous rulers had enjoyed under the combination of their different titles. The possibility to grant a privilege 'in particular' certainly created some room for the individual provinces to develop their own privilege policy but, in reality, most privileges were dealt with directly by the federal authorities.

The fact that the States General claimed the authority to issue privileges 'in general' suggests that, from 1588 onwards already, there was a clear concept of territorial unity. This contradicts the idea that the Dutch Republic was initially just a ragtag mess. Historians have a tendency to emphasize the independence of the Provincial States and the merchant cities. They then use this observation to argue that the 'modern' state does not necessarily have to be a centralized state and that decentralization "far from being the disruptive element in an otherwise 'modern' picture was a precondition for the extraordinary achievements of the Golden Age."[111] But the privilege acts tell a different story: they tells us that the States General, although in other domains nothing but a paper tiger, assumed central authority when it came to the introduction of new technology.

This is not to say that local authorities were bypassed; there was continuous consultation and it was precisely those fairly short lines between the different authorities that ensured that both applications and complaints were dealt with in a relatively short space of time (especially compared to today, when the time between applying for and being granted a patent can run into years). This can be seen as well in Figure 11, which shows the lapse of time between the application and the allotting of the final privilege both at the States General and the

110 "Ende alsoo dese Landen daeraen is gelegen dat dese Inventie ende konste nyet comme tot kennisse van vreemde, sondelinge deeser Landen vyanden [...]." Ibid.

111 Tracy, *The Founding of the Dutch Republic*, 314.

States of Holland. The cities formed the basis of the Dutch state structure and gave leadership to the central authorities. On the other hand, local power was kept in check by the corrective workings of central government. Good communication was essential for the smooth running of the system. To this end, both the language and the procedure for acquiring a privilege was highly standardized. And yet there were always exceptions in the Dutch *ancient regime* thinking about state management. It is precisely this tension that will be discussed in the next chapter, where we will cast our eye over the question of how privileges were administrated in theory and practice.

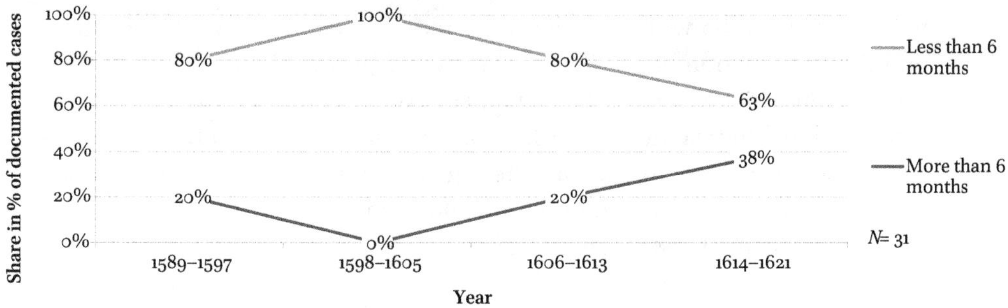

FIGURE 11 Total pendency of privilege applications in the Dutch Republic

CHAPTER 2

Administration and Management

This chapter sketches an overview of the administrative practices entailed in the procurement of a privilege. Since most of the privileges were granted by the States General in The Hague, the emphasis in this story will be on the central authorities; that said, the procedure at the States of Holland was in many ways pretty similar. Following the route of a standard application, attention will be given to the written request, the recommendation, the specification, and the examination of privileges. The second part of the chapter discusses the areas in which privileges were applied, exploring the various options that were available to inventors to cash in on their ingenuity. Finally, the conclusion makes an attempt to link the different elements together, enabling a more varied and deeper understanding of the role that invention privileges played in the Dutch system of state management.

The Application Procedure

The Written Request

The administrative procedure for obtaining a privilege was set in motion when an inventor filed a request with the States General in which he set out the reasons why he considered his invention worthy of privilege.[1] This request was often accompanied by a letter of recommendation written by those who wanted to implement the invention, and it frequently contained a sketch or a model that showed how the inventor planned to put his ideas into practice. The States General then examined and discussed the request *in camera*, sometimes demanding further clarification from the inventor and sometimes setting up a commission to inspect a prototype of the invention.

1 Privilege requests from before 1668 have been sporadically preserved in the archives of the States General. For the period 1581–1621, only some incomplete series for the following years remain: 1600–1601 (NL-HaNA, 1.01.02, inv.nr. 7474); 1603–1611 June (Ibid., inv.nr. 7475); 1611 July – December (Ibid., inv.nr. 7476); 1617–1618 June (Ibid., inv.nr. 7477); 1618 July – December (Ibid., inv.nr. 7478). Although the official procedure was only set in motion when the inventor handed in a petition, in some cases he had been asked by the authorities (both on an urban and on a provincial level) to develop a specific technology. In those cases, the act of petitioning was thus more or less just *pro forma*.

DOI:10.1163/9789004320420_004

Incoming requests all had a similar form and structure as determined by law. They began with a short description of the matter at hand, then proceeded to the reasons that would justify the privilege, before concluding with a formulation of the actual request.[2] Following that, there were two ways for a request to go through to the assembly of the States General: either through the registrar office, or by directly handing in the request to the chairman of the assembly. In most cases, the request first passed through the registrar office.[3] The registrar then added comments and gave his recommendation on the admissibility of the request by means of a so-called *apostille*, a special form of annotation in the margins of the request (see Figure 12).

Interestingly enough, inventors could hand in their request both at the registrar of the States General or at the registrar of the Council of State. The choice for either institution depended on the subject matter of the invention. Whereas the registrar of States General typically dealt with privileges concerning civilian life, the registrar of the Council of State was the right institution to turn to with applications concerning inventions in the field of military technology.[4] Another important element in the settlement of military inventions was that they were usually examined under the personal guidance of the Admiral General and Captain General of the Union, Stadtholder Maurice of Nassau.[5]

2 Merula, *Manier van procederen*, 293. Requests had to be drafted in Dutch: when Guido Baldo Bergamino petitioned for a privilege, it was adduced that he should draft his petition in Dutch and address it to Du Maurrier. RSG, 4:103 (20 April 1619).

3 For a detailed account of the registrar office in The Hague, see Van Riemsdijk, *De griffie van Hare Hoog Mogenden* (1885).

4 The registrar of the States General at the turn of the seventeenth century was Cornelis van Aerssen (1545–1627, in office 1584–1624), a former pensionary of the city of Brussels. Van Aerssen was a powerful man in the small Republic and known as a close ally of Oldenbarnevelt. Foreign representatives tended to approach Van Aerssen as a first point of call to access the higher echelons of those in power in The Hague and he played a central role in the sphere of internal affairs, too. As a registrar, Van Aerssen's duty was to reply to any incoming letters, and with that he could wield freely to accept or deny whatever he wanted. Although Van Aerssen was known to be a decent man, this form of freedom made some of his successors – especially his direct successor Cornelis Musch – susceptible to heavy corruption. Knevel, *Het Haagse bureau*, 57. On Van Aerssen, see Janssen, *Het stokje van Oldenbarnevelt*, 42; Den Tex, *Oldenbarnevelt*, 2:37; Fruin, *Geschiedenis der staatsinstellingen*, 156–158. The registrar of the Council of State, on the other hand, was Cristiaan Huygens Sr. (1551–1624), who was the forbear of an influential Dutch civil servant dynasty. Huygens Sr. had formerly been a member of the court entourage of William of Orange, before he took up his duties as registrar from 1584 until his death in 1624. On Huygen's family dynasty, see Stoffele, *Christiaan Huygens – a Family Affair*.

5 On het States army and Maurice's role as a military leader and strategist, see Puype and Wiekart, *Van Maurits naar Munster*; Rothenberg, "Maurice"; Stratmann, "De militaire hervormingen."

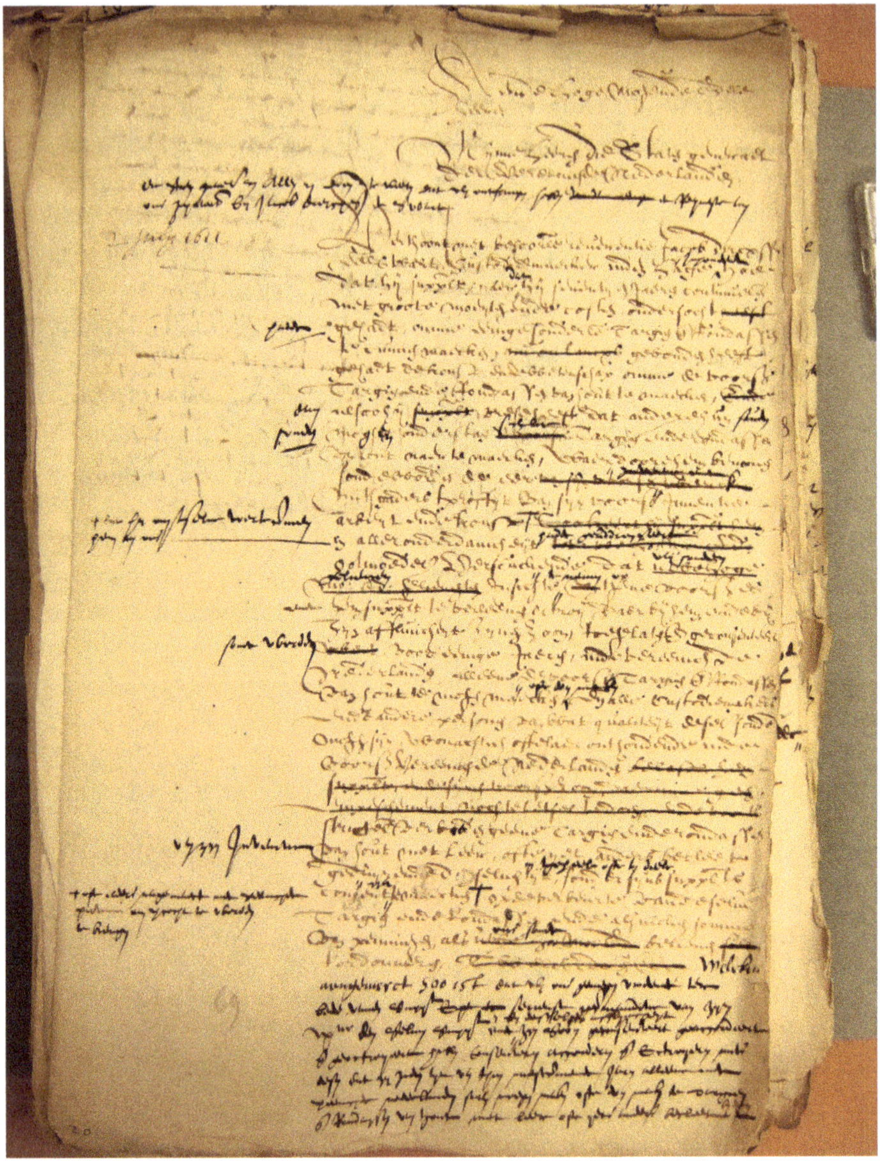

FIGURE 12 Annotations on a privilege request

Maurice not only examined inventions related to new bullets or weaponry, but also had a lively interest in corn mills that could help feed the army, inventions that saved on fuel, bridges, or machines that could drain the moats of cities under siege. To give just one example of Maurice's role, let us highlight the case of the inventor Bartholomeus Roelants, a tin caster from Delft, who invented a grenade that could:

> kill and eliminate [...] up to hundred men with one shot, of which [the inventor] had done a test at the direction of his Excellency in Den Bosch and targets around here in the presence of the Captain of the guard of his Excellency, Mr. Stevin, Verdoes, and the Generals of the Constable [≈ artillery master on ships, particularly of the VOC] who had been committed to this task by His Excellency.[6]

Other examples mention, as well, how privileges were granted after "wide and extensive communication with His Princely Excellency" Maurice of Nassau.[7] Thus, whereas the States General ultimately issued the privilege and was vested with the highest authority, the Council of State under the direction of Count Maurice was responsible for the settlement of privileges in the field of military technology.

In addition to the standard route of the registrar's office, inventors could hand in their request to the chairman of the States General.[8] He was the only one entitled to bring in topics directly. The chairmanship of the Assembly circulated among the seven provinces, changing every first day of the week. Nonetheless, the system was very fitful, which makes it difficult to determine the degree of direct influence by the chairman on the decision-making process in the Assembly. A comprehensive sample of two randomly chosen years shows that, of nineteen privilege applications that were filed in 1610–1612, twelve were dealt with under the presidency of someone who represented either Holland or Zeeland (see Table 1). It turned out to be impossible to provide any definitive evidence that it mattered who the chairman was.

Still, there are sporadic indications that the chairman did influence the decision-making process in the States General. When Thomas Grammaye, for

6 "[...] wel hondert man minder, oft meerder soude connen schadeloos ende om den hals brengen, daer van hy de proeve door last van syn Excellentie In den Bosch, ende doelen alhier in presentie van den Capitein van de guarde van Syne voors. Excellentie, Mr. Stevin, Verdoes, ende den Generaels van de Constapels (die daertoe by Syne Excellentie waeren gecommitteert) hadde gedaen." NL-HaNA, States General, 1.01.02, inv. no. 12298, fol. 356 [G55, 26 May 1601]. Roelants obtained a privilege for the duration of eight years.

7 NL-HaNA, States General, 1.01.02, inv. no. 12302, fol. 205v [G192, 16 December 1620].

8 Whereas incoming letters were registered by the date of reception (*receptum*), those presented directly in the meeting were dated using the term *exhibitum* or *lectum*. Incoming documents were kept in so-called *liassen* (meaning "bonded together"), which were ordered according to subject. On the back of a *lias* the registrar would write the name of the supplicant, the type of petition, and sometimes some additional information on the subject. Van Riemsdijk, *De griffie*, 120–25. Material that was considered too big to be included in the normal dossiers (such as models) were kept in separate boxes, now filed under *ingekomen delen en dossiers, afzonderlijk verzameld in series*. Van Riemsdijk, *De griffie*, 126–30.

TABLE 1 Requests filed during the years 1610–1612

No.	Date	Invention	Inventor	Archive SG	Duration (asked)	Remarkable details	Presidency
1	15-1-1610	Cannon balls and a mirror telescope	Johan Hoevelmans (from Frankfurt)	RSG NR, 13	Denied	– Van Dorth (Gelderland) and Warmond (Holland) receive commission to examine the invention together with some engineers and pyrotechnicians. – On 21 January and 1 April follows a negative report.	**Gheissen** [Gelderland]
2	16-4-1610	Drainage watermill	Piedro Sardy, engineer from Italy	fol. 163v [G 104]	10 (20)	– Recommendation from the ambassador of the French King.	**Magnus** [Zeeland]
3	24-12-1610	Mill operated without water or wind	Johan Orschou, millwright, and others	fol. 216 [G 105]	10 (16)	– Inventors were citizens of Amsterdam	**Poelgeest** [Holland]
4	8-2-1611	Printing patterns on textile fabrics	Jan Andriesz Moerbeeck and Co. from Amsterdam	fol. 229v [G 106]	5 (16)	– Two privileges on one day (see below)	**Magnus** [Zeeland]
5	8-2-1611	Horse-mill for forging, etc.	Samuel Marolois, mathematician	fol. 233v [G 107]	7 (10)	– Maralois' invention was made at the request of Pieter le Febvre and Thomas le Maire, merchants residing at Amsterdam, both being his brothers-in-law.	**Magnus** [Zeeland]

No.	Date	Invention	Inventor	Archive SG	Duration (asked)	Remarkable details	Presidency
6	12-2-1611	Method to build houses	Mario Brizzy, Italian nobleman from Sienna	fol. 232 [G 108]	7 (10)		**Magnus** [Zeeland]
7	8-6-1611	Lance-shields	Jacob Diercxz. de Swart, 'weaponry maker' (*custodiemaker*)	fol. 265 [G 109]	10 (some)	– Based in The Hague; later inventor in the service of Stadtholder Frederic Henry	**Magnus** [Zeeland]
8	9-7-1611	Determination of longitude	Thomas Leamer, English merchant in Amsterdam	fol. 266v	If useful, a compensation		**Duyck** [Holland]
9	3-9-1611	Medicine to be prepared from rosemary	Johan de Gantelmoes, French dokter	fol. 288v [G 110]	2 years	– Recommendation by Stadtholder Maurice	**Velsen** [Friesland]
10	1-10-1611	Economical oven	English inventor consortium	fol. 281v [G 111]	7 (20)	– Edmund Lassels, Andrew Palmer, and George Waymouth in the application (kept in archive) – Eduward Woodier, Edmont Lassels, George Waymouth, John Beare, John Waymouth, and Abraham Willelms in the privilege	**Oldenbarnevelt** [Holland]

TABLE 1 Requests filed during the years 1610–1612 (*cont.*)

No.	Date	Invention	Inventor	Archive SG	Duration (asked)	Remarkable details	Presidency
11	8-10-1611	Silk Production	Jan de la Bat, based in Middelburg	fol. 294v [G 112]	20	– Inventor based in Zeeland – Obtained permission for implementation on 31-5-1612 from Zeeland, but had a privilege from Holland and Friesland	**Magnus** [Zeeland]
12	21-12-1611	Making stone from clay	John Edrington, Thomas Michaelfield and Joseph Skelton	fol. 305 [G 113]	9 (21)	– Inventors are Englishmen based in Rotterdam	**Coenders** [Groningen]
13	11-1-1612	Ditcher	Aert Henricxz, resident of Delft	fol. 316v [G 114]	9 (35)		**Magnus** [Zeeland]
14	26-1-1612	Artificial marble	Hendrik de Keyser, Town architect of Amsterdam	fol. 320v [G 115]	9 (20)		**Haga** [Overijssel]
15	18-2-1612	Efficient mill	Herman Becx and Aernout Duyfkens	fol. 313v [G 116]	9 (25)	– Oldenbarnevelt was only president for that day, replacing a colleague; the inventors are from Amsterdam	**Oldenbarnevelt** [Holland]
16	31-3-1612	Drainage mill	Johan Thyll, former paymaster of the Company of Troopers of the Lord of Asperen	fol. 348v [G 117]	7 (20)		**Oldenbarnevelt** [Holland]

No.	Date	Invention	Inventor	Archive SG	Duration (asked)	Remarkable details	Presidency
17	21-4-1612	Corn mill, operated by wind	Andries Maerschalck and Nicolaes van Steenwinckel, residents of Middelburgh in the province of Zeelandt, and Frans Jansz. Pieck, resident of Rotterdam in the province of Holland	fol. 340 [G 118]	7		**Coenders** [Groningen]
18	23-5-1612	Longitude determination	Courdt Boddeker, citizen of Bremen.	fol. 342v	If useful, a recompense	– The delegates Bas and Symsz. (Holland) Joachimi (Zeeland) or "other members that wish to do so" will first speak with the inventor.	**Gerritsz** [Overijssel]
19	14-7-1612	Soot filter	Pieter Cool of Solingen, ex-smith, resident of the City of The Hague	fol. 358 [G 119]	7	– Res. Holland, fol. 165: The inventor had the support from Holland to get a privilege from the States General.	**Biesman** [Gelderland]

example, a scion of a powerful family in Guelders, requested a privilege on Saturday 27 July 1619 to protect an invention that would reduce the use of fuels, the chairman of the States General, a deputy for the province of Groningen, noted in the resolutions that the request was to be turned down.[9] After the weekend, on Monday 29 July 1619, the presidency changed, and a delegate for the province of Guelders took over the chairman's gavel.[10] The very same day, the request by Grammaye, rejected two days earlier, was approved without further objection.[11] There are more examples of this type that hint at the direct influence of the chairman: on Wednesday 13 March 1601, for instance, a representative for Holland was presiding over the assembly and three privileges for inventors based in Holland were passed in one day.[12]

One of the more striking charateristics of privilege applications was their personal tone. Petitioners regularly stated that their invention only came about as the result of blood, sweat, and tears. They were not afraid to portray rivals in the field as people who would do them a grave injustice if they were not protected by the law. The privilege application was in fact a supplication, not a normal request, and rhetorical language could help to persuade the authorities to take an inventor under their wing. The use of a submissive tone was for that matter not only reserved for inventors from a lower social class; it was used in equal measure by honorable gentlemen, and even by inventors who worked for the state. The famous engineer and former burgomaster of Alkmaar, Adriaen Anthonisz (1541–1620), by way of illustration, recounted in his petition that he had "brought to light after long practice, effort, and brain racking, a certain instrument that can be called an Astrolabium Armillare […] with which deviation of the magnet can be easily understood."[13] The inventor had "from

---

9 Dodt, *Archief*, 7:78. The chairman was Goosen Schaffer (1601–1637).

10 That day, the chairman's gavel was in the hands of Mr. Calfken, who was burgomaster of the city of Nijmegen and delegate to the States General for the province of Guelders from July-December 1619, and again July-October 1620.

11 NL-HaNA, States General, 1.01.02, inv. no. 12302 fol. 114v.

12 NL-HaNA, States General, 1.01.02, inv. no. 12298, fols. 343, 344, 346 [G52, G53, G54, 13 March 1601]. The chairman was Mr. Egmond.

13 "[…] na lange pratique, moeyten, ende hoofdbrekens gevonden ende in het ligt gebragt hadde sekere instrument dat genoemt moegen worden Astrolabium Armillare [sic!] daerby de sonnes hoogte volgt beschrijving […] afwijkinge van de magneet […] lichtelyk begrepen." NL-HaNA, States General, 1.01.02, inv. no. 12299, 111v [G77, 7 March 1603]. Despite the apparent ease of use, one needed a manual to understand how the invention worked. Anthonisz supplied one, obtaining a ten-year privilege for the protection of his invention "entirely or in parts, with or without the book." Adriaen Anthonisz was probably the most famous fortification engineer of the Republic. He was the first to apply the principles of the so-called 'Dutch fortification system', a variation of the *trace italienne*.

a young age practiced in geometry and astronomy" and "had brought to light many compasses, quadrants, and other instruments, useful for navigation."[14] It was only because he feared "that the new invention ... (provided by the gift of God) would forthwith be counterfeited and imitated, which would leave him frustrated from the fruits of it" that "he felt obliged to ask for a fifteen-year privilege."[15] He obtained a privilege for ten years on 7 March 1603.

The flattery of the subordinate party in the petition, and the haughty tone of the response, was not just a question of a difference in status: it also shows that the form of the request was determined by the superior party. By accepting the form as suitable, the subordinate party indirectly conceded that it recognized and respected the higher power. The difference between the duration of the legal protection requested by the petitioner and the significantly lower term of the privilege as awarded by the authorities, must be interpreted along similar lines. It was not simply a question of having a go. Nor was it that the authorities did not trust the petitioner or that the authorities carefully weighed up the decision concerning the duration of the privilege with a view to the monopoly position of the inventor and the accompanying price control.[16] In some cases the inventors did not indicate a duration for the requested privilege at all, but simply asked for protection "for a few years" (*voor eenige jaren*).[17] Giving the superior party room to make a decision was a way of paying them respect, and ultimately acknowledging their power. Figure 13 gives an overview of the difference between the number of years requested by the inventor and the eventual duration of the privilege as awarded by the authorities.

---

14 "[...] jonge leeftijd geoeffent inde geometrie ende atronomie, mitsgaders met compassen, Gratbogen, ende andere, mitsgaders instrumenten voor dienst en nut des zeevarts [...] in het ligt gebracht [...]." NL-HaNA, States General, 1.01.02, inv. no. 12299, 111v [G77, 7 March 1603].

15 "[...] (naer de gave van Godt verleent) dat vele ingenieurs haar profijt ende studie daarvan makende [...] by andere terstondt nagemaekt ende geconterfeyt soude worden ende hy daardoor gefrustreert soude blyven van vrugten syner voors. Inventie [...] genoodzaakt ons te verzoeken 15 nakomende jaren [...] int geheel ofte deel met boekxcks off sonder boekxcks na te maken [...]." Ibid. The privilege for "producing, publishing, and selling" (*maken uitgeven ende verkopen*) the invention had a validity of 10 years, and the fine was set at 100 pounds of xl groats for every counterfeited instrument. The invention could not "be imitated to be used or distributed in any manner" (*nagemaakt om gebruikt of gedistribueert te worden in enigerlei manier*).

16 That this is not the case can also be proven by the fact that the autorities also granted less years than originally requested by the inventor in case they determined the price for which an inventon was put on the market. Compare pp. 149–150.

17 NL-HaNA, States General, 1.01.02, inv. no. 12299, fol. 220.

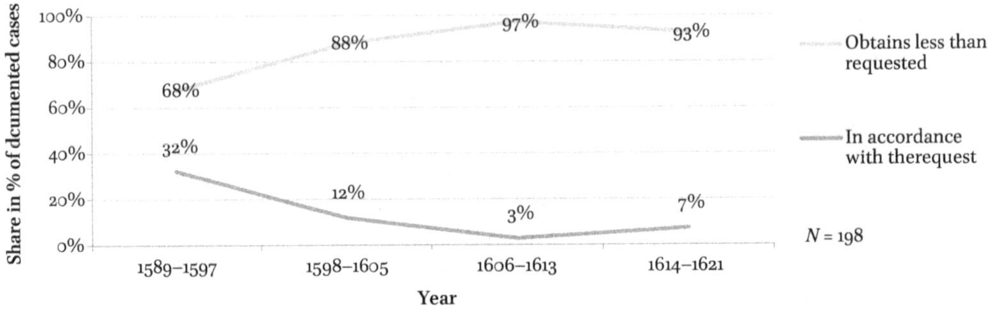

FIGURE 13 Discrepancy between the years asked and granted in terms of privileged protection

The Recommendation

Although, in theory, everyone was free to approach the authorities and present an invention, there were at all times many possible obstacles to obtaining a privilege. These obstacles ranged from unwilling administrators who delayed the application process to questions of legal representation. Women, for instance, were almost entirely absent in the official privilege documents, except when they acted as the widow of a deceased male inventor. The widow of a certain Jan Jansz Clock, for example, obtained a six-year privilege for an instrument to dredge under water. In this case, the water board of Rijnland had attested that the instrument worked to their satisfaction and that the widow had made an adequate test "with the help of her son (who had finished the aforementioned instrument that had been invented by his father and which had almost been ready)."[18] So, women only obtained the rights for the exclusive exploitation of inventions if those rights were appointed to them in one way or another through their male partners. A salient example in this context was the case of Gysbert Jansz Keyser, who in a moment of repentance, bedridden "because of bodily illness, yet in full possession of his faculties, memory and sense," declared in the presence of a notary that the privilege he had obtained from the States General actually belonged in equal part to his colleague Jan Hendricxz Soop, the "master of the glass furnaces of the city of Amsterdam."[19]

18 "met hulp van haren zoon (die t voors. Instrument by synen Vader geinventeert ende byna opgemaeckt wesende, voorts hadde opgemaeckt)." Dodt, *Archief*, 5:3 (10 August 1607, fol. 330).

19 "Keyser alhoewel sieck van lichame te bedde leggende nochtans sijn sinnen, memorie ende verstant, als dat uuytwendelijck gebleeck, wel gebruyckende, verklaart dat het

In an attempt to finally set things right (or rather as part of a clever business strategy) Keyser at last wanted to transfer half of the rights and revenues of the privilege to his partner, on the express condition that "if [...] Keyser passed out of this world his housewife would take his place as if he were alive."[20]

Conversely, there were elements that increased the chances of obtaining an exclusive privilege right, beginning with the submission of suitable expert testimonies that could confirm the utility and efficacy of a new invention. Those testimonies, which had to be given to a notary, were not necessarily provided by members from high society but rather by men knowledgeable of the topic in question (more information about the legal preconditions will follow in Chapter 4). As an illustration, the irrefutable benefits of a new type of copper mill for which a privilege had been requested were attested by:

> Leon de le Tombe, sworn in broker, about 51 years old, and Isaac Moors, coppersmith, [who] declare on behalf of the honorable Harman Becx and Aert Duyfkens that the copper [produced by the inventors Becx and Duyfkens] is as perfect as that from Aken and other places, where it used to be made up till now.[21]

---

octrooi dat hij 23 juli heeft geobtineert hem in feite maar voor de helft toekomt." Van Dillen, *Bronnen*, 2:56. [No. 104, 16 Augustus 1613]. For the privilege of De Keyser, see NL-HaNA, States General, 1.01.02, inv. no. 12300, fol. 417v [G124, 23 July 1613].

20 "[...] onder conditie off d'voors. Gijsbert Jansz Keyser deser werelt geraecte te overlyden, dat alsdan sijn huysvrou in sijn plaetse sal staen ende doen in allen schijn hy in 't leven waere, gelijck d'erffgenaemen van de voors. Soop naer sijn overlyden sijn plaetse meede (re)presenteren sullen." Van Dillen, *Bronnen*, 2:57.

21 "Leon de le Tombe, geswooren maeckelaer, out ontrent 51 jaeren, ende Isaac Moors, coperslaeger, ontrent 32 jaeren out, [...] ten versoucke van d'eerzame Harman Becx ende Aernoult Duyffkens, mede burgers alhier, [...] verclaert, getuyght ende geattesteert, hoe waer is, dat de coperwercken, soo keetels, beckens, bodems ende andere, die de producenten op d' voors. meulen sijn maeckende, slaende ende gietende, alsoock van root geel maecken, goet en schoon sijn ende soo volcommen in alles, als die van Aecken ende ander plaetsen, daer d'selve voor desen altijt gemaect plegen te werden, hier te lande commen. Voor redenen van wetenschap verclaerden sy, getuygen, [...] vercocht heeft tot glycken pryse, als hy andere, ja van Aecken ende elders gecommen waeren, verkocht hadde, [...]. Van Dillen, *Bronnen*, 2:17 (23 November 1612, no. 39). The Leon de le Tombe (also: Levy de Latombe, buried in Amsterdam on 6 August 1630) was a refugee from Antwerp. For more information about his highly interesting family network, see *The Montias Database*, Inv#. Lot 733.0272. Aert Duyfkens (also spelled Arnold/Aernoud/Aernout/Arnold Duyffkens) was a copper mechant from Hasselt; Herman Becx came from Cologne and was related to the family of Isaac Moors from Aachen. Klein suggest that the knowledge of this invention originally came from a master craftsman from Aachen who had been employed by the inventors shortly before. Klein, *De Trippen*, 325.

In addition to finding the right experts, having the right contacts in the right place could also help to secure a privilege. Inventors regularly made use of 'knowledge brokers' who would further their case with the relevant authorities.[22] Thus, it is known that the famous Dutch Golden Age poet and composer Constantijn Huygens Sr. (1596–1687), who served as secretary to three Stadtholders with an active career spanning most of the seventeenth century, helped a wide range of inventors, both foreign and domestic, to have their invention examined.[23] He interfered with the privilege for his son's pendulum clock[24] and was actively involved in a mission to help Galileo Galilei (1564–1642) obtain a Dutch privilege for his invention of finding longitude.[25] Less well known is that Huygens also discussed upcoming privilege applications relating to technological innovations with enemy authorities[26] and that he was actively approached by inventors to look after their affairs in the Republic. One of those inventors was engineer Andreas Reusner, who tried to canvass Huygens in order to obtain a privilege for his invention of a water work "which I am willing to show to you, but which is otherwise secret."[27]

Aside from the inventions that were recommended by brokers and experts in the field, a number of inventions were explicitly recommended by high-level stakeholders and interested parties (see Figure 14). One group that frequently

22 For a more in-depth discussion of knowledge brokers in the Dutch privilege system, see Buning, "Brokered Innovation", For an insight into the dynamics of brokerage practices on the Italian peninsula, see also the excellent study by Zanetti, *Janello Torriani*, in particular pp. 238–251.

23 About the role of Constantijn Huygens (1596–1687) as a middleman at the court of Stadtholder Frederick Henry, Prince of Orange (1584–1647), see Broekman, "Constantijn Huygens," in particular chapter 3, pp. 125–175. The States General occasionally called on Huygens as an examiner to give advice on upcoming privileges. For the advice by Huygens concerning a number of music pieces that were to be printed under privilege, see Groenveld and Hartog, "Twee musici." For a discussion of the role played by Huygens in securing a French privilege for the printed works of Descartes, see Clarke, *Descartes*, 138–141.

24 On Huygen's privilege, see for instance Iliffe, "In the Warehouse," in particular 41–52; Stoffele, "Christiaan Huygens," 111–18.

25 Huygens, *Briefwisseling*, 2:227 [No. 1536].

26 On 4 June 1643, Huygens wrote on behalf of Mr. Buyx to Guillaume de Bie, the registrar of finances in Brussels. He asked to support the privilege application of the inventor: "Il me dit que c'est un octroy qu'il veut demander de quelque sorte de moulins à eau, qui peuvent estre de grand service à desseicher maraiz." Huygens, *Briefwisseling*. 3:385 (no. 3266, Huygens to Guillaume de Bie, 6 April 1643). Huygens was related to both the inventor and the addressee (whom he addressed at other occasions as 'Monsieur mon cousin'). Other instances relating to inventions in the letters by Huygens include letters no. 3101, no. 3094 and no. 5276.

27 "Graag zou ik octrooi hebben voor een waterwerk, dat ik u wel wil laten zien, maar dat overigens geheim is." Huygens, *Briefwisseling*, 3:297 (17 June 1642).

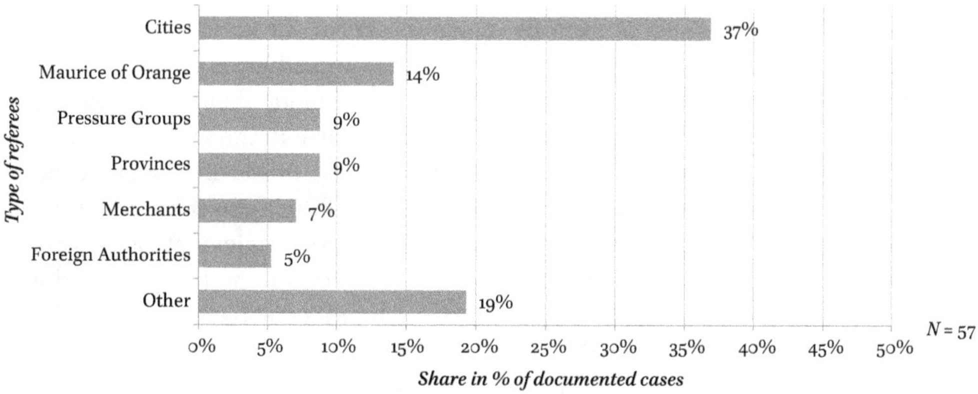

FIGURE 14 Categories of recommendation

reoccurred in this context were evidently the "merchants" (*Coopluyden*). Isaak Lieversz, for instance, obtained a privilege for making cinnabar on the basis of urine, showing a letter of recommendation from the Amsterdam magistrates as well as "the attestation of several respectable Merchants who traffic in the product."[28] There are numerous other examples of merchants receiving support from urban authorities. For example, when the inventor Isaac Burger obtained a privilege for a new type of economical peat oven, he submitted a recommendation letter from both the magistrates of Delft (dated 9 February) and Amsterdam (dated 6 February). The recommendation from the Amsterdam magistrates had been issued on the request of Burger, after the inventor had presented a letter in which a number of Amsterdam-based merchants indicated that they would be well served by the invention. On 7 March 1613, Burger signed a contract for the implementation of his invention with one of these merchants, Andries van den Broeck.[29]

Indeed, the vast majority of recommendation letters came from the urban authorities. These letters, as we have seen, were often a sign that the inventor had managed to sign a contract for the local implementation of his invention. This did not mean, however, that the privilege was then granted as a matter of course. When, for example, three inventors from Amsterdam presented a

28 "attestatie van verscheydenen treffelycke Coopluyden die daer mede handelen." NL-HaNA, States General, 1.01.02, inv. no. 12302, fol 16v [G158, 5 May 1617]. This general concern returns in several attestations.

29 RSG NR, 2:23; Dodt, *Archief*, 5:272. For the privilege, see NL-HaNA, States General, 1.01.02, inv. no. 12300, fol. 374v [G120, 13 February 1613]. Van Dillen, *Bronnen*, 2:32 [no.62] contains the contract for the exploitation of the invention. The interested parties were Andries van den Broeck, Nicolas du Gardin, Laurens Joosen Bake, Hans van den Berge(n), Reynier Verspreet en Abraham Anthonisz.

letter of recommendation from the burgomasters of Amsterdam to privilege "a certain art, invented by them, to print flowers and all kinds of colors on all kind of satin, silk and other textiles," the States General denied the request "because it is understood, that this invention has been in the world for a long time already."[30] A peculiarity of the Dutch state system was not just that the central authorities were restricted in the exercise of their power by the strong position of the various cities, but conversely that the individual cities were also accountable to the States and were not at liberty to act at will. Thus, the local and interprovincial authorities kept each other in check in a carefully constructed balance of power.

In addition to prospective local customers, foreign authorities and in particular ambassadors would also on occasion step in to support particular inventors. The Italian engineer Piedro Sardy, for instance, obtained a privilege from the Dutch authorities in April 1610 for his invention of a drainage mill driven by horsepower, which he had earlier "presented in France in the city of Paris."[31] The King of France had granted Sardy a twenty-year privilege for his invention; now the Dutch authorities decided to grant him a ten-year privilege as well. According to the Dutch privilege, however, Sardy "was held to leave his invention to the councils and cities who desire to use it for the sum of three hundred guilders once-only, and to individual inhabitants for the sum of fifty guilders."[32] Sardy was not best pleased with this decision, and after a couple of days the States General agreed:

> on serious intervention and recommendation of the Ambassador of the King of France [...] that the clause included in this privilege, which limited and restricted passing on the knowledge of the invention of the

30 "Gelesen eenen brief van burgermeesteren ende regeerders der stadt Amstelredam, van xx. deses, ten faveure van Jacob de Milder, Gerrit Lukas ende Wilhem Boer, inwoonders der voorsz stede, ten eynde hun soude worden geaccordeert octroy voor sekere jaeren, tot sekere conste, by hen geinventeert, omme op allerleye stoffen van satynen, seyde linten ende anders, te drucken, bloemwerck van alderhande coleuren, maer alsoo verstaen wordt, dat dese inventie lange inde werelt is geweest, is 't voorsz versouck afgeslagen." Dodt, *Archief*, 5:257 (27 March 1612).

31 "in Vranckryck ende gepresenteert hebbende binnen de Stadt van Paris." NL-HaNA, States General, 1.01.02, inv. no. 12300, 163v [G104]. Piedro Sardy (also spelled Pietro Sardi) later became known for his *L'artiglieria; divisa in tre libri* (Venice, 1621).

32 "[...] dies sal die Suppliant gehouden syn zyne Inventie over te laeten aen de Collegien ende Steden die deselve sullen begeren te gebruycken voor de somme van drye hondert guldens eens, ende aen elcken particulieren Inwoonder voor de somme van vyftich guldens eens." Ibid.

> engineer Piedro Sardy [...] to a certain sum [...] will be omitted, and not considered, thus that he can make his invention for the fare [he desires].[33]

High-placed foreign officials brokered deals for their clients to secure the most profitable exploitation rights for their inventions. Yet brokerage and courtiership certainly existed on a local level as well. The Stadtholder of Holland, for instance, repeatedly wrote letters of recommendation to support particular inventions. Besides letters concerning the maintenance of army material, some explicitly related to inventions that would enhance the prestige of his court.[34] One example was the invention by Passchier Lammertijn (see Figure 53), who obtained a twelve-year privilege in 1601 for the:

> Invention to weave in anything that can be written with a pen; an art in which His Excellency [Count Maurice] had found good contentment and pleasure, after having been shown a small [example] of his coat of arms [...].[35]

Although perhaps not as exuberant as in other countries, the court life of the Stadtholder clearly played a role in stimulating innovations in the Republic.

Because of the peculiar composition of the political landscape in the Republic, patronage practices extended beyond the strict boundaries of the Stadtholder court. The States General, as a collective, repeatedly acted as the benefactor of inventors too. One example of the many would be the case of Willem Kick (see Figure 15, more extensively discussed on pp. 201–202). After

33 "[...] op de serieuse Intercessie ende recommandatie van den heere Ambassadeur des Conincx van Vranckrycke, ende omme andere goede consideratien, geconsenteert ende geaccordeert [...] dat de clausule begrepen in dit Octroy, daerby den prys van het overlaten van de wetenschap van de Inventie van Piedro Sardy Ingenieur daer Inne verhaelt is gelimiteert ende gerestringeert aen seker somme In t selve Octroy ende oick anderssints gehouden sal worden als daervuyt gelaten, ende niet geconsereert, sulcx dat hy deselve syne Inventie tot syne costen sal moegen maken." Ibid. The privilege was dated 16 November 1610; the addition dated 18 November 1610.

34 About the court of Maurice, see Magreta, *The Development of Orange-Nassau Princely Artistic Activity*; Zandvliet, "Het hof van een dienaar." For patronage gestures in the Dutch privilege business, see also Vermij, "The Telescope at the Court." About the court of Stadthoder William Louis, see Waterbolk, "Met Willem Lodewijk aan tafel". About the court of his later successor, see the excellent study by Janssen, *Princely Power*.

35 "[...] Inventie, omme alle tgene men metter pennen scryven, oft trecken can, te doen weven, welcke conste hy Syn Excellentie int cleyne van desselffs wapenen, ende andere gethoont hebben, daerinne goet contentement, ende behagen genomen hadde." NL-HaNA, States General, 1.01.02, inv. no. 12298, fol. 346 [G54, 13 March 1601].

obtaining a privilege in 1609 for his invention of a new type of lacquerware, Kick found good customers among his Republican patrons. On 11 May 1612, it was "agreed [in the Assembly of the States General] to write out an order for Willem Kick, of twelve hundred guilders, for six dozen lacquered and gilded plates of all kinds, large and small, as well as a big gilded and lacquered *lict du champ*, which the High and Mighty have bought from him for the aforesaid price, among others to honor the Electress of the Palatinate."[36] There are many similar examples where the federal authorities decided to purchase inventions after privileging them. Perhaps the term 'Republican patronage' is appropriate in this context. Speaking about a privilege issued to the English inventor John Osborn for cutting whale bones "as if it was ebony,"[37] the official historian of the Republic, Nicolaes van Wassenaer (1571?-1630), described it in somewhat laborious terms as follows:

> The lofty members of the States [General], as fathers of their subjects, lovers of all subtle and admirable inventions, Patrons of all scientist, when it came to their attention that the above-mentioned inventor was dwelling in their domain, have granted him, in order to stimulate other speculators, an excellent privilege in the year 1618 to work these whalebones in their provinces without someone being allowed to imitate or distribute the same.[38]

---

36 "Is geaccordeert te depescheren ordonnantie voor Wilhem Kick, vande somme van twaelf hondert guldens, voor sess dousynen gelacte ende vergulde schotelen van elderhande soorten, groot ende cleyn, mitsgaders een groot vergult ende gelact lict du champ, dat hare Ho. Mo. van hem voor de voorsz somme hebben gecocht, omme onder anderen daermede oock de Cheurfurstinne te vereeren." Dodt, *Archief*, 5:275. For a more general account of gifts issued by the States General, see Sanders, *Het present van Staat*.

37 "alsoft die van massift ebbenhoudt waren". NL-HaNA, States General, 1.01.02, inv. no. 12302, fol. 54v (3 March 1618). See also below, at pp. 86, 114. The invention by Osborn later led to a number of conflicts, especially when on 25 April 1620, Rooswinckel obtained a privilege for a method to work baleen that was completely new and "nothing in common but was entirely different from the invention by Jan Osborn" (*heel verscheyden van de de Inventie van Jan Osborn Engelsman*). NL-HaNA, States General, 1.01.02, inv. no. 12302, fol. 172 [G186, 25 April 1620]. The final solution was that the two inventors were forced to work together and a new privilege was issued on both their names. NL-HaNA, States General, 1.01.02, inv. no. 12302, fol. 174v [G187, 26 May 1620].

38 "De Edele Hoogh-Mogende heeren Staten, vaders van hare ondersaten, beminders van alle subtyle en prijsweerdige inventien, Maecenates van alle geleerden, verstaende dat de voors. inventeur in hare dominie plaets hielt, hebben hem, om andere speculateurs te prickelen, een treffelijck octroy vergunt in den jare 1618, dese baleynen alleen in haer provincien te mogen wercken, sonder dat ymant hem die mocht namaken of distribueren." Wassenaer, *Historisch verhael*, VIII, fol. 87, as cited in Muller, "John Osborn," 202.

Just as monarchs did so elsewhere in early modern Europe, the Dutch Republican authorities presented themselves as the Maecenas of the arts. Figure 16 shows an example of the work by Osborn; it is a medallion (1626) made of whale bone depicting Stadtholder Frederick Henry, Prince of Orange, who succeeded his half-brother Maurice in 1625. The members of the States Assembly included an addendum to the ten-year privilege, stating that the inventor was "to please, accommodate, and serve everybody for a small profit, without exploiting the good people [≈the upper classes]."[39]

Aside from political authorities and merchants, other pressure groups, such as trade companies and artisan associations, occasionally recommended specific inventions as well. Trading companies were, for instance, directly involved in the invention of a reliable method for finding longitude at sea.[40] Proof that artisan organizations spurred on innovation can be found in the 1591 case of an inventor in the city of Haarlem, who was "requested by several bleachers to produce for them a type of water mill."[41] Not every artisan association was a guild.[42] Yet the idea, once widespread, that invention privileges functioned

---

39 "[...] des sal de voors. Jan Osborn gehouden syn een yeder te gerieven accommoderen ende te dienen voor een cleyn gewin, sonder de goede luyden te beschatten." Ibid.

40 See also pp. 148, 179–186.

41 "[...] versocht zynde by eenige bleyckers om voor hen alsulcken instrument ofte forme van watermolens te maken ende leveren op heurluyder bleyckeryen ten dienste ende gerieve vandenselven [...]." NL-HaNA, States General, 1.01.02, inv. no. 12298, fol. 26 [G11, 21 February 1591]. The twelve-year privilege for the exclusive production of this new type of water mill was granted for twelve years on the strong recommendation of the burgomasters and city council of Haarlem. RSG, 7:506.

42 For a discussion of definitions, see De Munck, Lourens, and Lucassen, "The Establishment," 67. At the risk of oversimplification, one could say that a craft guild was a governement-licenced group of artisans that was organized around their use of the same raw material. This form of organization resulted in situations (such as in the city of Utrecht) where saddlers and bookmakers ended up in one guild, even if they had not much in common in daily economic life (note that I talk about craft guilds; other professional guilds, such as the schoolmaster's guilds followed a different dynamic.) One often had to pass an examination to become a member of a guild and it was obligatory to become a citizen of the city where one was domiciled. Once a member, the guild executed the role as, in a way, acting manager by providing things such as widow pensions, funeral insurances, etc. Guilds also had strong religious overtones; for the elaboration of this theme, see Thijs, "Religion and Social Structure." It is important to note that there was never something like a 'national policy' regarding the guilds; guilds were very much an urban affair and practices differed considerably from place to place. In the Zaansteek, for example, one of the most important industrial centers of the Dutch Republic, there never were any guilds at all. In cities where the guilds did make headway, certain crafts were left 'free' and worked outside the guild system. In Utrecht, for instance – a city that knew some of the strongest guild organizations – there were no guilds for shipmasters, fullers, or dyers; in

FIGURE 15 Wooden chest, attributed to Willem Kick, *c.* 1620–1625

FIGURE 16 Baleen portrait medallion of Frederick Henry of Orange

essentially as a counterbalance to the suffocating power of the guilds is in no way supported by the sources.[43] Some inventors appear to have been guild members too.[44] Moreover, guilds were involved by the authorities in the assessment of new inventions.[45] This could explain, perhaps, why guilds in the Dutch Republic did not oppose the introduction of new inventions produced under privilege. Another reason could be that there was more than enough work for those working within and outside the guild, so there was no reason to revolt against the introduction of novel techniques. I return to this issue more extensively in Chapter 5, when I explore the relationship between invention privileges and ideologies of profit.

Amsterdam, there were no guilds for millers, sawyers, or masons. Posthumus, *Neringen*, 9. Also, the 'guild-obligation' (*gildendwang, zunftzwang*) was for instance not enforced everywhere or in every part of a certain craft.

43 Those guilds, supposedly being a typical phenomenon of the middle ages, had controlled the market through craft secrecy and thus had "thwarted" any technological improvement up to the sixteenth century. Doorman, *Patents*, 14. See also Landes, *Revolution in Time*, 210. For this reason, political leaders took invention privileges as a means to 'open up knowledge' and to speed up innovation. In reality, things were not so simple. Although it is known that the guilds occasionally did oppose technological change, this fact is all too easily overgeneralized. For a more thorough analysis of the public resistance against the introduction of novel technologies, see below, pp. 216–226. It should be noted that the importance of the guild system for the development of knowledge economies has been thoroughly re-evaluated over the last twenty years. For a concise overview of this topic, see Epstein and Prak, "Introduction." Two crucial articles for the Dutch context are Lucassen and Lourens, "Ambachtsgilden in Nederland"; Lourens and Lucassen, "De oprichting en ontwikkeling." For a 'summary' in English, see De Munck, Lourens, and Lucassen, "The Establishment." The study of the guild system in the Republic has been somewhat hampered by the fact that only very few documents survived. In particular, data regarding guilds that were dissolved before 1791 (the year in which guilds were officially suppressed in the Netherlands) has proven very hard to retrieve. Thus, it remains unclear how guilds were distributed over the Republic, how many members they had, or even what the exact occupations of these members were. All in all, it appears that the guild system was relatively weak in the United Provinces until the second half of the seventeenth century, i.e. when the system of invention privileges got into its stride. Lucassen and Lourens, "Ambachtsgilden in Nederland," 44. As Nicolaas Posthumus formulated it, "as industrial development advanced, the guilds tagged along" (*Al naar mate de industrieele ontwikkeling voortging, volgden de gilden*). Posthumus, *Neringen*, 10. There are comparisons to be made with the situation in England at this juncture, where guilds also grew stronger after the Middle Ages. See Berlin, "Guilds in Decline?," notably 229. Berlin did not see any clear opposition between technological innovation and the guild system either, by the way.

44 See also pp.148–149.

45 In the case of an invention relating to the casting of lead in thin sheets, for example, proposed by a couple of plumbers, the pewterers and other masters of the tinsmith's guild first had to be heard before granting a privilege. RSG, 10:821.

Patterns: Drawings and Models

Aside from the recommendations and testimonials, inventors were often obliged to provide drawings and models of their invention. As Figure 17 attests, it was particularly in the early days of the Republic that many requests were accompanied by such specifications. Often, it was the type of invention that determined what type of specification was desired by the authorities. When an invention related to a project that envisioned the application of a new method rather than a new machine, text-based clarifications were generally considered sufficient. If the inventor wished to incorporate the use of mechanical instruments, small-scale drawings or models of the invention were required as well. These models were usually made from wood or copper.[46] According to a privilege act granted to a couple of brothers from the city of Utrecht, for example, the authorities were convinced of all the advantages of a newly invented watermill, since these advantages "could be seen in a certain small copper model made by the suppliants and brought along to that end [to the Assembly, to convince us]."[47]

Aside from the difference between textual and visual illustration, another distinction with regards to the type of specification was dependent on the tangibility of the invention. It appears that, when an invention was not yet reduced to practice at the time of the application, drawings and descriptions were particularly desired; whereas in the event that an invention was reduced to practice at the time of the application, preference was given to examining its functionality by means of witness statements and through 'ocular inspection' by selected experts.[48] In most cases, the authorities made use of a combination of different forms of evidence, as in the example of Albert de Veer, whose invention of a water pump was "revealed by the respective figures, as well as by witness statements, that he had attached to his afore-mentioned Supplication."[49] Moreover, the use of a drawing did not exclude the use of a

---

46 Examples of wooden models can be found in NL-HaNA, States General, 1.01.02, inv. no. 12298 [G39, 24 February 1598] and NL-HaNA, States of Holland, 3.01.04.01, inv. no. 5028, fol. 759 [H12, 14 December 1594]. To my knowledge, no contemporary examples of these models have survived.

47 "Al t'welck met suffisant ende vast werck soude connen bevonden werden, ende Wij eensdeels souden connen sien aen seecker clein model van coper gemaeckt by de supplianten tot dien eynde overgebracht." Doorman, *Octrooien*, 163.

48 I shall return to this issue in more detail in Chapter 4.

49 "[...] sonder aende selve water-meulens eenige veranderinge te maecken als beneden in de waterasse blyvende het bovenwerck altemael in zijn geheel *na uuyt wysinge van de respective figuren, ende getuygenissen*, die hy by zyne voors. Supplicatie hadde gevoecht, zukx dat men daerdeur het water in veel meerder overvloet, ende hoochte mochte [...]." Emphasis added. NL-HaNA, States General, 1.01.02, inv. no. 12298, 258 [12 December 1598].

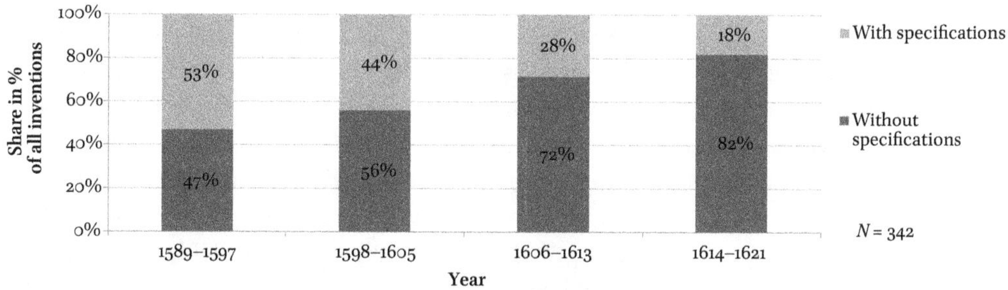

FIGURE 17 Specifications offset against the total number of applications

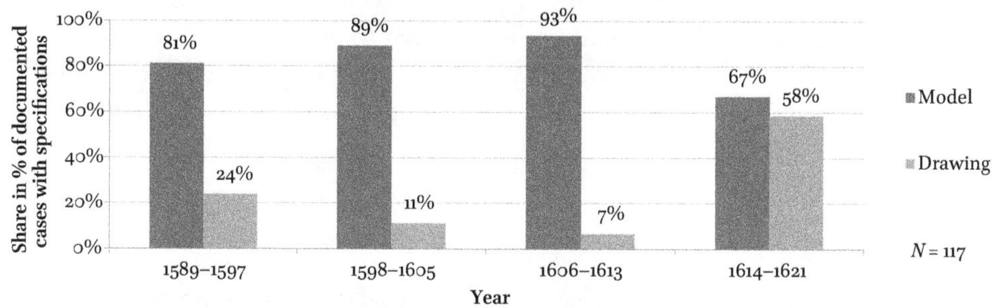

FIGURE 18 The type of specifications used

prototype model (see Figure 18). The inventor Jacob Claesz Schot, for illustration, had devised a water mill with eight basins that could elevate an enormous amount of water; he had submitted a drawing of his invention, yet in addition "had done a satisfactory test with a model using only four basins, appearing from the certification, and [...] with reference to the good knowledge that Lord of Brederode has obtained from the work."[50]

The increasing use of drawings and models in the context of the privilege system seems to have been closely related to concurrent developments in the world of architecture.[51] From the fifteenth century onwards, in imitation of Italian examples, Flemish patrons began demanding design drawings

50 NL-HaNA, States General, 1.01.02, inv. no. 12302, fol. 260v-261 [G210, 21 April 1622]. Jacob Claesz Schot. "met een modelle alleen van vier backen eene genoegsame preuve hadde gedaen, blyckende by de Certificatie, ende refererende den Suppliant hem wyders tot de goede kennisse die den heere van Brederode van t selve werck is hebbende, als t'effect van dien gesien hebbende."

51 The coming into use of early modern "privilege specification" is, instead, extremely difficult to pin down. None of the contemporary sources I have set eyes on, dealt with the issue specifically. Moreover, there is hardly any substantial historiography associated to the phenomenon. Notable exceptions are: Biagioli, "Patent Specification"; Davies, "The Early History"; Popplow, "Why Draw Pictures of Machines?" For machine drawings in the

and models from their architects in order to gain a clearer understanding of what to expect when giving commissions.[52] These drawings and models were called *patroenen* ('patterns'), which later became the common name for privilege drawings and models as well.[53] The use of these patterns in the world of architecture was closely related to the emergence of a new building culture in the Renaissance, when patrons became increasingly involved in the process of negotiation: patterns served to convince the patrons and to provide them with an impression of the envisioned construction.

Drawings and models in the privilege business served a similar purpose. They were not a blueprint allowing "any person skilled in the art" to reproduce the invention, but rather something with which to negotiate with the higher authorities.[54] This explains why the visual information in machine drawings contradicts our expectation that "one drawing of one machine should also

---

early modern period, see also Ferguson, *Engineering and the Mind's Eye*; Hall, "The Didactic and the Elegant."

52 One of the first to write about the use of architectural models in the process of negotiation between architects and patrons was the famous theoretician Leon Battista Alberti (1404–1472). He declared to "highly commend the ancient Custom of Builders, who not only in Draughts and Paintings, but in real Models of Wood or other Substance, examin' d and weigh' d over and over again, with the Advice of Men of the best Experience, the whole Work and the Admeasurements of all its Parts, before they put themselves to the Expence or Trouble." Alberti, *Ten Books on Architecture*, 86. Considering that Flemish architects carefully studied the architectural theories that came from Southern Europe, the adaptation of models in Northern Europe probably should be ascribed to the influence of Italian architects. Miedema, "Over de waardering," 72ff. See also Briggs, "Architectural Models." Alberti, who is mostly commemorated for his work on perspective, also had one or two things to say about inventions. In his *Ten books on Architecture* the word 'invention' appears about fifty times. Often, it is used in the sense of original 'idea' (e.g. page 36: "though the Invention and Design of the Architect *Policrates* must needs have been wonderful") but mostly it meant something like 'mental ability', as in the following example: "But after having made Experiment of every Thing, the Wit and Invention of Man has found out nothing yet more convenient than Tiles of baked Clay." Alberti, *Ten Books on Architecture* (London, 1755), 206, 366.

53 Meischke, "Het architectonisch ontwerp," 168, 170–72. Goldtwaite speaks of a similar ambiguity of the words *modello/ disegno* in Italian. Goldthwaite, *The Building of Renaissance Florence*, 396.

54 Nowadays, the specification is a *conditio sine qua non* for obtaining a patent. The US constitution declares that: "The specification shall contain a written description of the invention, and of the manner and process of making and using it, in such full, clear, concise, and exact terms as to enable any person skilled in the art to which it pertains, or with which it is most nearly connected, to make and use the same, and shall set forth the best mode contemplated by the inventor of carrying out his invention." 35 U.S.C. 112 Specification. For more background, see Rankin, "The 'Person Skilled in the Art'."

have one viewpoint."[55] The point was to "show the whole device without any distracting settings, and use as many viewpoints as necessary to convey precise information."[56] On the other hand, specific information (such as dimensions) was usually missing in the drawings. It has been argued that this was also because inventors wanted to secure their position as experts.[57] The lack of information functioned in this case as a form of security: it ensured that the inventor was still needed if the project were ever reduced to practice. One might add that drawings and models had an important function in the process of tinkering. The Italian architect Leon Battista Alberti (1404–1472) had elaborated on this aspect in his famous treatise on architecture, specifying that:

> ... the making of curious, polish'd Models, with the Delicacy of Painting, is not required from an Architect that only designs to shew the real Thing itself; but is rather the Part of a vain Architect, that makes it his Business by charming the Eye and striking the Fancy of the Beholder, to divert him from a rigorous Examination of the Parts which he ought to make, and to draw him into an Admiration of himself. For this Reason I would not have the Models too exactly finish' d, nor too delicate and neat, but plain and simple, more to be admired for the Contrivance of the Inventor, than the Hand of the Workman. Between the Design of the Painter and that of the Architect, there is this Difference, that the Painter by the Exactness of his Shades, Lines and Angles, endeavors to make the Parts seem to rise from the Canvass, whereas the Architect, without any Regard to the Shades, makes his Relieves from the Design of his Platform, as one that would have his Work valued, not by the apparent Perspective, but by the real Compartments founded upon Reason. [58]

55 Cf. McGee, "The Origins of Early Modern Machine Design," 56.

56 Leng, "Social Character, Pictorial Style," 94.

57 Popplow, "Why Draw Pictures of Machines?," 34. Cf. McGee, "The Origins of Early Modern Machine Design," 59. It should be noted that Popplow and McGee are speaking about early modern machine drawings in general, not about 'privilege specifications' only.

58 Alberti, *Ten Books on Architecture*, 86. Compare at this juncture the remarks by Simon Stevin, the famous scientist-engineer and fervent inventor, who claimed "that one draws in advance several forms [...] in order for the people who have to come to an agreement with one another may understand each other well, and from that point onwards decide what is best. These forms are made in two ways, firstly in flat on paper, and then physically from compost, wax, wood, or another material" (*datmen van te vooren verscheydenen formen teeckent* [...], *op dat de persoonen die daer af te spreken hebben, daer soodanighe formen malcander wel verstaen meughen, ende eintlick het beste daer af besluyten. Deze vormen worden gemaeckt op tweedeley wijse, eerst int platte op papier, daer na lichaamelick van poteerde, was, hout, of ander stof*). Stevin, *De Stercktenbouwing*, Leiden 1594, p.7; as cited

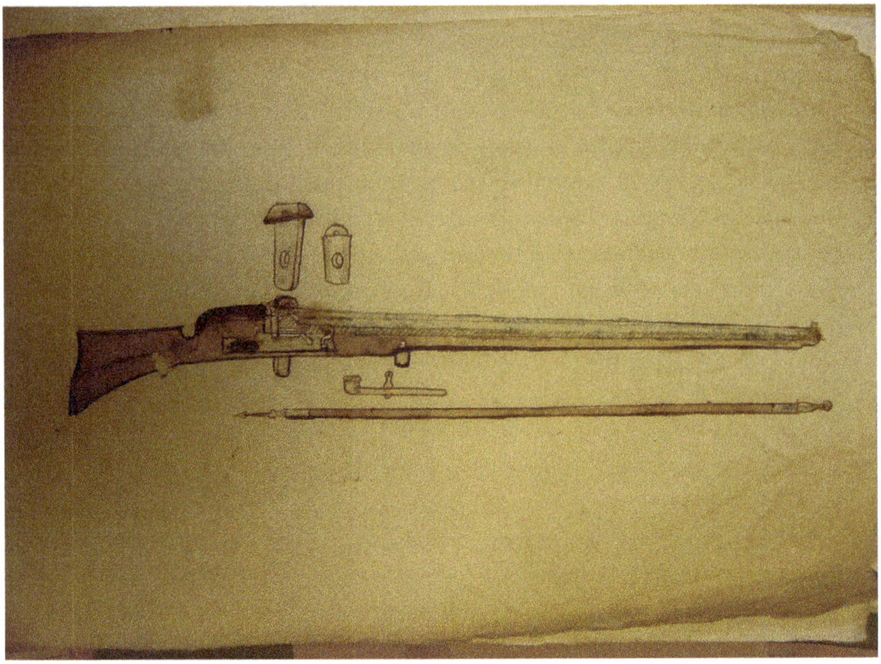

FIGURE 19 The privilege specification by Andries Scholten for a new type of drilled barrel, forged from solid iron, 1600

Privilege models effectively fulfilled the same function as architectural models. Having this contextual background in mind might help us to appreciate why the drawings that accompanied privilege requests did not reach the level of technical sophistication that one expects nowadays from patent specifications drawn up according to specific standards (see Figure 19, see also book cover).

Presenting a model, whether on paper or in 3D, unmistakably contributed to the credibility of a proposal to implement a specific invention.[59] But in some cases, a model was desired *after* a privilege had been granted and written out. These cases reveal the secondary function of the privilege specifications, namely to serve as a measure against which to test the exact thrust of the privilege in later court cases. As an illustration, one could highlight the privilege of a couple of carpenters who sought for a monopoly for the exploitation of their

in Taverne, *In 't land*, 424, n.32. See on this point the excellent essay by Tieskens, "Omdat teykeningen."

59 Examples where models were presented at the assembly of the States General are NL-HaNA, States General, 1.01.02, inv. no. 12300, fol. 426v [G125, 6 September 1613] and NL-HaNA, States General, 1.01.02, inv. no. 12301, 46v [G132, 11 June 1614]. Doorman remarked that the habit to provide drawings decreased from 1635 onwards only to increase again after 1771. Doorman, *Octrooien*, 21. I have not been able to verify this observation.

treadmill (for lifting heavy loads) and a pump. The privilege literally stated that it would only take effect "if a model of the invention is delivered within a month to be kept in order to prevent disputes."[60] It seems that the States General even created a special storage room in its quarters where they kept the models of new inventions. The master carpenter Anthoine van den Vate, at least, requested the placement of the model of his invention in the "peek room" (*pinckcamer*) in the quarters of the central authorities in The Hague.[61]

The Dutch authorities were particularly keen on the production of visual clarifications. If the inventor did not submit a drawing or a model together with the request, he was asked to do so in the near future. If the inventor failed to do so, the authorities would usually stop handling the request. In other cases, the validity of the privilege was made dependent on the production of a model. The engineers Pierre Mybois and Joseph Skelton, for example, having invented a new type of horse mill, were "held to make their inventions at a small scale within the next three months, and at a large scale within a year, or else would lose the effects of this privilege."[62]

This last fragment of text shows that there were actually two stages of disclosure. At first, when the inventor filed a request, descriptions were kept very vague and generic. Only after the privilege had been made out would the inventor give more information away, for instance by producing a scale model. It was also at this point that drawings sometimes became more schematic in nature, and more details were added, such as dimensions. Particularly in the resolutions of the States of Holland, one comes across quite accurate sketches in the early years that visually support the issued privileges (see Figure 20 and Figure 21). Also known in this context are the beautiful "patterns" produced

---

60 "midts dat de voorsz. inventien zyn nieuwe inventien in dese Landen by anderen noyt ghevonden of gepractiseert, daer vanse binnen een maendt een model van elcks sullen moeten overleveren om bewaert te worden tot voorkominge van disputen." NL-HaNA, States of Holland, 3.01.04.01, inv. no. 5050, fol. 16 [H40, 17 January 1618].

61 RSG, 1:288 (2 October 1588). Compare at this juncture the remarks by Guiseppe Ceredi, who recounted that in Venice: "I have seen many other models of machines that have never been reduced to practice; and in particular in the secret rooms of the Ufficio de' Provveditori di commune in Venice: where everyone who is convinced to have found some ingenious and beautiful thing takes his models in order to obtain a privilege." Ceredi, *Tre Discorsi sopra il modo d'alzar acque d'Luoghi bassi*, 19.

62 "dies sullen de voors. Supplianten gehouden syn de voors. heure Inventien Int cleyne te maecken binnen drye maenden naestcommende, ende int groote binnen een jaer, by verlies van het effect van dit Octroy." NL-HaNA, States General, 1.01.02, inv. no. 12302 fol. 2v [G149, 2 January 1617]. For similar examples, see NL-HaNA, States General, 1.01.02, inv. no. 12302, fol. 9 [G153 11 February 1617] and NL-HaNA, States General, 1.01.02, inv. no. 12302, fol. 7 [154, 13 February 1617].

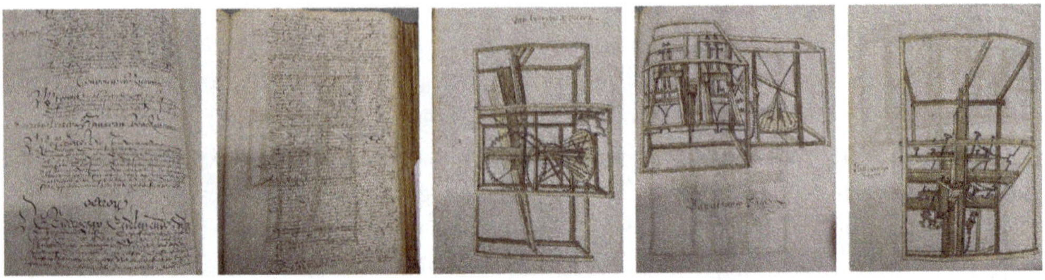

FIGURE 20 Privilege for Frans van Bouckelwaert's invention of a horse-mill for wood-sawing, 1594

FIGURE 21 Privilege for a new dredger invented by Jan Simonsz van der Goude, 1595

by Simon Stevin, which most closely resemble modern construction drawings (Figure 22).[63] However, Stevin was the exception rather than the rule, and most drawings were clearly of a less abstract level (cf. Figure 23 and Figure 24). Why, by the way, from 1602 the habit fell out of custom at the States of Holland to depict drawings directly with the resolutions is not entirely clear, but it is likely that the drawings were kept separately from then on, as was already the custom at the States General, and that they were therefore not preserved.

63 This specification supporting the patent of nine different inventions is kept in a separate quire in the National Archives: NL-HaNA, Kaartcollectie Binnenland Hingman, 4.VTHR, inv. no. 4795. It is in my opinion likely that the document was not a patent application, but a further specification after the patent had been issued. On this point, also compare Stevin's 1586 patent, where his rights are made contingent on the inventor "exhibiting the portraiture of his instruments" (*sal alhier exhiberen de pourtraicture van zijn instrumenten*), later on to be attached later to the patent. Stevin, *Principal Works*, 5:27.

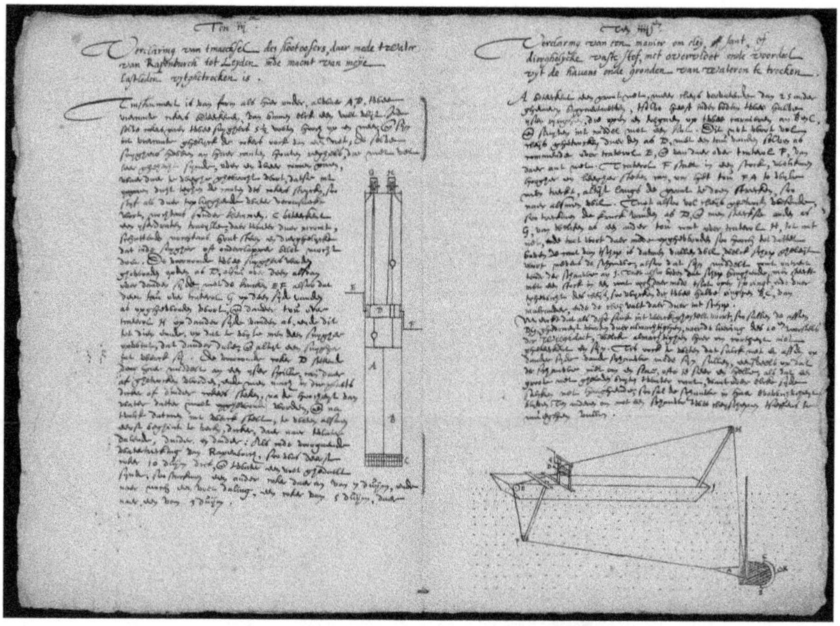

FIGURE 22 Specification for a set of inventions made by Simon Stevin, 1589

The reason for the initial restraint on the part of the inventor was probably a fear that too much openness would lead to dissapointment. For instance, the inventor who had "been asked by several bleachers to make [...] a water-mill and to deliver it to their bleaching fields [...] had not been willing to contract this assignment [...] unless he would first be granted a decent form of privilege."[64] In another example, the States of Holland granted city carpenter of Delft, Cornelis Dircksz Muys, a privilege:

> to practice and bring to light the new invention [...] which the Supplicant had invented and conceived [...] taking into consideration that the Supplicant has spent much time and effort to invent the Instrument; and he fears that, when he would want to use it in some public Works or Mills, it would we readily imitated and counterfeited by others [...].[65]

64 "[...] versocht zynde by eenige bleyckers om voor hen alsulcken instrument ofte forme van watermolens te maken ende leveren op heurluyder bleyckeryen ten dienste ende gerieve vandenselven, tzelve niet en hadde willen aennemen alshem oock niet geraden zynde t enware dat hem by ons ierst verleent ware behoorlyke acte by forme van octroy [...]." NL-HaNA, States General, 1.01.02, inv. no. 12298, fol. 26 [G11, 21 February 1591].

65 "[...] ende gemerckt hy Supplianten veel tydt ende arbeydt tot het inventeren van 't selve Instrument besteedt heeft, ende beduchtende is, dat indien hy 't selve in eenige publycque Wercken of Molens soude willen gebruycken, terstont by eenige nagemaeckt ende gheconterfyt soude werden, waer door hy de vruchten van syn arbeydt ende kosten in

FIGURE 23
Privilege for a new type of windmill for cutting wood, invented Franck Jansz van Hoorn, 1595

FIGURE 24 Privilege for a new type of drainage device invented by Cornelis Dircksz Muys, 1589

The privilege functioned as a means of a security for the inventor: it was a hedging instrument intended to persuade the inventor to publicize his invention.

The limited amount of disclosure in the initial stages of negotiations for a privilege supports Mario Biagioli's argument that there could not be such a thing as a 'patent bargain' in the early modern period, since the purpose of specifications was not to inform the public at large.[66] Whereas the modern patent system hinges entirely upon the idea that inventors receive temporary monopoly rights in exchange for making detailed information about their invention freely available, we see that before the democratic revolutions in France and America:

> the privilege was not a contract involving people acting as political representatives. Patent officials may have "represented" the prince or sovereign power that hired them, but those political powers did not represent the people. As a result, the privilege was not a contract between the inventor and the public (through an elected government), but rather a material gift from one person (the prince or a *persona ficta* like a city) to another person (the inventor). The very notion of the modern patent bargain would have been unthinkable outside of a political regime based on representation.[67]

Inventions, to be sure, "would be of great benefit" to the common good when experimented upon and put to use.[68] But not everyone had access to information about the invention. Thus, even if a concern for disseminating novelties

---

't practiseren ende vinden van de voorsz. Kunst gedaen, Verliesen, ende van 't effect van dien. gefrustreert soude werden, ende hy Supplianten met het selve Instrument gaerne 't gemene Landt soude dienen; versoeckende daeromme tot dien eynde consent ende octroy […]." NL-HaNA, States of Holland, 3.01.04.01, inv. no. 346, fol. 348 [H7, 25 May 1589]. Another example was that of Josua van Sunderen, who demanded the security of receiving one-third of any future gains before revealing his invention to bring profit to the country. RSG NR 6:26 (24 January 1623). The request was denied, and the proposal was not followed up.

66 Biagioli, "From Print to Patents," 152. The 'patent bargain' is the expression commonly used to decribe the exchange between the inventor and the state in terms of a social contract: inventors receive temporary monopolies in return for the disclosure of their invention by means of patent specifications that are accesible to everyone (and which eventually fall into the public domain).

67 Biagioli, "Patent Republic," 1159.

68 The terminology was used, for instance, in the following example: "[…] indien […] geexperimenteert ende in gebruyckt ware, tselve bevonden soude worden tot groots voordeel." NL-HaNA, States General, 1.01.02, inv. no. 12299, fol. 75 [G73, 10 July 1602].

was an important constituent, the logic of the privilege regime was still far removed from its modern counterpart.

The discrepancy can be further illustrated by highlighting the case of the English inventor John Osborn, who obtained a Dutch privilege for the making of medals from whalebones (see also pp. 72–74, 114). Osborn got into a conflict with his competitors Thomas Comes and Richard Osborn (his brother) shortly after filing a complaint against them for infringement. The two copycats demanded access to the complaint in order to comprehend in what way they had infringed upon the privilege.[69] However, after notary summons, John declared that: "They have been denied [insight into] the copy of the [complaint] by the gentlemen of the States themselves in order to prevent any lawsuit and [they were] to settle amicably."[70] The authorities were clearly reluctant to share relevant information and strictly supervised the diffusion of knowledge.

When it came to the diffusion of technical details about a privileged invention, the authorities maintained a tight grip as well. There were numerous non-disclosure agreements among private individuals.[71] But there are also plenty of examples where the state strove for secrecy when it came to new inventions.[72]

---

69 "om daeruuyt te sien in wat manieren [...] 't verschil geconsidereert [...]." Van Dillen, *Bronnen*, 2:528 [no. 929, 9 April 1624].

70 "De copye van de requeste is hen van de heeren Staten selven geweygert te geven om processen te voorhoeden ende om metten anderen in 't minnelijck te accorderen." Van Dillen, *Bronnen*, 2:528 [no. 929, 9 April 1624].

71 For examples of contracts involving secrecy, see Van Dillen, *Bronnen*, 2:207 [no. 340]; Van Dillen, *Bronnen*, 2:711 [no. 1191]; Van Dillen, *Bronnen*, 2:252 [no. 412, 2 November 1617]. One contract, related to the invention of a copper mill, for instance, spoke of an agreement between the inventors Harman Becx and Aert Duyfkens and the carpenter Pieter Pleunisz. The carpenter would get a daily wage, as well as 300 carolus guilders "in which regard Pleunisz has expressively promised not to disclose, teach or reveal the way or the means, either in part or as a whole, to no-one in the world, whoever it is" (*In regard van dewelcke* [...] *Pleunisz expresselijck belooft de maniere noch gelegenheyt van dit werck in 't geheel noch in in deel aan niemant ter werelt, wie hy oock sy, te ontdecken, leeren noch openbaren*). The carpenter was also not allowed to make similar objects, nor around these parts or abroad. Van Dillen, *Bronnen*, 1:723 [no. 1210, 3 December 1611]. For the privilege, see NL-HaNA, States General, 1.01.02, inv. no. 12300, fol. 313v [G116, 18 February 1612]. Harman Becx and Aert Duyfkens were known investors in innovative technologies. Another one of their projects was a cotton mill; on 1 December 1611, they closed a deal with the Englishman Edmont Web not to stand in each others way when it came to printing textiles. Van Dillen, *Bronnen*, 1:722 [no. 1209]. For this project Becx and Duyfkens worked together with Jan Andriesz Moerbeeck, an arms dealer who obtained a privilege to found cannons (see p. 114).

72 About the complex issue of secrecy in the Dutch Republic, see Davids, "Public Knowledge"; Zandvliet, *Mapping for Money*, 91–96; Friedrich, "The Importance." On the issue of

This mainly concerned inventions in the area of military technology. The most famous case was probably that of the telescope, when the inventor Hans Lippershey was asked by the States General "to promise that he will not divulge his invention to anyone else for some time."[73] In another example, when three citizens of Amsterdam obtained a thirty-year privilege for the production of a new type of war vessel, the members of the States General considered it "important that strangers, and especially enemies, did not find out about the details of this invention and art."[74] The central authorities therefore authorized the magistrates of those places where the boats were made to extract an oath of secrecy from the craftsmen involved in the construction.

Cost and Taxes

Privilege applicants were usually expected to bear the cost of the examination process. But, other than that, the cost of obtaining an invention privilege in the Dutch Republic seem to have been rather low. Under Burgundian-Habsburg rule, as well as in other sovereignties such as England or France, it was common to levy a tax on the revenues that would accrue from a privilege.[75] The inventors Andrien de Fonteny and Pierre Marentin, for example, had to pay a tax to the Habsburg treasury for their invention to prepare oil from cereals:

> in recognition thereof [= the privilege] the said suppliants will be held to pay to our profit the twentieth *denier* [≈ coin] of the profit which will proceed their aforementioned product in the hands of our beloved and loyal councillor and receiver general of our finances: Lieven Wouters.[76]

---

secrecy more generally see also Vermeir, "Openness versus Secrecy?"; Vermeir and Margócsy, "States of Secrecy."

73 Van Helden, *The Invention of the Telescope*, 36. We will return to this case below, pp. 91–92. In another example, the States General expressively agreed with an inventor that the specification of his invention "would be kept secret" (*sal secreetleyk bewaart worden*). RSG, 9:355 (9 April 1596).

74 "Ende alsoo dese Landen daeraen is gelegen dat dese Inventie ende konste nyet comme tot kennisse van vreemde, sondelinge deeser Landen vyanden, authorizeren hare ho. Mo. Magistraten" NL-HaNA, States General, 1.01.02, inv. no. 12302, fol. 207 [G192, 16 December 1620].

75 For the issue of rents in English Kingdom, see Davies, "The Early History," 108–9. Davies is not convinced, however, that rents ever formed a strong motive for the award of invention privileges.

76 "Pourvue que en recongnoissance ducelluy [= le privilege] lesdits supplians seront tenuz payer a nostre prouffit le vingtiesme denier du prouffict que procedera leurdit artifice es mains de nostre aime et feal conseiller et receveur general de nosdits finances Lieven Wouters." Doorman, *Octrooien*, 73 [K12, 15 January 1568]. The *receveur general* supervised

Under the brief reign of Leicester in the Republic, it remained common to pay a yearly tax in recognition for the privilege rights. The inventor Simon Stevin, for instance, was "held to pay a yearly fee of 3 pounds to the County of Holland in recognition [of the privilege for his invention of a more efficient mill]."[77] But with the *coup d'état* by the Republican States General, tax obligations suddenly disappeared from the privilege acts.

The change is hard to account for. Taking into consideration that Holland had a reputation for "being the most heavily taxed nation in Europe,"[78] it is rather unlikely that the Dutch suddenly pressed for a free market system identified by a low tax burden. A more likely explanation for canceling the rent obligations is that, after the departure of Leicester who had been assisted by his own Council of Finance, there was simply no longer an office with central authority to collect taxes in the Republic.[79] Tax collection then shifted (even more markedly) to a local level, whereas something like a 'central office' controlling the taxing practices (the *Generaliteitsrekenkamer*) would only be institutionalized in 1608.[80] By that time, new practices in the privilege business may have already settled down. It remained one of the most characteristic features of the Dutch system that no taxes were raised from the central government on the profits made on the basis of an invention privilege.

This did not mean that privileges came for free. For a start, applicants had to pay money to the Keeper of the Seal for authenticating a privilege. In an attempt to take up as many additional positions as possible, Johan van Oldenbarnevelt had at some point assumed the post of Lord Keeper for the province

both the revenues and the expenses. The office had been instituted by Philip the Bold (*Philips de Stoute*; 1384–1404) in 1387. De Schepper, "Beleid en bestuur," 28.

77 "Mits dat hij [= Stevin] gehouden werdt in erkentenisse van desen te betalen jaerlicx tot proufytte van de Graeffelicheyt van Hollandt drie ponden te xl grooten [...]." The privilege had a duration of 18 years, so the total of tax would be 54 pounds of xl groats. The privilege had to be presented at the Rekenkamer van Holland "to be registered" (*omme aldaer geinterineert ende geregistreert te worden*). Stevin, *Principal Works*, 5:28.

78 't Hart, *The Making of a Bourgeois State*, 137.

79 Cf. Tracy, *The Founding of the Dutch Republic*, 277. There were two other Audit offices in the Northern Netherlands in the last decades of the sixteenth century; the *Gelderse Rekenkamer* in Arnhem (erected in 1559) and the *Hollandse Rekenkamer* seated in The Hague (erected in 1447). But they became less and less important in this period and from 1588 onwards their tasks were temporarily taken over by the Council of State and the States General. For an concise and extremely useful introduction to the Dutch tax system (with particular reference to the province of Holland), see 't Hart, *The Making of a Bourgeois State*, 119–157 (Chapter 5).

80 Fruin, *Staatsinstellingen*, 201.

of Holland.[81] Over the period 1602–1619, he levied six pounds of 40 groats per privilege.[82] Whether Oldenbarnevelt also managed the costs for the seal of the States General remains, alas, unclear; most of the financial records of the States General have been lost and there is no information in the official documents about the costs for a seal issued by the federal authorities.

Furthermore, the official documents did not list all the costs involved in the entire process of obtaining a privilege. Everyone knew that in the political jousting in The Hague all kinds of 'douceurs' had to be presented.[83] Bribery was such a familiar phenomenon that the States General decided at some point to hang a plaque (1588) at the entrance hall of their colleagues from the Council of State:

> written in Dutch, French, English and Scottish, which will warn everyone, that they shall not present, give or promise, any gifts or presents to [the members of the Council of State], their wives, children, or family, or anyone else, either directly or indirectly, under penalty of a fine [...].[84]

Whether the injunction had any effect is difficult to say, but corruption seems to have been a recurrent problem. In 1651, the States issued another placard prohibiting "to present, give, or promise [...] any gifts or presents, no matter how small, including beverages or comestibles, to obtain [...] any office, benefice, *privilege*, resolution, or arrangement."[85] Yet I have not come across any evidence of bribery that directly related to invention privileges.

---

81 Invention privileges enter the account books from 1602 onwards. But not all of the privileges can be traced in these accounts. The privilege granted to Isaac Burger on 5 March 1603, for instance, cannot be found in archive of Oldenbarnevelt (3.01.14, inv. no. 1060).

82 The Dutch pound of 40 groats (*groten*) was the common money-of-account in the Republic. One Dutch pound was equal to the Dutch *guilder*, which could be divided into 20 *stuivers* or *schellingen*; each *stuiver* was sub-divided into 16 *penningen*. Payments were often made using the Flemish pound, which was an actual coin (and the money-of-account used in Zeeland). One Flemish pound equaled six guilders, or 240 groats, or 120 stuivers. Van Gelder, *De Nederlandse munten*, 32. Unless otherwise indicated, I mean Dutch pounds when I use "pounds" in the text.

83 Knevel, *Het Haagse bureau*, 132.

84 "[...] gestelt in de Nederlandtsche, Fransche, Engelsche ende Schotsche Talen, daer by een yeder gewaerschouwt sal worden, dat sy aen de voorsz. Persoonen, derselver Huysvrouwen, Kinderen, Familien, ofte yemandt anders van haren't wegen, eenige giften ofte geschencken sullen presenteren, geven noch beloven, directelijck noch indirectelijck, op de peynen ende amende [...]." Cau and others, *Groot placaet-boeck*, 4:123. Article 30 of the *Instruction for the Council of State*, 12 April 1588.

85 "[...] te presenteren, geven ofte belooven, [directelick ofte indirectelick, oock niet by eenige onderhandelinge, koop, permutatie, oft andersints,] eenige giften ofte geschencken van

Examinations

Once an inventor had handed in all the proper documents related to his invention, the authorities would discuss the possibility of issuing a legal privilege *in camera*. In some cases, the inventor immediately obtained a temporary right for the exclusive exploitation of his invention. In other cases, the authorities would decide that more research was required before a privilege could be dispatched. In a later chapter, I will scrutinize the legal theories that underpinned the privilege system, discussing both the type of evidence used in this context as well as how examinations were carried out in practice. In this section, I want to highlight some of the actors involved in the process of examining a privilege application.

Firstly, there were the delegates in the (inter)provincial Assembly. The composition of the Assembly changed regularly, so it is difficult to identify who was responsible for what. There were no fixed commissions to examine inventions at the beginning of the seventeenth century, even if some representatives returned on a regular basis in the official documents. In particular, three representatives from Zeeland (the former attorney Johan Van der Warcke, Casper van Vosbergen, a Member of the High Court of Holland and Zeeland, and Dr. jur. Jacob Simonsz Magnus)[86] seem to have been intensively involved with the assessment of new technologies on which a privilege was claimed, but this impression could also be distorted because only the representatives for Zeeland in the States General were elected for life. A number of other long-serving officials regularly return in the examination process as well. Among them, the quartermaster-general of the army of the States, Simon Stevin, and Joris de Bye, the Treasurer-General, who was expected to give his opinion on the financial aspects of future privileged monopoly grants.[87]

eenige dingen, hoe kleyn die souden mogen wesen, oock van Dranck ofte Eet-waren, om t'obtineren, [ofte gheobtineert te hebben, voor hem selven ofte yemandt anders, directelick ofte indirectelick,] eenige Ampten, Officien, Beneficien, Octroyen, Sententien, Resolutiën ofte Dispositien, op eenigerhande saecken, ofte om de expeditie van dien, onder wat pretext dat oock soude mogen wesen." Van Aitzema, *Historie*, 375. This text was part of the oath that an office holder had to take when taking up a public office within the Generality. For the official form, see also Van Aitzema, 464.

86 For biographical and bibliographical information on these actors, see *Repertorium van ambtsdragers en ambtenaren 1428–1861*.

87 By the orders of the Council of State all expenses were carried out by the *thesaurier-generaal*. In the period under consideration, the position of treasurer-general was held by Joris de Bye (from 1587 to 1628). Stapel, "Joris de Bye." The office of treasurer-general was important, if only because the Treasurer also had an advising vote regarding finances in the Council of State. For the period under consideration, the position of receiver-general

In addition to the members of the States General, the Stadtholder of Holland, Maurice of Nassau, also played a key role. As we saw earlier, Maurice was largely responsible for the assessment of inventions made in the field of military technology in his capacity as Captain General. One of the most famous inventions ever examined in this setting was probably that of the refracting telescope—an invention initially intended to be used on the battlefield, even though it was later made famous by Galileo as a scientific instrument par excellence. Maurice was directly involved with the early attempts to monopolize the use of the new instrument and he probably made some suggestions for technical improvements too.[88] In any case, he was already in possession of a telescope when on 2 October 1608, the German inventor Hans Lippershey (1570–1619) approached the States General with the request to privilege his newly invented instrument "for seeing things far away as if they were nearby."[89] The States General decided that, prior to dispatching the privilege, each province should delegate a person to test the new instrument "on the tower of the quarters of His Excellency."[90]

A few weeks later, Jacob Metius (after 1571 -1624/1631) appeared on the scene. Metius had somehow got wind of the pending application and claimed that one could see a distant object just as clearly with his instrument (which he had already been working on for two years) as with the instrument presented by Lippershey. All this, according to the States' resolutions, could be furthermore confirmed "according to the judgment of his Excellency [Count Maurice] himself and of others who had tested the respective instruments against each other."[91] Apparently, the tower of the Stadtholder headquarters had been turned into a testing ground to compare the different telescopes.[92] After exam-

was held by the Doubleth family, i.e. Philips Doubleth (*c.* 1560–1612, in office 1588–1612), Jan Doubleth, 1580–1650, in office 1612–1619), and Johan Doubleth (1618–1629).

88 Vermij, "The Telescope at the Court," 83.

89 The genesis of the telescope has been told so many times that it seems unnecessary to repeat it here in detail. For an excellent introduction, see Van Helden, *The Invention of the Telescope*. See also Van Helden, Gent, and Zuidervaart, *The Origins of the Telescope*. On the monopolistic tactics of Galileo in gaining credit for his telescopic discoveries, see Biagioli, *Galileo's Instruments*, in particular 116–134.

90 Van Helden, *The Invention of the Telescope*, 23. The Province of Holland sent no less than six representatives from different towns.

91 Van Helden, 39.

92 The Stadtholder's Quarters (*Stadhouderlijk Kwartier*) had moved from Delft to The Hague in 1584. The so-called Maurice tower (*Mauritstoren*) was ready for use shortly before 1600. It was situated on the northwest corner of the *Binnenhof*, where the States of Holland and the States General held their Assemblies. The tower became the centre of Maurice's Court. He had several private rooms in this residence, including his bedroom and a dining

ination, the States decided not to privilege either the invention by Metius or Lippershey, but instead to buy six of Lippershey's telescopes for 1000 guilders a piece for army usage. The example of the telescope offers a fine illustration of how interprovincial authorities, local representatives, and Stadtholder worked together within the framework of the examination process to determine what was best for the Generality.

The fact that the States asked for counsel before issuing a privilege proves that power was negotiated rather than imposed. Representatives in the Assemblies as well as local authorities affected by the new privilege gave their advice as vested parties. As we have seen, the interests of guilds and other artisan collectives were also taken into consideration in this context. But there is little evidence that the guilds were directly involved with the technical appraisal during the examination process as well.

Guilds were, however, not the only type of professional organization in the Dutch Republic. Another, and perhaps even more important, organization for industrial production was the so-called *nering*, succinctly defined by Nicolaas Posthumus as an "organization of house-industrial producers completed with industrial entrepreneurs" organized on the basis of the finished end product and not – as with the guilds – around the use of raw materials.[93] Like the guilds, *neringen* were recognized as legal entities with their own government, seal, and capital. But unlike the guilds, a *nering* did not have the power to prescribe its own statutes.[94] The city government would issue laws (*keuren*) that covered any part of a certain trade (such as working hours or the techniques that one was supposed to use). The governing body of the *nering* then looked after the

---

room and a fixed guest room for his cousin the other Stadtholder William Louis. Beyond that, there was also a library, a room for maps, an armoury and a room where jewelry was kept. About the *Mauritstoren* as the centre of the princely court, see De Vos, "Propaganda," especially 129; Zandvliet, "Het hof," 44.

93 "[...] organisaties van huisindustrieele producenten met als aanvulling loonindustriele ondernemers." Posthumus, *Neringen*, 20. In contrast to the guilds, where one had to pass an examination to become a member, one automatically became a member of a *nering* by being involved in a certain trade in a city. In the Leiden textile industry, for example, textile merchants (*drapiers*) took care of the import and delivery of raw materials that were later treated by artisans (often in their houses) and then redistributed to other retailers who finally distributed the commodity. These actors were all part of the same *nering* because of their involvement in the textile industry. Posthumus, 24. Yet not everybody involved in the Leiden textile industry was organized in the *nering*; the cloth-shearers, for instance, kept working in guild context. As Posthumus himself indicated, this shows some of the shortcomings of the definition. Posthumus, 38.

94 A *nering* could be compared to what has been has been identified as 'sectoral guilds'. Epstein and Prak discuss some examples of sectoral guilds in the silk industry in Bologna and Lyon. Epstein and Prak, "Introduction," 33.

implementation of such laws.[95] One of the most important tasks to be carried out by the governors was the inspection of the quality of delivered materials.[96] It was in this capacity of recognized experts that they were repeatedly called in to access the quality of novel inventions too (for an illustration of the governors' working methods in the Leiden textile industry, see the case of Pieter Kayseel on pp. 204–205). The authorities attached a great deal of importance to the opinion of the *nering*. Thus, indirectly, the cities had a major influence on the examination and implementation of privileged inventions because of the control of the *nering*.

Urban authorities were involved with the examination of privileges in other ways as well. A recurrent problem in the history of Dutch water management was, for example, refreshing the water flowing through urban canals. Especially in summer those canals were tainted by a stench that was associated with the spread of contagious diseases, such as the plague. Each town endeavored to come up with its own solution, which usually involved a method for circulating the water. Local circumstances were crucial for finding a suitable solution and the urban authorities relied on a wide range of examiners to assess the possibility of implementing privileged technologies.[97] Privileges were granted centrally so as to encourage inventors to come up with new ideas for regulating urban water facilities; the available knowledge could then be implemented or not on a local level.

Largescale drainage projects with local relevance were virtually always carried out in coordination with interprovincial authorities as well. One example was the 1613 invention of a "windmill that could be operated by horse, if need be, pumping up large amounts of water [...] as [confirmed by] the model examined by the delegates of the States of Holland and West-Friesland."[98] The inventor, Michiel van Elderhuys, obtained his ten-year privilege under the

95 Posthumus, *Neringen*, 27. The *nering* was managed by several superintendents (*superintendanten*) and governors (*gouverneurs*): the superintendents were selected from among the last magistrates (*schepenen*) in office and were charged with general supervision, whereas the governors formed the executive committee.

96 This inspection usually took place in the market hall of each *nering* (of which Leiden, for instance, had nine). Posthumus, 22. There, governors would inspect the material that was put on sale according to quality standards that had to comply with the standards imposed by the statutes that had been set by the city government.

97 For a fuller description of this issue, as well as the role of privileged technology in this context, see Buning, "Stench and the City."

98 "windtwatermuelen, oock desnoot synde met de peerden [...] synde het model daervan door Gedeputeerde van de E. Mo. Heeren Staten van Hollandt ende Westvrieslandt gevisiteert ende voorgoet geapprobeert ende aengenomen." NL-HaNA, States General, 1.01.02, inv. no. 12300, fol. 379v [G122, 5 April 1613].

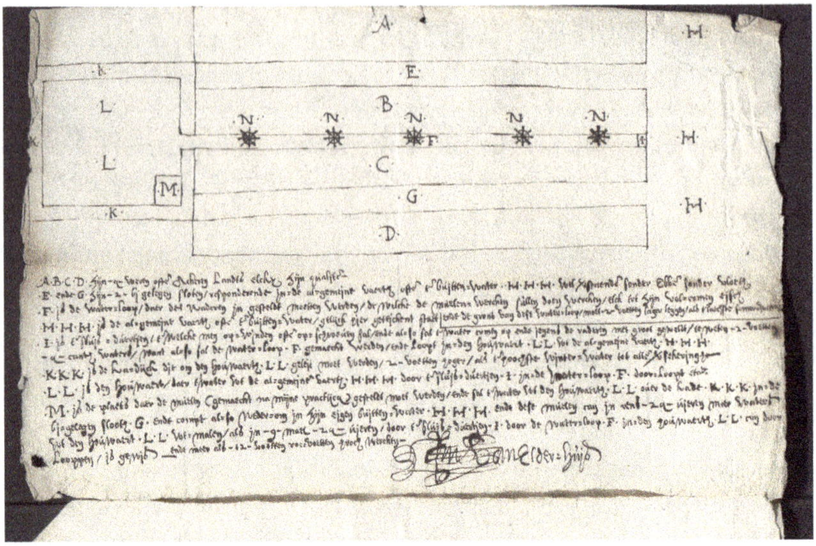

FIGURE 25 Project sketch by Michiel van Elderhuys, *c.* 1620

strict condition that he would "communicate his invention for a civil price" to those who desired to use it.[99] He then offered to implement his mills to dredge a moat around the military stronghold 's-Gravenweerd (see Figure 25). After extensive communication with the Council of State and the Treasurer-General, the States General decided the project was too risky.[100]

Also in the case of the various land reclamation projects for which the Dutch would become famous, continuous consultation was demanded, for instance with the water boards and other cities represented in the Provincial States (if only because drainage activities required the expropriation of land as well as the adjustment of municipal borders).[101] Since the cities and other local

99 "Ende dat d'voors. Elderhuys gehouden sal syn syne Inventie voor eenen moderaten ende civilen prys te communiceren den ghenen die dat van hem sullen begeren." Ibid.

100 RSG NR, 4:608 (9 oktober 1620); RSG NR, 4:608; RSG NR, 4:667 (7 December 1620); RSG NR, 4:681, 28 December, final descision). A later request for compensation (*traktement*) by Elderhuys is denied; RSG NR 5:148, 12 May 1621 (and again on 15 May). Instead, he obtained 20 guilders for his services. In 1617, Elderhuys had made the decision to travel to England to sell off his invention, obtaining a privilege reviewed and approved in person by Attorney-General Sir Francis Bacon (1561–1621). Pastorino, "Weighing Experience," 58.

101 For a further discussion of the impact of these 'negotiation arenas' on the development of technological knowledge, see Buning, "Stend and the City". For more background on the use of technology and water management, see also Zeischka, *Minerva in de polder*. The Water Boards (*waterschappen*) were charged with regional water management in the Republic. They were powerful independent bodies entitled to levy taxes and to administer justice.

authorities ultimately abided by the decision taken in the Provincial States, water management issues involving privileges in that way indirectly strengthened the Assembly's 'institutional capital'.[102]

Rewards

When an inventor came forward to apply for a privilege grant, there were basically three outcomes. The first was that the inventor would obtain the privilege; the second that he did not; and the third that the decision was postponed, because the inventor had not provided sufficient disclosure, for instance, or because there were objections from the individual provinces and cities. In addition to these three outcomes, however, there was a whole set of other possibilities that closely related to the process of obtaining a privilege. It is to those possibilities that we now turn.

The first option was that an inventor who applied for a privilege was taken into the service of the States. In some cases, this meant that he would not obtain a privilege. This happened to Johannes Muller, for instance, when he petitioned for a privilege for a new type of corn mill; the States General decided, for reasons not further defined, "not to agree to a privilege [...] but instead have the suppliant employed into the service of the land."[103] The States agreed to pay the inventor a reasonable salary to produce the mills exclusively for the benefit of the country "and not for anyone else unless with express permission."[104] In other cases, the inventor was taken into the service of the States after obtaining a privilege. When, for instance, on 16 May 1593, the inventor Jooris Florisz

102 Cf. Fransen, *Dijk onder spanning*, 390. The relationship between drainage projects and state formation is somewhat of a topic in itself, even though it intersects with privileges for technological inventions. For a recent case study focusing on the relationship between knowledge and political power, see Ash, *The Draining of the Fens*. For a comparative perspective, see for instance Van Cruyningen, "Dealing with Drainage." Also relevant to a better understanding of the triangular relationship between state formation, bureaucratic expertise, and the management of natural recourses is Appuhn, *A Forest on the Sea*; Popplow and Meyer, "'To Employ Each of Nature's Products'"; Warde, "Law, the 'Commune'."

103 "De Staten Generael der Vereenichde Nederlanden hebben naer voorgaende deliberatie niet goetgevonden te accorderen het octroy [...] op seecker molen van granen sonder steenen te breecken, maer hebben den suppl. in plaets van dien in dienste van t lant aengenomen [...] des sall hy Suppliant de voors. Molens daerop by hem octroy is versocht alleenlyck op redelyck salaris vermogen ende moeten maecken sulcx begeert wordende ten dienst van t land ende aen egeen anderen als met expres consent." Doorman, *Octrooien*, 196 (6 April 1634).

104 Ibid.

obtained a twenty-year privilege for the production of a new type of oil that would prevent corrosion, the States commissioned the treasurer-general Joris de Bye to negotiate with Florisz how much he wanted in return for maintaining the country's ordnance with the invented oil.[105] On 21 October, the treasurer-general reported that Florisz wished to be paid 2400 florins monthly, which was such an enormous amount that the proposal was simply neglected. But a couple of months later, on 3 December 1593 to be exact, Florisz was taken into the service of the States for 20 Flemish pounds per year to keep the ordnance of 1000 men "clean and free of corrosion."[106] Contracts of this kind were not awarded often, and mostly only when military interests were at stake.

The second option, for when an inventor came forward to offer his invention, was that he received an amount of money in the form of a 'recognition' (*vereering*). Sometimes, this recognition was part of the entire privilege process. Willem Kick, for example, who had already obtained a privilege for porcelain production, received:

> as a form of honor, for the dedication of his invention to the High and Mighty, in recognition of the costs and labor he had put into the invention on behalf of the country, the sum of 320 carolus guilders.[107]

In most cases, however, the recognition substituted a privilege. The best way to illustrate this is perhaps by drawing attention to the case of a certain Frenchman by the name of De Valet, who approached the States of Holland "indicating that he could make ships sail much faster by a secret art, requesting privilege thereof or, on the war ships, the 20ᵉ penning of the Prince [= a 5% sales tax]."[108] The inventor was first sent to the Admiralty to demonstrate his

105 RSG, 8:74. The negotiations started on 15 October 1593. "[…] seker olie om alle geweren mette selve gestreken zynde schoon ende suyver te houden, alwaert datsse van zout water ende andere stercke materien besmet waren geweest […]." NL-HaNA, States General, 1.01.02, inv. no. 12298, fol. 75v [G15, 16 September 1593]

106 "uuyten roest zuyver ende reyn." RSG, 8:78. Another example was that of Hans Lucas van Persijn, who was promised an official post in one of the cities (either for himself or for one of his children) in case it was possible to reduce his invention to practice. RSG NR, 4:348 (27 November 1621). Yet another example was that of the carpenter Willem van den Houff, who had invented a cart and who wished to become a clerk or a supervisor of some sort. RSG NR, 4:581 (27 July 1622).

107 "Is Wilhelm Kick […] toegeleet tot eene vereeringe, voor de dedicatie van deselve syne inventie aen haere Ho.Mo. in recognitie van de costen ende arbeyt, by hem aen deselve inventie tot dienste van het land geëmployeert, de somme 320 carolus guldens." Dodt, *Archief*, 6:79. 2 August 1618. For more information about this case, see pp. 72, 201–202.

108 "…geeft te kennen dat hij door zeckeren const de scheepen veel harder can doen zeylen, versouckt daervan octroy ofte opte oorlochscheepen den 20en penning van de prinse."

invention. During the discussions in the Assembly of the States of Holland, the delegate for the city of Medemblik jotted down in his notes that "if the art is deemed fit he will be either granted a privilege, if such is possible, or at least an honest recognition done by the land."[109] In short, a recognition and a privilege were based on the same principle.

Furthermore, there are plenty of examples where the States General acted as a Maecenas for inventors without discussing the option to issue a privilege for exclusive market control. Willem Swart, for instance, obtained twenty florins for showing off a porcelain harpsichord on which he had worked for thirteen years.[110] Ghysbert Doncker "alias the Philosopher" (*alias de Philosoof*), on the other hand, obtained 74 florins in recognition of his *perpetuum mobile*.[111] Another example was that of Thomas Duyn, who received a total of 138 pounds "for a certain invention to harm the enemy."[112] Apparently, Duyn's invention, for which he never obtained a privilege, concerned a type of submarine. The Council of State noted in a resolution from 17 March 1604 that:

> On the basis of the report by Mr. Boner, who has been in contact with Thomas Duyn, the Englishman, who continues to make a ship-like instrument, which will carry its load under water, and which cannot be damaged by canons, it has been agreed [...] to employ his person [= Duyn] to have the boat made, disbursing him on top of his pension twenty-five stuivers a day for the duration of the construction, and an honest recompense of twenty-five guilders in case his instrument will prove useful after examination, taking into consideration all of the efforts he has made, and costs he has incurred.[113]

Veenendaal-Barth, Smit, and Vree, *Particuliere notulen*, 1:508. [No. 2273, 27 September 1622]. The Admiralties were to do additional tests. De Valet had first approached the States on 19 September, when Amsterdam; Hoorn; Enkhuizen and Medemblik were commited to examine the invention. Ibid, 1:484 [No., 2191, 19 September 1622].

109 "[...] ende ingevalle de const goedt bevonden werdt dat hem alsdan off octroy zal werden vergunt, so sulcx practicabel es, offte ten mintsten een eerlijcke recognitie bij 't landt ghedaen." Stellingwerf, *Particuliere notulen*, 1:508.

110 Dodt, *Archief*, 6:373 [21 Aug 1615].

111 Dodt, *Archief*, 7:33 [11 May 1617]. The amount was augmented after a couple of weeks "for good reasons" to 100 florins.

112 NL-HaNA, States General, 1.01.02, inv. no. 12503, fol. 61v. On 26 May 1601, Thomas Duyn receives 120 pounds and on 24 April 1602 he recieves another 18 pounds for "a certain invention to harm the enemy" (*zekere inventie om den vyant afbreuk te doen*).

113 "Op het Rapport van de Heere Boner in communicatie geweest hebbende met Thomas Duijn, Engelsman, die hem vervolcht te macecken een instrument schipgewijs, 't welck zijne inhebbende last onder 't waeter soude voeren ende met gheen canon beschadigt kunnen wordden, Is goetgevuonden [...] dat men zijne persoen aenneempt omme tselve schips instrument te doen maecken ende dat men hem des daegs geduijrende den tijt van

Recognition was not always a form of random power display: it could also be a form of primitive 'right'. In the proceedings of the Leiden Court of Justice, for example, one finds documents relating to the inventor Reynier Passchier, who had requested the members of the Court of Leiden to visit his new mill in accordance with the ordinance published by the bench of burgomasters and aldermen of Leiden "that those, who would desire to erect a fulling mill in or around this city would obtain a certain recognition."[114] Another example, where a recognition in the form of a reward functioned as an incentive for innovation was the competition held to find a solution to the problem of determining longitude at sea. On 1 April 1600, the States General decided to hold a competition offering 5,000 guilders to the first inventor who managed to disclose a reliable method for determining longitude. Several inventors then promptly came forward to offer their solutions, among whom for instance Galileo Galilei (this case will come up in Chapter 4 more extensively).

Holding competitions or giving away invention privileges was, however, not the only means for authorities to stimulate technological progress or to favor emerging industries. Another possibility was for the States to concede specific tax cuts. The entrepreneurs Dirck Jansz Hooft and Bartholomeus Munter, for example, who had erected a watermill (without privilege) near the city of Schiedam for processing copper materials, requested on 12 October 1613 "freedom of export tax (*licenten*) for all incoming materials, being copper and calamine [= a zinc type of mineral], as well as tax freedom for all commodities produced from the same materials."[115] In their justification of the request, the States used a type of language very similar to that used to justify a privilege. They agreed to favor suppliants for five years, taking into consideration "the large costs that they have made already to erect the aforementioned mills" and "in order to attract the business to this country."[116] The very same day, Antony

tmaecken desself over zijne psoen zall toelegghen vijff ende twintich stuuers ende tselve instrument bij proeve goet gevonden zijnde zall men hem Thomas Duijn een eerlicke recompensie gheven naer zijne alreede gedaene moeijten ende costen toegeleijt, vijff en twintich gulden [...]." *Navorscher* X, 35. Note that this submarine has not been descibed in any of the existing literature in the field.

114 "[...] in aansluiting met de ordonnantie door dit college gepubliceerd, dat degeenen, die eenen volmolen binnen offte omtrent deeser stede soude begeeren te stellen, zeeckere vereringe van U EE. Soud genieten." Posthumus, *Bronnen* 4:24 [No. 29]. For the use of fulling mills, see Munro, "Industrial Energy," 245–62.

115 "[...] vrij licent van alle incommende materialen, als rauw cooper ende calmynen, mitsgaders het vuytgaen vande gemaeckte wercken derselver stoffe." Dodt, *Archief*, 5:289–290.

116 "omme de nering in dese Landen te trecken [...] ten aansien vande groote costen, die sy alreede tot het maecken vande voorschreven molens [...] aangewendt hebben." To "prevent all fraud" (*omme alle fraulden voortecoomen*), the suppliants declared that each

Slicher and Harman Becx, who had "erected a mill in Amsterdam similar to the former," requested and obtained the same tax advantages.[117] Contrary to Hooft and Munter, Harman Becx had obtained a nine-year privilege for a new type of copper mill on 18 February 1612.[118] The example serves to illustrate how difficult it is to distinguish invention privileges from other means of supporting emerging industries, except for the fact that only an invention privilege contained the explicit clause that imitation of a specific technology was not allowed without consent and some compensation.

War, Trade, and Industry: Areas of Usage

It was in the movement of administrative papers that some projects were allowed go ahead whereas others did not. And it was by supporting specific projects, and by rejecting others, that the Dutch authorities shaped the production of knowledge.[119] The state could foster innovation, just as much as it could block the development of new ideas. Furthermore, the authorities influenced what type of knowledge was considered desirable by supporting specific research fields at the expense of others. It is worth, if only for that reason, to examine if there was a pattern in the kind of projects that the Dutch supported by means of privileges.

Not every request that came in was automatically granted. The auhorities turned down roughly 20%, either after examination, or because they considered at the meeting that the invention did not meet the stadards of privilegability. It could be that the technical basis was considered insufficiently substantiated, that the invention was not as new as the inventor claimed, or that the authorities did not consider the invention to benefit the common good, for example because it put people out of work (see pp. 219–220). Figure 26 provides a detailed overview of the various reasons for rejection.

---

mill would need hundred thousand pounds of calamine and copper. The freedom of tax would last for the duration of five years. Dodt, *Archief*, 5:289–290.

117 Dodt, *Archief*, 5:289–290.

118 Becx had originally obtained his privilege in companion with Aert Duyfkens, another merchant with several stakes in new inventions.

119 The argument presented here dovetails with the literature on paper tools and technologies. See, for example, Becker and Clark, *Little Tools of Knowledge*; Gitelman, *Paper Knowledge*; B. Jardine, "State of the Field"; Krajewski, *Paper Machines*; Yale, *Sociable Knowledge*; Bittel, Von Oertzen, and Leong, *Working with Paper*. See also the more recent attempt to place bureaucratic knowledge production within the framework of historical epistemology: Felten and Von Oertzen, "Bureaucracy as Knowledge."

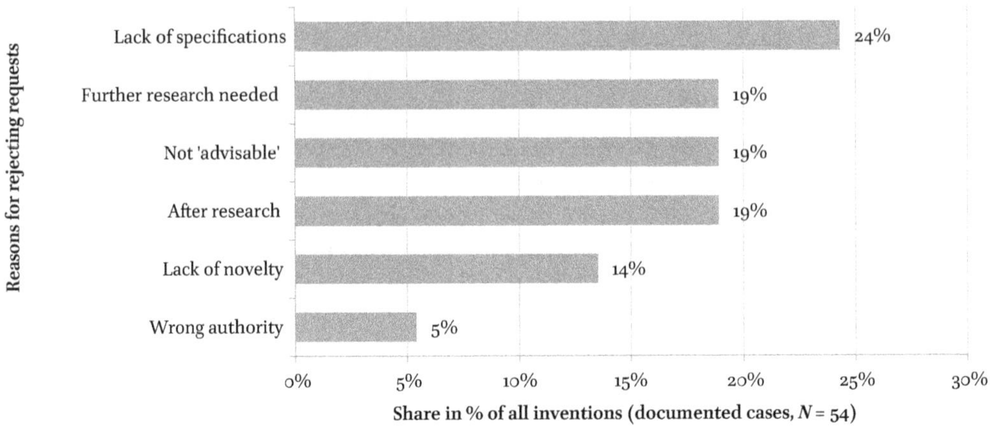

FIGURE 26 Reasons for rejecting privilege requests, 1589–1621

Administrative prerequisites as well as the screening of projects were to hold off entirely nonsensical projects, or projects that would undermine the credibility of the system. However, the vast majority of requests was granted. One striking fact that keeps coming back is the generosity with which the authorities rewarded those who attempted to find a solution for a specific problem. Inventors were seldom if ever shown the door and even the most unusual inventions received due attention. There were very few preconditions regarding the subject matter of an invention; an invention, for example, did not have to correspond to prevailing scientific theories (as is the case today). Never say never, seems to have been the motto; an unusual solution might just be of defining significance. In the event that an inventor returned repeatedly to bang on about a privilege, then, at a certain point, enough was enough. But even unsuccessful proposals were regularly rewarded. By way of illustration, it was decided to pay George Robbert of Pfeilffwergh 5 to 6 guilders per month on top of his normal salary, for his proposal "to enable sandy soil to yield decent fruits by preparing the seed", in case he decided to enlist in a military company in the Republic, so that he could meanwhile bring his invention to perfection.[120] But if the inventor wanted to leave, he would get 40 to 50 guilders "for his travel expenses" (*totte costen van syne reyse*) nonetheless. Similar indemnities were

120 "... een seer hoochnuttelyck arcanum ende conststuck ... door den ackerbouw, gelyck vuyt byleggende natuurlycke conste ende bewys te sien is, namentlyck alsoo, dat een gantsch magere sandige acker, door preparatie des saets alle jaer treffelycke vruchten dragen sal, alsoff men den acker met de allerbeste mist toegemaeckt hadde ...." Dodt, 7:11 (27 July 1617).

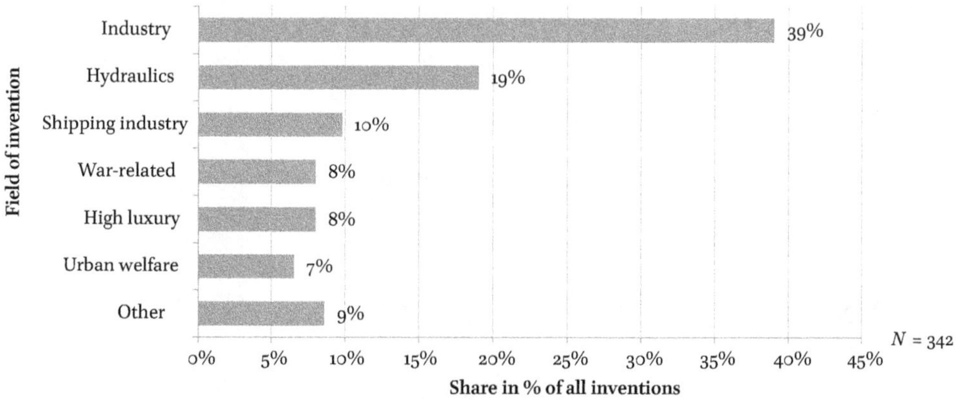

FIGURE 27 Invention privileges in the United Provinces by field, 1589–1621

disbursed on a regular basis. It seems the authorities thus wanted to promote an image of the United Provinces as being open to new ideas.

As Figure 27 attests, most of the privileges related to inventions in the field of military technology (such as cannons and shot), hydraulic equipment (pumps, dredging equipment), industrial technology (industrial mills, furnaces, stoves, and so on), and urban welfare (soot filters, water supply systems, and suchlike). There was also practically a separate category in which privileges related to expensive luxury items produced for the higher echelons of society. In 1618, for example, the stonemason and lens grinder Davidt Hermansz Nieman obtained a privilege to print "beautiful histories" on marble "as everyone may desire to his entertainment."[121] The famous Amsterdam sculptor Kasper Panten, to give another example, obtained a three-year privilege "to solely produce and cast any work and pattern that he will invent and model in the meanwhile."[122] The

121 "[…] toetssteen, Item Marmersteen, ende andere specien van Steen, die hart syn […] door de Meulen soo bequamelyck konde bereyden, dat men daervan sal connen maecken sonder yet te loor te gaen, Calumnen, Taeffels, daer Inne letteren, advisen, ende *soodanige schoone historien, als een yder tot syn vermaeck sal begeren*, Ja al wat men op Papier can prenten, ende dat soo suyver, ende schoon dattet een Glans heeft als een Spiegel, Item tot vorderinge van t'gebou friesen, ende ciraet in d'huysen, ende plaeten om in Gevels te setten daer op groote verheven Gulde letteren wapenen, ende alles anders tot yegelycx plaisier, t'welck alle in hem selffs goet blyft, ende onbederffelyck doordien den steen in sich selffs is ende onvervalst blyft […]." NL-HaNA, States General, 1.01.02, inv. no. 12302, fol. 82v [G172, 13 September 1618]. Emphasis added. On the patent of Nieman, see also Hurx, "The Most Expert in Europe."

122 "[…] alleene te mogen maecken ende gieten het werck ende patronen by hem daerentusschen te inventeren ende boutseren." Dodt, *Archief*, 7:13 (30 August 1617). Initially, Panten (*c.* 1585–1630) had requested a privilege "for life, for all the inventions that he has already found, as well as those he will still discover related to the art of casting stone, and sculpt-

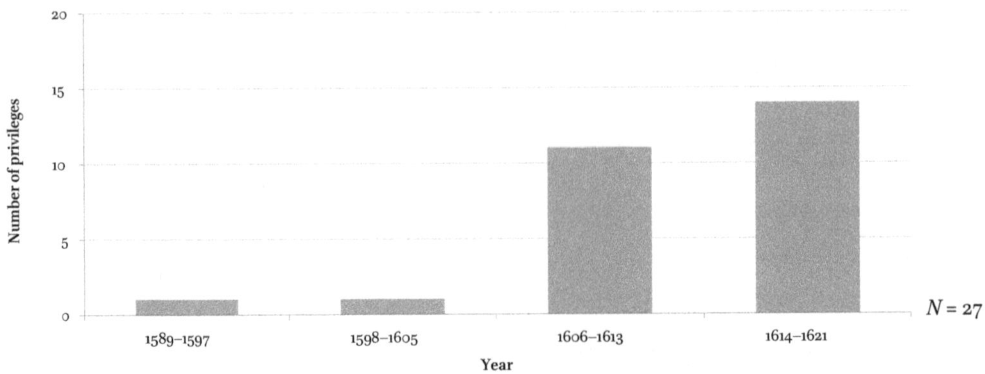

FIGURE 28 Number of inventions that related to high-end luxury items

number of privileges issued for the the esthetically pleasing, or 'high luxury' goods, increased over time, as Figure 28 can attest. Possibly, this increase was adjunct to the outcome of the Truce Troubles and the expansion of the court of Maurice (who began using the title 'Prince of Orange' February 1618).[123] It is also possible that there was a more general desire for luxuries in a period of relative peace that coincided with the Twelve Year Truce. In any case, most of the examples that relate to exclusive luxury items fall outside of the period we are dealing with here, i.e. they relate to the accession to power of Prince Frederick Henry of Orange (1584–1647) after the death of Prince Maurice (1625). Not long after, one finds automatic organs in the privilege rolls and things such as a harpsichord "that plays by itself, without the help of anyone" (1626)[124] or a book wheel "on which six books in folio format could be placed and used simultaneously" (1630).[125]

---

ing posts, columns, figures, chimneys, mantelpieces, etc." (*by hem al reede gevonden, ende die hy noch vinden sal, syn leven lanck geduyrende, van seeckere conste van steengieten ende boetseren van posten, colommen, figuren, schoorsteenen, schoorsteenmantels, enz.*) Dodt, *Archief*, 7:13 (21 August, 1617). But the States had insisted that Panten should first show some of the inventions and decided to issue the exclusive rights only for a limited period of time. For Kasper Panten's (also Kasper Panthen, Caspar van Panten) later career in Sweden, see Wrangel and Beets-Damsté, *De betrekkingen*, 37.

123 Taking on power, Maurice had not obtained the title 'Prince of Orange' because of the existence of an older brother who had been taken hostage by Spanish forces and who was raised Catholic in the Spanish courts. Only when the older brother Philip William of Orange (1554–1618) passed away in February 1618, did Maurice officially change his title to 'Prince of Orange.' About the complex issue of Maurice's titles, see Zandvliet, "Het hof van een dienaar," 44.

124 Doorman, *Octrooien*, 166 [G259, 18 February 1626].

125 Doorman, *Octrooien*, 179 [G304, 9 January 1630].

Yet a persistent taste for luxury was certainly not the only motivator of novelties. War was equally important. The relatively high number of innovations in the field of military technology must of course be seen as a logical result of the ongoing Dutch Revolt. Not only did government expenditure on military apparatus reach a record high, but also the many trade relations were characterized by the idea that "trade cannot be maintained without war, nor war without trade."[126]

The inventors of weapon machinery were rarely unknown previously to the authorities. The primary producers of the material during the early years of the Republic consisted of a relatively small group of entrepreneurs with close connections to political power.[127] What these entrepreneurs had in common as that they repeatedly sought for privileges to secure their investments and were closely involved with the import of raw materials needed for the production of war machinery. Jan Andriesz Moerbeeck, for example, one of the arms dealers from the inner circle, addressed the States General on 21 June 1619 on behalf of his company, requesting a privilege to produce cannons with imported material.[128] He would supply the raw materials and wished in addition to be exempted from paying any import tax.[129] The States General initially decided to consult with Prince Maurice, as was standard when it came to the maintenance of the military apparatus. The advice of the Captain General was probably favorable but has not been recorded. The entire issue was then raised again in the States General on 29 June. This time, the representatives for the province of Holland indicated that they first wanted to consult their 'principles' (*principaelen*) before making a decision about the privilege.[130] A month later, the States General decided that the Board of Admiralty of Amsterdam should be asked for advice as well. Since the Admiralties were responsible for collecting import tax in the Republic, the delegates at the general assembly

126 Jan Pietersz Coen, Governor-General of the Dutch East India Company, 1614. As cited in Parker, *The Military Revolution*, 132. The Dutch East Indies Company, for illustration, reserved sometimes up to 70% of its annual budget for military activities. The total expenditure on fortifications, for illustration, steadily rose from 3.2 million in 1591 to 13.4 million in 1632, and to 18.8 million in 1640. Parker, *The Army of Flanders*, 16–17.

127 De Jong, *Staat van oorlog*, 87.

128 Dodt, *Archief*, 6:392. Jan Andriesz Moerbeeck was also involved in the exploitation of an invention related to cotton printing, together with Harman Becx and Aert Duyfkens. Van Dillen, *Bronnen*, 1:700 [no.1186]. For more details, see page 86, note 71. Besides being a business man, Moerbeeck was also a writer and informal advisor to the WIC. In 1624, he published a pamphlet entitled "Why the West India Company should attempt to Conquer Brazil from the king of Spain, without delay." Moerbeeck, *Redenen*.

129 Dodt, *Archief*, 7:70.

130 Dodt, *Archief*, 7:72.

presumably wanted to get a second opinion about the financial consequences of the privilege.[131] In any case, the advice of the Admiralty was decisive and on 13 July 1619, Moerbeeck & Co obtained a twelve-year privilege to produce cannons.[132] The requested tax exemption was denied.[133]

Inventions produced for military purposes were not just related to the production of weapon machinery. For instance, there is the example of a corn mill used to feed the army (see pp. 145–146), but also plenty of other examples including bridges, boats, and ovens that served the needs of the military. Moreover, the threat of war motivated the production of instruments later used for different purposes, the most famous example of which was probably the telescope. The refracting telescope was, for that matter, not the only type of telescope on the research agenda of the States General. A medical doctor in Geertruidenberg, Thomas Raephout, petitioned for example for a privilege to protect another "invention with which one can see openly what happens within a radius of fifty miles."[134] Raephout's request was discussed in a meeting of the States General on 7 August 1615, when the States decided to commission its deputies Mr Grave and Mr Goyer (from Zeeland) to "hear the supplicant in the presence of his Excellency [Count Maurice]."[135] A day later, the commissioners reported that they had spoken with Raephout. Apparently, his invention was a kind of telescope that involved a mirror construction, but the commissioners "had not understood or noticed any substance in his proposal."[136] Their advice was not to engage further with Raephout's proposal, and the States General sent their agent Valkenburg to inform the inventor about this decision.

---

131 About the admiralties and taxes, see De Vries and Woude, *The First Modern Economy*, 98. From 1597 up to 1795 there were five admiralties. Of the five admiralties that constituted the standing navy in the republic three were in Holland (Rotterdam, Amsterdam, Hoorn/ Enkhuizen), one in the province of Zeeland (Middelburg) and one in Friesland (Dokkum, later Harlingen). Their coming into existence was a complex story of inter-city and inter-provincial rivalries. For an excellent overview, see 't Hart, *The Making of a Bourgeois State*, 39–43.

132 NL-HaNA, States General, 1.01.02, inv. no. 12302, fol. 129 [G178, 13 July 1619].

133 Dodt, *Archief*, 7:76. I have found no further evidence of the implementation of this project.

134 Another example of a mirror telescope can be found in the case of Johan Hoevelmans (from Frankfurt) from 1610. See Table 1.

135 "[…] te hooren op zyne aengediende inventie, daermede dat men vuyt zeecker instrument, by hem gevonden, kan sien, wat opentlyck vyftich mylen in't ronde geschiet, ende haere Ho. Mo. vande apparentie vande waerheyt rapport te doen." Dodt, *Archief*, 7:372 (7 August 1615). NL-HaNA, States General, 1.01.02, inv. no. 40, fol. 176.

136 "maer nyet en hebben kunnen begrypen ofte vermercken, dat in syn voorgeven fundament is […]." Dodt, *Archief*, 7:372 (8 August 1615). The States nevertheless decided to accord Thomas 10 guilders for his maintainance.

Aside from the production of particular products, another side effect of military innovation was that it helped legitimize particular sciences that had been left somewhat at the fringes of society. The German Doctor of Medicine Jacob Peutius, for example, whose ideas were similar to those of the famous German-Swiss occultist Paracelsus, obtained a privilege in 1597 for "multiplication of bread in a city under siege [and] to make good and healthy drinks in similar conditions, as well as to make salt and to find saltpeter."[137] Not coincidentally, saltpeter was one of the core ingredients for the production of gunpowder (the others being sulfur and charcoal). Thus, military interests within the privilege system plausibly contributed to the relatively high status of alchemy in the Republic.[138]

Still, privileges for army material or for luxury goods would never dominate the market. Much more often, privileges were used to stimulate the production of hydraulic equipment, such as sluices, pumps, dredging equipment, etc. For a country so governed by water and regularly plagued by floods, the frequent appearance of these technologies is not surprising. What is instead remarkable, in the period under consideration, is the seemingly omnipresent urge to drain various lakes and turn them into land. Although this development had started earlier in the sixteenth century, it only really found its stride at the beginning

137 "[...] van het multi-pliceeren des broots in een belegerde stadt om in gelycke occurentiën goeden ende gesonden dranck te maken, item sout te maken ende salpeter te vinden." RSG, 9:355. Jacob Peutius had studied in Leiden and Hardewijk, and stood in contact with the famous reformer Théodore Bèze (1519–1605). The German doctor had first filed his request on 2 April 1596, when the States decided to commission Junius, Van de Warcke, De Bye, and Doubleth to negotiate with the supplicant. The commissionars reported back on 9 April, when it was decided that Peutius should submit his inventions in writing. On 13 April, Peutius was promised 100 guilders upon delivery of all his inventions. On 21 July, he was personally present in the States General with a sample of his inventions. Peutius demanded a monthly stipend, and on 31 August, Sille, Cuyck, and Rengers were committed to investigate the matter further. Peutius wished to be "treated" (*getracteert*) as an engineer, yet it is unclear if he managed to obtain this title. On 31 December 1596, he did obtain 100 guilders, and he was asked to come again to the Assembly of the States General to discuss the details of his privilege. Information on this case can be found in RSG, 9:354–55. On 25 January 1597, Peutius finally obtained a fifteen-year grant of privilege; a clause that forbid the import of similar products produced abroad was, however, denied. RSG, 9:703. On 6 september 1603, Peutius received another 40 pounds for medicine that he had prepared for the city of Oostende. NL-HaNA, States General, 1.01.02, inv. no. 12548.119.

138 Even scientists defending what we came to call the 'mechanical worldview' strongly supported the business of alchemy. Among them, the military engineer Simon Stevin, who advocated: "the marvelous and very unusual subject of alchemy, unknown among the Greeks, which recently began to make its appearance again, by means of which people are able to inquire into the nature of substances, to their profit, in another way than was possible for them to understand without this great science." Stevin, *Principal Works*, 3:605.

of the seventeenth century. The most famous example of these large-scale land reclamation projects was probably the *Beemsterpolder* (1612). This polder was drained with a series of mills designed and 'patented' by state engineer Simon Stevin; the hydraulic engineer Jan Adriaansz Leeghwater (1575–1650) supervised the milling, using Stevin's inventions just after the privilege had expired and the technology had entered the public domain.[139]

Another important area of privilege usage was shipping. This was a broad category that not only included the invention of boats and navigation techniques, but a number of other issues as well. The surgeon Adriaen Sare Romen, for example, obtained a privilege because he had managed to produce:

> a new substance of pitch [...] to daub ships that sail on the East and West Indies, as well as other far-away places. The strength of the invention is that water cannot damage these ships, that the sun cannot melt it, that the worms cannot eat it, and that fire has no grip on it.[140]

Many inventions in the shipping industry were directly linked to the discovery of the New World and the exploration of new trading zones. So, there was a certain overlap between inventions for shipping and inventions produced for

139 For a fascinating study of 'conservative innovation' during the reclamation of the Beemster polder, see Fleischer, "The Beemster Polder." It is worth noting that during a meeting held on November 13, 1607, the chief administrators of the Beemster considered new innovations from various inventors. The examination of the various inventions (some of which had already been patented) was carried out by the members of te water boards (*hoofdingelanden*) themselves together with licensed mill masters (*molenmeesters*), on the condition that "when the said inventions are found on further examination to be very uncertain" (*alzoo de genoemde uitvindingen bij nader onderzoek zeer onzeker bevonden worden*), one would stick to the "ordinary old works" (*ordinaire oude werken*) so that the project would not be delayed. Bouman, *Bedijking*, 59–60.

140 "[...] seecker substantie van peck, by hem geinventeert, omme daermede te besmeren de schepen, varende op Oost- ende Westindien, mitsgaders andere verre gelegen plaetssen, daervan de cracht is, dat het water die schepen nyet en kan beschadigen, dat de sonne het voorsz. peck nyet en kan doen smelten, die wormen met en kunnen dooreten, ende dat het vier ofte vlamme daorop nyet en kan vatten ofte branden." Dodt, *Archief*, 5:291. NL-HaNA, States General, 1.01.02, inv. no. 12300, fol. 432–432v [G126, 29 November 1613]. Sare Romen was a surgeon in the service of the States General. His invention had been "tried during a long voyage to St. Thomas and Brasil on a certain ship named the Greyhound (*Hasewindt*) under our armada, of which squire Pieter van der Does had been the Commander" (*geprobeert was aan seecker Schip genaempt de Hasewindt gedaen hebbende de lange voyage op St. Thome ende Brasilien onder onse Armade waerover Commandeur was Joncker Pieter van der Does*). The ship had shortly afterwards sailed to Venice to transport corn. The Dutch West Indies Company later established a post on St. Thomas in 1657, and colonized Brazil between 1630 and 1654.

military purposes. In 1620, for illustration, the brothers Harman and Dirck Warnartsz together with the investor Hendrick Frist, obtained a privilege to build a multifunctional amphibious assault ship. The supplicants explained that the new invention served "to put people on land, to find and discover unknown and foreign Coasts, Rivers, Bays, and sand bars."[141]

The 'global dimension' of many inventions was not limited to shipping issues. Another issue that emerged was to process materials imported from the New World. One of the first invention privileges ever granted by the independent Dutch authorities, for example, related to a new type of mill that served to cut "all types of hardwood such as Brazilian [wood]."[142] Yet another example was the furnace invented by the protestant minister Jan van Linden de Jonge, who described his invention as:

> an instrument that makes all salty, filthy, and unfit waters sweet, pure, and tasty, by using the art of distillation, which is very useful for those who take to sea, who make long journeys, and who otherwise would contract diseases and ailments because of a lack of sweet water [...] it is not only fit for drinking and cooking, but also to treat scurvy [...].[143]

These types of inventions only made sense in the light of long-distance trade. In the traditional histories of modern patent law, however, the importance of cross-continental encounters is often obscured, because of the use of a

---

141 "[...] mitsgaders, omme te doene vliegende tochten, ende volck te Landen, omme alle vreemde ende uytheemsche kusten, Rivieren, Bayen, ende droogten op te soecken, ende te ontdecken [...]." NL-HaNA, States General, 1.01.02, inv. no. 12302, fol. 205v [G192, 16 December 1620]. The privilege was granted "after wide and extensive communication with His Princely Excellency" who in his capacity of Admiral had supervised the examination.

142 "[...] alderhande hert hout als Presily ende ander can breken [...]." NL-HaNA, States General, 1.01.02, inv. no. 12298, fol. 3 [G3, 30 January 1589]. The suppliant was Albert (also Elbert and Ellert) de Veer, secretary of the city of Amsterdam. Besides, he was the owner of a inn just outside the *Regulierspoort* and he held a exclusive privilege in Holland for breaking flax. Elias, *De vroedschap van Amsterdam*, 1:51–53.

143 "Instrument, waermede men met seer cleyn vier in merckelycke quantiteyt alle soute, vuyle, ende onbequaeme wateren soet, reyn, ende wel smakende conde maken, door die conste van distillatie, seer dienstelyck wesende den genen die op schepen varen, die lange reysen doen, ende anders door gebreck van soete wateren in verscheiden sieckten, ende ongemacken commen te geraecken, ende dat tvoors. gedistilleert water niet alleen goet is om drincken, ende koken, maer oyck met cleyn middel bereyt conde worden om het scheurbuyck ende andere sieckten wech te nemen." NL-HaNA, States General, 1.01.02, inv. no. 12298, fol. 310v [G49, 12 June 1600]. The installation of an efficient distillation plant on board of ships remained a problem for seaborne empires until a satisfactory solution was found in the nineteenth century. Cf. Goodman, *Power and Penury*, 137–41.

particular set of categories. For example, the invention of the mill to cut Brazilian wood would be categorized as an 'industrial windmill', whereas the invention of the furnace would fall under the category 'heating equipment'. The importance of global trade is therefore consigned to the background, or does not figure at all. If, instead, one would use a different type of catagories – asking the question 'what inventions served for' instead of 'what they are' – a very different picture emerges: as Figure 29 attests, a careful estimation shows that on average ca. 10% of all inventions indeed belonged to the category of 'global trade'.

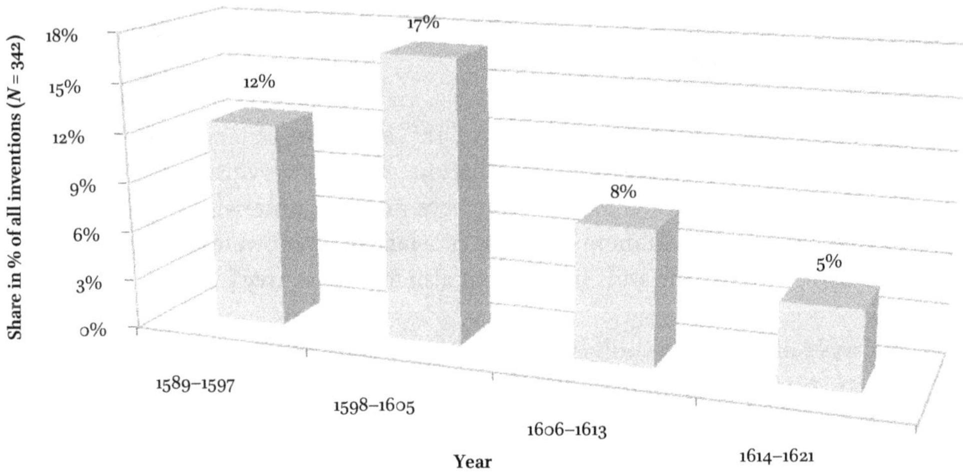

FIGURE 29 Percentage of inventions related to global trade

The problem of categorization returns in other cases as well. Pauwels Auleander, for example, received a privilege for his invention of an instrument "highly needed and useful to extinguish fire, which started both inside and outside the houses."[144] The invention was clearly produced with a view to reducing some of the risks of living in the city. But the fire extinguisher was also useful for "artificial rain fall" (i.e. irrigation) as well as for war and navigation purposes. Another example was that of Pieter Kayseel and François Verhage, who had invented a mill driven by man power (instead of wind or horse power). The mill "occupied little space" and worked "in such a way that the sound will not be heard more [loudly] by the direct neighbors than any other

144 "[...] ten hoochsten noodich ende proffitelyck was tot uuytblussinge des brants, soo wel van binnen als van buyten de huysen opgegaen [.........] artificiellen regen te doen vallen [...]." NL-HaNA, States General, 1.01.02, inv. no. 12301, fol. 47 [G132, 11 June 1614].

manual work."[145] Obviously, this mill serving the production of "cloth and candles as well other things" was particular useful in the cities, and it could either be classed as contributing to 'urban welfare' or to 'industry' (it has now been classed as a 'multifunctional industrial mill'). In short, there is no escaping a certain interpretation when categorizing the use of inventions and, although the percentages in the different tables give a good impression, they should be taken *cum grano salis*.

Nevertheless, by any account, there is no doubt that 'industry' was by far the biggest category among the privileged inventions. This category consisted of various branches, including multifunctional industrial windmills, furnaces, water mills, wood production, copper production, but also the production of porcelain, hemp, and paper. Figure 30 gives an overview of these different industry sectors.

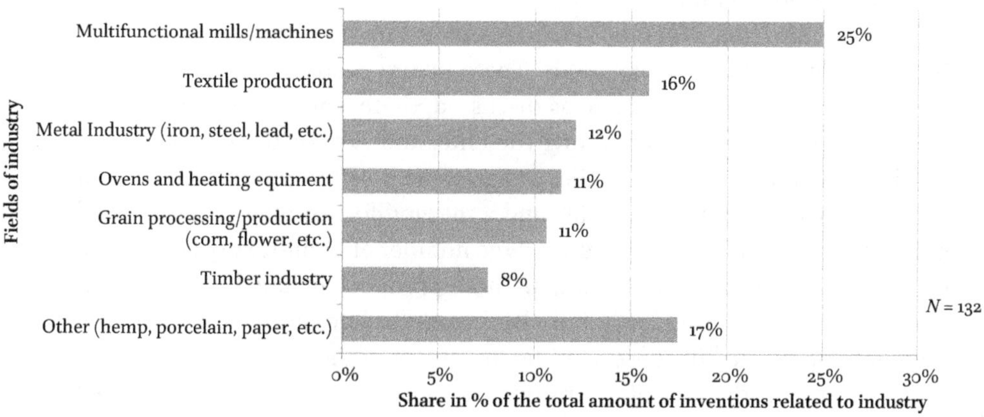

FIGURE 30 The type of privileged industry, 1589–1621

Some of the categories in Figure 30 are again closely connected to other categories. The production of lead and iron, for example, serviced particularly the machinery of war, since it was used for the production of bullets, amongst other things. Boulting machines, on the other hand, clearly belonged to the category of 'grain processing' as did corn mills, but some multifunctional machines appeared to be useful for grinding corn as well. One example in this context was the invention of:

145 "In soodanigen stillicheyt dat t'geluydt van dien by de aldernaeste gebuyren nyet meerder can gehoort worden als andere gemeene hantwercken [...]." NL-HaNA, States General, 1.01.02, inv. no. 12301, fol. 196 [G136, 21 May 1615].

> a mill, which would turn by one man [running] in a wheel and which, using two water wheels, would steadily throw out over eight tons of water, to five feet high. The water that had been thrown out could in turn make a corn mill turn without wind, being as well able to render dry, not expensively, large amounts of water, lakes and other place filled with water—of which [the inventor] had made an experiment on a small scale.[146]

In fact, most of the industrial mills were not made for the production of a specific product only; they were meant to generate energy more efficiently. Since the construction of economical ovens and heating equipment was equally driven by a desire for the efficient use of energy, one could perhaps speak of a sort of 'energy revolution' that unrolled in the era under consideration.[147] What sparked this revolution were the growing energy needs to produce goods for domestic production.

Underpinning the different catagories in Figure 30 runs the issue of an increased demand for 'luxury goods'. These luxury goods can be divided into two groups.[148] On the one hand, as discussed, there were a number of items produced for a select group among the elite, and not for the masses. Many textile inventions belonged to this category, such as the production of gold and silver wire, mills to weave silk, and a method for weaving Turkish camel hair. On the other hand, there were a large number of 'semi-luxury goods' destined to be sold in a broader market. These products included paper, porcelain and pottery, soap, and salt, but also strings for lutes and hat makers. The frequent appearance of these semi-luxury goods in the privilege documents seems to support the idea of a consumer revolution that supposedly unrolled in the sixteenth and seventeenth centuries—an idea that has been developed

---

146 "[…] werck van een molen, t welck door t gaen van een man In een Radt soude omgaen ende met twee Schepvaten gestadelyck over d'acht tonnen waters soude connen vuytwerpen, tot vyfftalff voeten hooch, Welck uuytgeworpen water, wederom een Corenmolen sonder wint soude connen doen omgaen, wesende mits d'oncostelycheyt van twerck bequaem omme groote wateren, plassen ende andere plaetzen met water vervult drooch te maecken, daervan hy de prouff int cleyne gemaeckt hadde." NL-HaNA, States General, 1.01.02, inv. no. 12299, fol. 336v [G100, 6 November 1607].

147 See also Chapter 5.

148 The distinction I draw here runs somewhat parrelel to the distiction between "Old" and "New Luxury" as drawn by Jan de Vries. De Vries, "Luxury in the Dutch Golden Age." De Vries suggested that New Luxury was rather urban than courtly and was rooted in a growing economy built around sociabilty and comfort rather than distinction and display. I see more continuity between the old and the new, and emphasise rather the dissemination of products.

in particular by Joan Thirsk.[149] Thirsk embedded invention privileges in the wider context of state-sponsored projects to spur the development of a consumer society in early modern England. In the Dutch Republic, many inventions responded to a new type of consumerism too. Dierck Jansz Spiegel, for example, adduced in his privilege application that his new method of making pottery would come "to the relief of the poor community" (*tot soulagemente van de schamele gemeente*).[150] Archeological research has shown that people of moderate means were probably indeed the client base of Spiegel; his products (see Figure 31) are mainly found in the barnyards and refuse pits of the average urban dweller.[151]

FIGURE 31
Red glazed ceramic pot, trademarked by DCS (Dierck Claesz Spiegel), 1603

Now, what about the effectiveness of invention privileges? Which kind of projects "made it" and which did not? It is difficult to give a straightforward answer to this question, not only beause of a lack of data but also because of the large degree of variation in this regard. Some of the inventions made under the privilege scheme were highly successful and decisive in fabricating

149 Thirsk, *Economic Policy and Projects*.
150 NL-HaNA, States General, 1.01.02, inv. no. 12299, fol. 77v (20 July 1602).
151 See the excellent essay by Ostkamp and Venhuis, "'tot Soulagemente'," 49.

the image of the Dutch as a highly innovative bunch of people, setting the pace for technological development in the first half of the seventeenth century. Examples include the wind-powered sawmill invented by Cornelis Cornelisz of Uitgeest, which not only defined the Dutch landscape but also spurred on industrial development in the Zaanstreek for centuries to come.[152] Other inventions were less successful, either because schemes did not work out as planned or because there were problems in terms of distribution and implementation. A good example was the privileged water mill invented by Simon Stevin, which he exploited at various locations together with Johan de Groot (father of the famous jurist Hugo Grotius). One of those locations was IJsselstein, where the polder administrators were highly dissatisfied with the performance of the mill, felt cheated, and finally sued De Groot and Stevin. The inventor concluded, after inspection, that the mill was poorly constructed, that the millers were incapable, and that sabotage had been at play.[153] There are even stories about inventors who ended up in jail on allegations of fraud.[154] Yet, as Joan Thirsk rightly remarked in her analysis of privileged projects in early modern England: "Scandals about projects ... were but the scum on the surface of a healthy current of water that flowed strongly underneath, and it is this current which we must follow."[155] Indeed, for my purposes, the exact success rate of inventions is less important than the awareness that that privilege system opened up new entrepreneurial possibilities and incentives that changed the nature of desirable knowledge in the field of technology.

Infringement and Fines

The sanction for infringement on invention privileges usually consisted of the confiscation of any counterfeited products or instruments plus a monetary penalty. Penalties fluctuated between 100 and 1000 pounds, which was an amount that typically had to be distributed in equal parts between the local prosecutor, the inventor, and poor relief.[156] Yet other allocation arrangments

152 Pieters, *Uitvinders in Nederland*, 53–55; Kaptein, "Een unieke economie." The mill was not an entirely new invention but an amalgamation of techniques that were already in use elsewhere. Kaptein and Schotsman, "Alkmaar as bakermat", 204.

153 Devreese and Vanden Berghe, *"Magic is No Magic,"* 86. There are some 70 documents related to this case that was litigated in the years 1590–1595.

154 For a later example, see Buning, "Brokered Innovation."

155 Thirsk, *Economic Policy and Projects*, 11.

156 For comparable clauses in the English system of patents, see Duncan, "Monopolies under Elizabeth I, 1558–1585," 262.

were also possible. In one case, the monetary penalty was to be paid in equal parts to the inventor and to the States Auditor. In another case, the fine was to be paid to the inventor, the prosecutor, and the declarant. There seems to have been no clear system of awarding punitive damages from the powers above.

The number of infringement cases that eventually led to guilty verdicts was limited. When lawsuits did emerge, they were usually about the presumed novelty of an invention, or about the geographical scope of a privilege, in which case pecuniary penalties were of lesser importance. Violations of the privilege were mostly settled under the watchful eye of "arbitrators."[157] The inventor Willem Kick, for instance, sued his companion Pieter Jansz when making a surprise visit to his house, where he found several objects gilded and varnished according to his privileged method. The inventor was accompanied by a notary, who reported the event and claimed that "the objects were made villainously and against his privilege. They were henceforth forfeited in addition to a penalty of 100 Flemish pounds."[158] Eventually, it was agreed that the objects would temporarily remain in the house of Pieter Jansz, until the two men had agreed upon a settlement under the guidance of a number of "good men" (*goede mannen*).[159] The temporary agreement was made on condition that the defendant would not change or touch anything in the stock of counterfeit products and that he "would also indicate and reveal several other people, who were infringing the privilege of the plaintiff and made similar works."[160]

In the event that a legal conflict got out of hand, it was generally resolved by local and provincial courts, and in the final instance by the High Court of Holland and Zeeland.[161] In some cases, however, the States General preferred to keep the "determination of the punishment to themselves," which meant that it would deal with the issue whenever problems arose.[162] But even when

157 Arbitrator was a contemporary term in this context; they were to hear quarrelers so as to deal with the dispute "in a friendly manner" (*Het zijn Rechters ofte Scheyd-Mannen, buyten de Vyer-schale, met gemeyn verdag van den twistenden ghekoren, om haer-lieder verschil, ofte twist, in 't vrundelijk af te doen*). Van Aller, *Generale regulen*, 81.

158 "d' voors. Goederen gemaect waren in villipendie ende tegen sijn octroy dat daeromme d'selve met noch 100 ponden Vlaems daerenboven waren verbeurt [...]". Van Dillen, *Bronnen*, 2:375-376 [No. 643, 20 Sept 1620].

159 Ibid.

160 "oock soude aenwysen ende openbaeren verscheyden ander persoonen, die des requirants octrooy infringeerden ende gelycke wercken maecten." Ibid.

161 For the history of the highest judiciary power during the Republic, as well as a very useful guide to its archives, see Bailly and Verhas, *Hoge Raad van Holland, Zeeland en West-Friesland*.

162 "[...] dan by aldien daerop tusschen beyden questie geryst, sullen haere ho. Mo. deselve questie decideren, gelyck sy sullen bevinden te behooren, nae de gelegentheyt der saecke."

cases were dealt with in the usual way, the States General often defined the lines along which conflicts were to be settled. When Henrick LeMaire, for example, demanded the cancellation of a privilege granted to Richard Fouler, the States General decided his request be sent to the magistrate of Delft for advice.[163] When Louwys Lorrel complained about a privilege granted to his rivals Aert Duyfkens and Steven Versteech, the States General decided that "in as far as the suppliant is obstructed to exercise the before-mentioned art [of printing flowers and patterns on satin and other material] he shall have the right to oppose [the privilege] in front of a competent judge."[164] Time and time again, the States General was the first point of contact.

Challenging infringements could nevertheless be a complicated affair. One inventor who claimed that his rights had been infringed upon, let slip that he would not "seek to obtain any other privilege any time soon, because it is but a continual battle, where one always has to stand with a sword in hand."[165] He was not alone in encountering difficulties when defending his rights. John Osborn, for example, who had obtained a ten-year privilege for his method of working whalebones in 1618, complained to the States General that those who infringed upon his privilege could not be sued, because they claimed "to keep no works among them, and pretend not to be the proprietors of the instruments that are needed [to produce] those works, or obstruct [him] with opposition and legal processes."[166] His privilege was thereupon amended in 1625, and it was agreed from then on "that over and above the penalties to be paid

---

Dodt, *Archief*, 4:14; NL-HaNA, States General, 1.01.02, inv. no. 12301, fol. 212v [G138, 6 June 1615]. Another example where the States General "kept the judgement of issues that will [arise] from this privilege or because thereof" (*kennisse ende Judicatuyre van questien die op dit octroy off ter cause van tselve*) can be found in NL-HaNA, States General, 1.01.02, inv. no. 12301, fol 136v [G134, 21 January 1615].

163 Dodt, *Archief*, 7:61.

164 "Is verstaen, by sooverre als den suppliant verboth werdt gedaen int exerceeren van de voors. konste, dat hy hem daertoe met rechte sal mogen opponeren voor zynen competenten rechter." Dodt, *Archief*, 7:67 (9 October 1619). Lorrel complained that Duyfkens' invention was not new, and that his privilege should for that reason be nullified.

165 "Doch ick sal my nu wel wachten, van eenigh dingh Octroy te halen, want 't is maer een geduerigen strijt, men diende wel altijdt met 't swaert inde handt te staen, en noch soude men niet beletten, dat sulck volck onder den duym doen souden [...]." Jan Verqualje, *Uytvindingh* (Amsterdam, 1661). Preface, without page numbering.

166 "[...] ende dat sy luiden geen wercken by haer houden, ende pretexeren off geen eigendom te hebben aen de instrumenten tot deselve wercken nodich, off d'executie van de selve den supplt. met processen ende oppositien belettende [...]." Doorman, *Octrooien*, 163 [G247, 15 February 1625]. Original privilege NL-HaNA, States General, 1.01.02, inv. no. 12309, fol. 54v [G167, 3 March 1618].

according to the privilege, all the instruments such as presses, molds, models and patterns serving or suitable for the pressing of baleen will be confiscated."[167]

Occasionally, the authorities intervened as well when Dutch merchants ran into trouble abroad for infringing on privilege rights that applied locally. In 1585, for illustration, the young Stadtholder Maurice of Nassau urged Leicester to send a letter to the Queen of England, because eight barrels of steel had been impounded by the Merchant Adventurers in the name of a twelve-year patent for exclusive steel production granted to a certain Robert Beale, a "scribe in the Privy Council."[168] Maurice defended the interests of a company of Dutch merchants who complained to the States General that the seizure was in conflict with the general trade agreements between England and the Netherlands.

Conclusions

This chapter began by charting how bureaucratic practices shaped projects and inventions. In doing so, a number of interesting issues emerged. First of all, it appeared that there was a clear division of tasks when it came to the settlement of inventions in the field of military technology: whereas the States General was vested with the highest authority, and ultimately granted the privilege, the Council of State was largely responsible for the examination and settlement of privileges in the field of military technology. In his capacity as Captain General of the troops of the Republic, it appears that Stadtholder Maurice played a much more important role than hitherto assumed in promoting innovation. Often, he was personally involved in the examination process of particular technologies and he kept a close eye on their further development.

Another thing that came to light is that cities, although extremely important at every stage of granting privileges, were ultimately subject to the consensus that emerged from the (inter)provincial assemblies. The federal authorities eventually paid the piper to call the tune. In fact, exclusive privileges were just one option among many to encourage innovation: the States repeatedly allotted rewards, bonuses, and other financial advantages to encourage inventors too. One remarkable facet is how generous and open-minded the authorities

167 "[...] boven de penen in denselven octroy begrepen oock ten behoeve van de voorsz. Jan Osborn sijne erven oft sijns actie hebbende sullen verbeurt zijn alle Instrumenten van perssen, formen, modellen ende patronen tot het drucken van baleinen dienende one geapproprieert sijnde [...]." Ibid.

168 "clercq in den Secreten Raede." RSG, 5:51.

were in their negotiations with the different actors. This was no doubt intended to contribute to the public perception of the Republic as a desirable location in which to implement new ideas.

Finally, this chapter has discussed the areas of privilege usage. It showed how many inventions could be classified under multiple categories and that global trade was much more important for Dutch innovation than usually assumed. Nonetheless, it was not war or shipping that drove the authorities to push innovation by means of exclusive monopolies; by all accounts, it was the promotion of local industry that was at the center of privileged innovation. As a reference point, the system of invention privileges confirms thus the idea that "for the Dutch, commerce was king, and in Holland commercial interests effectively replaced *raison d'état*."[169] The next chapter will explore in more detail how state officials, merchants, and inventors indeed found each other in an emerging market of ideas.

169 Braudel, *Civilization and Capitalism*, 3:205.

CHAPTER 3

Merchants of Ideas

People of all kinds and origins encountered one another because of the privilege system. These encounters occurred in both a very concrete as well as in a more abstract sense. In the concrete sense, people with different backgrounds and from various social standing were brought face to face with one another. In a more abstract sense, the privilege system functioned as a place of encounter where different ways of understanding technological inventions could fuse. This chapter elaborates on how it was the language of profit that brought inventors, merchants, visionaries, and administrators together.

The Dawn of Projects

Although some authors in classical antiquity may have seen a correlation between the notion of priority and the ownership of ideas, the possibility of upholding the rights to a new idea in a courtroom was a typical consequence of Renaissance thought; with the emergence of a new concept of individual authorship, ideas gained a new legal and economic status too. Yet claiming ownership over ideas is not a straightforward issue, if only because ideas are intangible and allow for simultaneous use. One cannot establish the boundaries of ideas by literally 'taking the idea' from another person.[1] Moreover, ideas continuously build upon each other, which makes it hard to define where new ideas begin and old ideas end.

Indeed, modern patent law has worked its way around these problems by acknowledging that not all the elements of novel ideas are necessarily novel of themselves. When inventing a new type of bridge, for example, an engineer is permitted to incorporate common knowledge and make use of known materials, without his invention losing its novelty. What matters is that the inventor applies any familiar knowledge in such a way that others could not have predicted the result. This is what the law calls the 'inventive step' or 'non-obviousness', which is a crucial requirement nowadays if inventions are to be eligible for patent protection.[2] If a person with ordinary skills can deduce an invention

1 Drahos, *A Philosophy of Intellectual Property*, 22.

2 The expression 'inventive step' is preferred in Europe, whereas 'non-obviousness' is used in the legal system of the United States. The standard view is that the doctrine of non-obviousness

 | DOI:10.1163/9789004320420_005

on the basis of what is already known to the public, for instance by making a minor improvement to an existing machine, the invention will not be considered worthy or innovative enough for legal protection. What matters, in terms of claiming ownership over new ideas, is that the inventor puts the different elements in a new order and thus makes them appear in an altered (and new) fashion.[3]

The thought that ownership might follow from a combination of existing things did not originate with the development of intellectual property. Roman law had, for instance, recognized a "mode of acquiring property that involved the addition of value to property through labor or the addition of new materials."[4] It was called the principle of accession. Accession was based on the idea that an "accessory thing when annexed to a principle thing becomes part of the latter, and thereupon and thereby becomes the property of the owner of the principal thing."[5] A classic example was the situation where someone had built a house on landed property owned by someone else. The land in that case was considered more important than the house in question and, in consequence, the owner of the land (principal) became the owner of the house (the accessory).[6] The acquisition of new property by means of accession only came into play after an intimate and inseparable union of the accessory with the principle. The Romans had even used the expression of "making a new thing" (*novam speciem facere*) in this context. But even though it is tempting to see in these discussions within Roman property law a forerunner to our notion of intellectual property, there was at least one fundamental difference: Roman law had never brought the production of materials into the context of legal action *in a stage where it did not have a tangible form.*

---

only developed in the course of the nineteenth century in the framework of rapidly developing economies that were based on technological advancement; the genesis of the non-obviousness doctrine in the US is usually traced to the Supreme Court's decision in *Hotchkiss v. Greenwood* 52 US 248 (1850). For a discussion of the role of the inventive step in the privilege system, see Buning, "Making Things New."

3 This is obviously a somewhat simplified rendering of all the complexities that are involved in the determination of the inventive step. Especially in the case of biological patents, the distinction between products of nature and inventions is sometimes difficult to make. In the case of DNA patents, for instance, what matters is the degree to which sequences are "lifted" from their natural state. On these issues, see Beauchamp, "Patenting Nature"; Kevles, "Patents"; Kevles, "Inventions."

4 This paragraph is strongly indebted to Long, "Accessio". See also Van der Merwe, "Nova Species," notably 102–108.

5 Brown, *A New Law Dictionary*, 9. In Dutch law, the principle of accession was called '*natrek*'. Van Leeuwen, *Rooms-Hollands-Regt*, 171–75.

6 Nowadays, lawyers use the term 'specification' for this process.

As we have seen, early-modern authorities repeatedly issued privilege rights prior to the actual reduction to practice of an invention. The exclusive exploitation rights were an enticement to invest, rather than a reward for an invention that had already proven its utility. The Dutch authorities often granted the inventor a limited amount of time (mostly about six months to a year) to crystallize his ideas, or else his privilege rights would expire.[7] This formula allowed the inventor enough time to look for investors, who were prepared to put in the money to realize the project in exchange for a percentage of the profit that could be made on the basis of the future monopoly rights.[8]

The privilege thus became something like a hallmark showing that the authorities had given their approval for a particular project. For this reason, it was in their interest to carefully examine the proposals put forward by different inventors. *Not* granting a privilege did not mean, however, that the authorities considered a project of no importance, or that the implementation of an invention was forbidden. When Pieter Silverer from Trieste, for instance, requested a privilege for the duration of 100 years to provide all the cities in the Republic with fresh and clean water, "not only for the brewers but also for other Burghers in their houses, playhouses and elsewhere," the States of Holland decided that:

> the Suppliant will be allowed to use his art in these Lands at his own risk, cost, and peril, and for the profit and gain as the Suppliant will be able to negotiate with the Inhabitants of the Land, without that the States [...] could accord a privilege as requested.[9]

Inventors were allowed to set their inventions to work without a privilege, as long as they respected local legislation. On the other hand, the authorities took the position "that there are no rules that oblige someone to accept the invention of the supplicant against his will."[10] The privilege was nothing more than

7 This was not unique to the Dutch system: also in Venice, for instance, it was common to require from the inventor to "*fare la experientia*" within a year. Berveglieri, *Inventori stranieri a Venezia*, 27.

8 Thus, inventors might adduce that they were "poor artisans" (arme ambachtsluyden) and that they hoped that the privilege would open the possibility that the project would be allowed to proceed without them being ruined. NL-HaNA, States General, 1.01.02, inv. no. 12298, fol. 335 [G50]. For other examples, see also Davids, "Patents and Patentees," 273–75.

9 NL-HaNA, States of Holland, 3.01.04.01, inv. no. 5677, fol. 295 (12 March 1590).

10 Philip Brugman had complained to the States General about infringement upon his privilege that had taken place in Leiden. The States decided to draw up a report: "ende gehoort

a form of endorsement by the States to assure potential investors a reasonable return on their investments.

The authorities did insist on concrete results. If an inventor did not manage to reduce his invention to practice within the time-scale foreseen, his rights would be invalidated. Figure 32 attests how the authorities explicitly mentioned the existence of an "exploitation clause" in the privilege acts from ca. 1606 onwards. The data suggests that, over the course of time, the Dutch increasingly included such a clause to prevent inventors from blocking the market based on their rights. At the same time, the data shows that the authorities became much more lax about demanding the careful documentation of projects in the making. It may be that they increasingly abandoned the idea of "specifications" in order to consciously push for speculation on new ideas; after all, little would be lost if an inventor who had gone to the trouble of obtaining exclusive rights was not able to reduce his ideas to practice, since his rights were automatically invalidated should a successful outcome not be achieved.

Even if many invention privileges related to projects, the legal understanding of 'invention' was profoundly characterized by materiality.[11] Privileges dealt with physical objects only, or with methods leading to concrete results, which is in stark contrast to modern patent law and its emphasis on the inventive step. Although, in theory, it must be possible today to reduce the inventive step to practice, the inventive step itself is intangible and not identical to the physical object of the invention. In the early modern period, such a distinction was not made: all that counted was that the invention would result in something tangible.[12] As Mario Biagioli put it:

---

zynde het rapport van de gelegenheyt der saeken [...] is verstaen dat den suppliant egeen redenen en heeft van clagen, diewyle neyt en blyckt, dat hem binnen Leyden zyne voors. inventie wert nagemaect tegen het voors. octroy, en dat daertoe geen redenen staet, dat dat men yemandt constringere tegen zynen danck des suppliants inventie aen te nemen." Dodt, *Archief*, 7:86 (5 October 1619). For a transcription of the privilege, as well as the efforts by Leiden not to prolong the privilege when it expired in in 1624, see Posthumus, *Bronnen*, 4:112-116 [no. 100 and 101]. For a copy of the privilege in the State Archives, see NL-HaNA, States General, 1.01.02, inv. no. 12302, fol. 36 [G163, 23 September 1617]. The inspection required for the report was carried out by the Court of Leiden together with two governors from the *hooftneringe* (cf. 92–93). The privilege indeed was not prolonged when it expired, but the inventor got exclusive permission from Amsterdam to press woolens nevertheless. Van Dillen, *Bronnen*, 2:578 [no. 1022, 12 July 1625].

11 See also Biagioli, "Patent Republic," 1144; Pottage and Sherman, *Figures of Invention*, 25, 52. The general agreement is that the distinction between idea and embodiment only emerged after the breakdown of the privilege regime around 1790, and notably with the development of written specifications.

12 To say it differently: in the early modern period, one had an intangible (privilege) on something real (a physical object), whereas a patent is an intangible (right) that has bear-

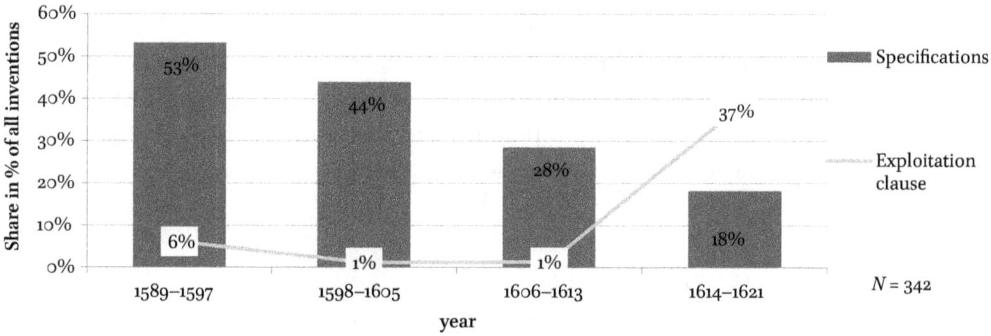

FIGURE 32 Exploitation term offset against the number of specifications

> [People might have had ideas about their inventions in the early modern period but] whatever went on in the inventors' minds (or in the books they might have written about their inventions) was perfectly irrelevant to the working of the privilege system. Privileges rewarded working machines in specific places, period. They were technology transfer tools, not instruments of intellectual property. I mean this not only in the historical and empirical sense that privileges were introduced to facilitate the immigration of machines, manufactures, and skilled labor into the confines of the privilege-granting nation, but also in the logical sense that, as a regime, the privilege had no need for the idea of the invention (as a category), nor did it have a conceptual or legal space to represent it even if such an idea were to be found in the inventor's mind. In this sense, there was just nothing intellectual about the privilege. It was all about locality, materiality, and utility.[13]

It is indeed beyond reasonable doubt that the early modern privilege system left little room for a noncommittal interpretation of the 'reduction to

ing on an intangible (the inventive step). For an excellent exploration of the genesis of invention in modern patent law, see Pottage and Sherman, *Figures of Invention*. By the way, this is not to say that early modern inventions could only occur in one specific form for that reason. This is exemplified, for instance, by the patent of Cornelis Diksz Muys (June 9, 1583) for certain instruments with cutting chisels, which states that the inventor "may increase or decrease the number of iron chisels and increase or decrease them in height and thickness according to the requirements of the location where, and of the instruments in which, the inventor wishes to use them" (*in den getale van Ysere Beytels sal mogen Vermeerderen of verminderen ende in de hooghte ende dickte verlichten of verswaren na den eysch van de plaetsen ende van den Instrumenten daeraen hy Suppliant 't selve sal willen ghebruycken*).

13 Biagioli, "Patent Republic," 1145–46.

practice' clause in comparison to the modern patent system. Even if an inventor obtained his privilege in a conceptual stage, he was then obliged to reduce the invention to practice within a reasonable time period. In contrast to the present day, actual reduction to practice was therefore an absolute prerequisite if the privilege were to have any legal validity.[14]

Venture Capital

Privileges were conceptualised as a reward for services provided (see also chapter 4 on legal theory). In some cases, the reward was issued for a finished product. In other cases, the object of the invention still had to be built. In both scenarios, however, privileges functioned as an inducement to attract investments necessary for the implementation of inventions. How this market of venture capital worked in practice can be illustrated by examining the contract between the inventor-engineer Isaac Burger, who had obtained a privilege for his invention of a new type of oven, and Philips Ouseel, who was the merchant-representative for an Amsterdam-based company that refined Scottish coal.[15] On 8 June 1623, the two parties agreed by means of a notary contract to split the profits made on the exploitation of the invention equally. Isaac Burger had sold his rights to the privilege on the condition that Ouseel would furnish the lump sum of 200 Flemish pounds "in order to prepare and finish the work [...] without enjoying interest here upon [...and...] in order to establish the invention in other places in these United Lands."[16] Other contracts followed

14 Today, patent law makes a distinction between the 'actual reduction' and the 'constructive reduction' to practice. The 'actual reduction to practice' requires a presentation of the invention in a physical or tangible form, whereas the 'constructive reduction to practice' occurs when a patent application on the claimed invention is filed. Actual reduction to practice has fallen out of use nowadays. Instead, patents are judged on their constructive reduction to practice. Filling in the original application is in that case evidence of the conception of the invention. Hence, the inventor need not provide evidence of any actual reduction to practice when relying on the content of the patent application: the functionality of the patent does not have to be proven in a physical reality. USPTO, "MPEP."

15 For the privilege, see NL-HaNA, States General, inv. no. 12302, fol. 338v [G223, 27 June 1623].

16 For the contract, see Van Dillen, *Bronnen,* 2:494: "[...] om 't werck te bereyden ende clear te maecken ende als het werck in orde gestelt is ende bevonden wert goede avance te sullen geven, te tellen 200 ponden Vlaems uyt sijn eygen beurs, sonder interessen hiervan te genieten. [...] om 't werck in andere plaetsen van dese Geünieerde Landen op te stellen ende in treyn te brengen [...]." Ouseel was bound to raise his investment to a total of 600 pounds, and it was agreed that within two to four years Ouseel would withdraw his invested capital.

the same logic: the inventor was the person who had devised the machinery, and merchants made the invention possible.

Some merchants resurfaced on various occasions as central players who provided the capital for the materialization of inventions. One of those merchants was Pieter Denijs, a brewer in the inn *'t Rode Hart* in Amsterdam and someone with several stakes in different projects.[17] One of these was the commercial exploitation of a drainage mill that would pump up water more effectively. On 16 October 1621, the inventors Abraham Jansz Segers and Pieter Sturck had obtained a privilege from the States General to build a new type of mill.[18] On 12 February 1622, Sturck sold his share of the privilege to Segers. The sales agreement stipulated that Sturck would receive 1000 guilders emanating from the first revenues, if the mill raised a specific amount of water (3 feet water per square meter, and six feet high). If the mill did not manage to raise that amount of water, and if Segers decided to continue the experiments nevertheless, the contract would be considered void.[19] Four months later, on 16 June 1622, it appeared that Segers had entered into a contract with one Henrick Frits, who was yet another investor with multiple stakes in different projects. According to this contract, Frits had obtained a one-third share in the privilege on condition that he furnished a "considerable sum of money."[20] Then the company borrowed 1275 guilders from Pieter Denijs.[21] The shareholders apparently had trouble reimbursing the money. On 13 August 1622, they signed a debt declaration in which they declared that they still owed the brewer-projector 300 guilders.[22] On 22 April 1623, it was decided to move to a different type of payment. Denijs obtained, by means of a notary contract, the "free and full ownership" of the privilege. The contract further stipulated that Denijs would:

> advance all the expenses [...] to take down the mill here, transport, bring and erect [it] in Frisia in a certain polder, located on the Southside of Worckum on the side of the sea, where he will [put it to work to] drain the polder according to a certain contract. [23]

---

17 For some other projects, see Van Dillen, *Bronnen*, 2:452 [no. 781, 10 June 1622]; Van Dillen, *Bronnen*, 2:676 [no. 1212, 20 may 1629]. The first project concerned a method for pulverizing hard stone, the second the melting of iron.

18 NL-HaNA, States General, 1.01.02, inv. no. 12302, fol. 244v [G203, 16 October 1621].

19 Van Dillen, *Bronnen*, 2:436-437 [No.752, 12 February 1622].

20 "een merckelycke somme van penningen." Van Dillen, *Bronnen*, 2:454 [No.786, 15 June 1622].

21 Ibid.

22 Van Dillen, *Bronnen*, 2:454, note 3 [No.786, 13 Augustus 1622].

23 "Denijs sal den gemelten meulen met asistentie van andere arbeyders, by denselven Denijs dartoe te huyren off bestellen, alhyer opbreken, transporteren, brengen ende

Segers would execute the work in return for one-third of the profit, but not before the prolongation of the invention privilege was secured, "since the time to test the privilege has expired." On 26 October 1623, Pieter Sturck was authorized to petition for the prolongation in the name of Segers. He promised to carry out this assignment "as customary, usual and according to the way of the chamber of the High and Mighty"[24] and indeed obtained a privilege on 7 November 1623, again with a one-year exploitation period, for the duration of twenty years.[25] The eventual outcome of the project is, alas, unknown. Yet the example of the drainage mill shows how investors and inventors cooperated closely in a rich and emerging market of ideas.

Investors and future shareholders could become aware of new inventions in various ways. Although if there is no evidence that a public promulgation of the privilege was a necessary prerequisite for legal validity, there are a few examples where the text of a privilege seems to have circulated on behalf of the authorities.[26] One example was the privilege for the water pump invented by Arent Smith, which was printed on broadsheet (see Figure 33). Another example was the privilege issued to a certain Samuel van Pietsen for the invention of waterproof linen (this pamphlet even ended up in the collection of papers of the English intelligencer Samuel Hartlib, together with a handwritten transcription into German).[27] Although difficult to interpret, this kind of "notices"

rechten in Vrieslant op seeckere bedyckinge, leggende aen de zuytsyde van Worckum aen de zeekant [...] om d'selve bedyckinge drooch te maelen volgende seecker contract [...] alle de oncosten sal verschieten [...]." Van Dillen, *Bronnen*, 2:486 [No.854, 22 April 1623].

24 "[...] omdat wesende den tijt t'octroy te proeven geexpireert ende derhalve vernieuwt dient te syn. Sturck sal doen nae costume, usantie ende maniere der camere van heure HoMo behoort te presenteren." Van Dillen, *Bronnen*, 2:509 [no. 892, 26 October 1623].

25 NL-HaNA, States General, 1.01.02, inv. no. 12302, fol. 364 [G227, 7 November 1623].

26 A privilege, in the end, was a law—and new laws demanded public promulgation to become legally valid. The custom was that bailiffs sent out a copy of a new law to all the jurisdictions, who in turn carried out the promulgation. The new laws were then proclaimed out loud in church on specific hours, whereupon the inhabitants of a place were held to be knowledgeable of the new law, and to respect it. Merula, *Manier van procederen*, 56; Van Leeuwen, *Rooms-hollands-regt*, 23. Thus, inhabitants could no longer claim ignorance of the law. Cf. Van Aller, *Generale regulen*, 235–238.

27 The invention of *lijnwaed* (waterproof linnen) got ample attention in its time. Simon de Vries, for instance, extensively praised the usefulness of the invention, which served not only to make watertight tents and suitable drinking vessels, but also as a canvas for painters whose paintings could now "last for hundreds of years, without cracking or flaking, even when hung against damp walls" (*konden dueren honderden van Jaeren, sonder te barsten, of af te schilferen, al hingense tegens vochtige Mueren*). The most important property of the new linen, however, was that both the rich and the poor could be warmly clothed, since the material retained body heat; this was useful for soldiers and fishermen, but also for the common people (and especially housewives) who could thus move more easily without falling ill. De Vries, *Historische oeffeningen*, 204–210.

FIGURE 33
Privileged water pump, invented by Arent Smith

may also have functioned as a form of advertisement; even if specific information about the invention was not disclosed in the privilege act, it alerted interested parties to the fact that an invention in a particular field had been made.

The second way for investors to discover the existence of a certain invention was through proper advertisements. Later in the seventeenth century, these advertisements would appear in the newspapers that were slowly creating a new market; before that time, they were probably displayed in inns and churches, as well as around markets and in specific shops.[28] Adverts of this kind usually drew attention to the existence of a prototype, with the aim of finding entrepreneurs who wanted to commercially exploit the invention. One example was the 'announcement' of a self-driving wagon, with a handwritten notice that the new machine could be seen in operation in Leiden (see Figure 34). Another example, from somewhat later in the century, was a leaflet engraved by Caspar Luyken for the invention of a drainage mill (see Figure 35).

---

28 For more detail on advertisements and the Dutch newspaper business, see Der Weduwen, *Dutch and Flemish Newspapers*; Margócsy, *Commercial Visions.*

FIGURE 34 Advertisement for the invention of a self-driving wagon

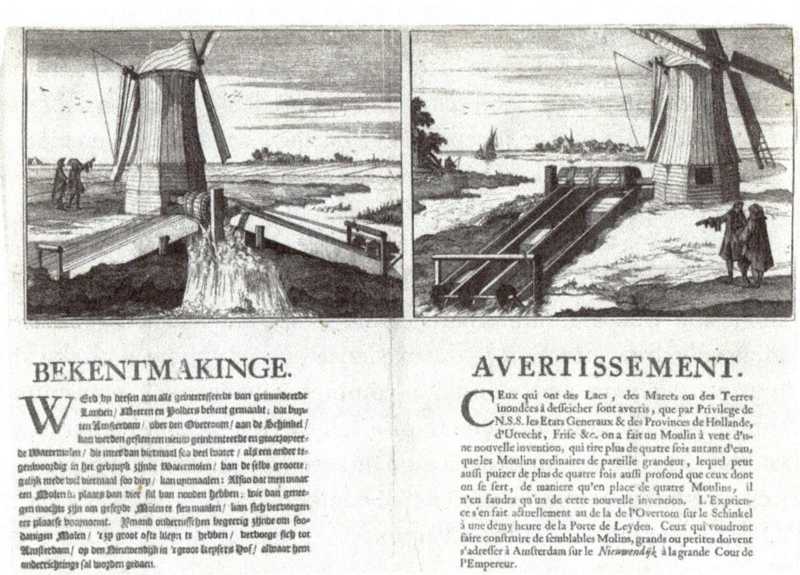

FIGURE 35 Advertisement for a new type of water mill

Although it is not entirely certain how these announcements were viewed and used, their textual content gives us insight into local practices regarding the diffusion of technical knowledge. Let us look, for illustration, at the leaflet that advertised the invention of a drainage mill (Figure 35). Drafted in French and in Dutch, the text of the flyer spoke of a prototype that could be visited by potential buyers. The mill was for sale either "small or big", where "small" possibly meant "in the form of a model" that would allow to rebuild the invention at another location. Whereas the Dutch text described the location of the prototype test mill as being "outside of Amsterdam, across the Overtoom [= a ship portage], along the Schinkel [= a canal]," the French text spoke of the same location as being "at half an hour from the city gate of Leiden."[29] The second difference is that the French text spoke of a privilege granted by the "N.S.S. of the States General & of the Provinces of Holland, Utrecht, Frisia & c.," whereas the Dutch text simply mentioned that the invention had been privileged, without mentioning any specific authorities. Clearly, both texts addressed a different audience. The Dutch text extended to Dutch-speaking merchants, who mostly resided in Amsterdam and would probably be sufficiently assured knowing that the invention was privileged. The French text, on the other hand, was aimed at an international (French-speaking) elite. International entrepreneurs, who might have had a potential interest in investing in the implementation of an invention, would also have wanted to know where exactly the privilege was valid, and thus the text offered additional assurance that the invention was properly privileged by the relevant authorities.

Once an invention was successfully implemented, local artists would sometimes launch out on a panegyric. A well-known instance was the laudatory poem written by Hugo Grotius on the occasion of Count Maurice's trip over the beach at Scheveningen, in the winter of 1601, in a new type of sailing chariot that could reach a speed up to 40 km/hour (Figure 36).[30] Remarkably enough, the invention was never privileged. One possible reason was that a very similar invention had been made shortly before, i.e. a sailing chariot that slided across ice as if over sea. This invention had been privileged, and was given full laudatory praise in an engraving (Figure 37) by Cristoffel van Sichem (1546–1624) as yet another proof of Dutch "ingenuity" (*cloeckheyt*).[31] Once one disposed

29 See text in Figure 35. Interested buyers were told to address themselves to the *groot Keysershof*, a renowned inn in Amsterdam, to obtain further information.

30 Devreese and Vanden Berghe, *Magic is No Magic*, 46.

31 On 27 January 1600, the States of Holland issued a ten-year privilege to Adriaen Terrier, living outside of Haarlem. In order to obtain a privilege, Terrier had submitted a drawing that clearly showed the mast and the staysail, lenght and depth, "without however depicting the leeboards or the rudder, which contained most of the art and the secret

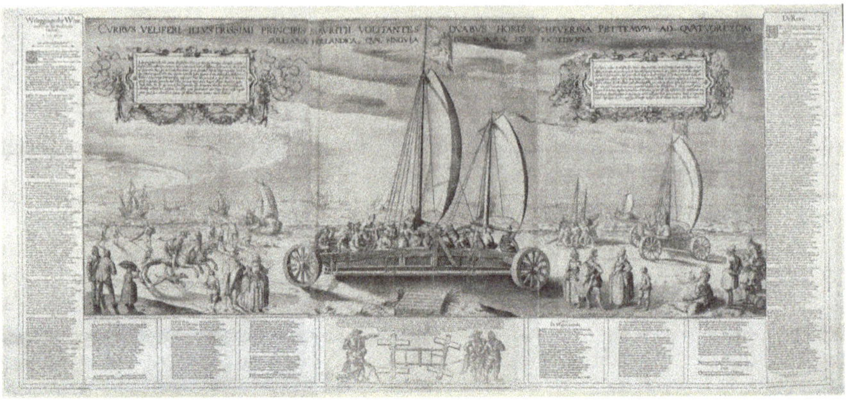

FIGURE 36 The sailing chariot invented by Simon Stevin

FIGURE 37 A new invention of a sailing chariot

over a sailing chariot that moved on ice, it must not have been too difficult to replace the blades by wheels, and to have the same machine move on sand.

of the matter" (*sonder nochtans daer inne te hebben doen uytbeelden de Swaerden noch oock het Roer, daer inne de meeste konste ende het secreet van de saecke is gelegen, door desen getransfixeert*). Christoffel van Sichem (I) shortly afterwards also produced the diagram of Stevin's invention of the sailing-car. Rijksmuseum, RP-P-OB-80.564. It has also been suggested that Stevin could have been inspired by the Chinese sailing wheelbarrow, described in Jan Huyghen van Linschoten's *Itinerario*, published 1596. Duyvendak, "Simon Stevin's 'Sailing-Chariot.'" One explanation does not necessarily preclude the other.

The Applicants

Aside from the inventors of lesser origin, who have been highlighted in the previous paragraphs, there was also a group of affluent inventors (arms dealers, merchants, and so on) who wanted to strengthen their market position by procuring a monopoly. Moreover, there must have been inventors for whom the financial reward was of lesser or no importance, and who simply wanted to be recognized for their capabilities by the highest authorities. This paragraph attempts to break down these different 'inventor types' further by dealing successively with the occupation, the geographical origins, and the place of residence of the privilege applicants. It ignores the question whether the person who pretended to be the inventor was indeed the 'true and first inventor' of a particular product or technique in the Dutch Republic but starts from the assumption that the applicant and the inventor can be considered as equivalent.[32]

Occupation

The inventors who presented themselves at the doors of the States General and the States of Holland came from virtually every layer of society (see Figure 38). Nevertheless, most of them were artisans. One reason for the prominence of these artisans might have been the foundation of a new type of school for the formation of engineers. Paralleled by developments at the University of Franeker,[33] the University of Leiden agreed in 1600 "that here in the university arithmetic or counting and surveying would openly be read in proper lower German to particularly encourage those who want to practice engineering."[34] Unfortunately, there are only a few remaining documents that give insights into

32 The reason for doing so is not only the unreliability of the sources on, but also the ambiguous status of, the 'inventor' in the early Republic; see also pp. 6, 194–197.

33 In 1600, the first full-time chair for mathematics in the Republic was established at the University of Franeker. It was held by Adriaan Adriaansz Metius (1571–1635), who taught in the vernacular, and who was at several occasions directly involved with the privilege business. Dijksterhuis, "Duytsche Mathematique."

34 "[...] dat in de universiteijt alhier opentlijc soude gelesen worden in goede nederduijtsche tale de aritmetique ofte het tellen ende het landmeten tot bevordering bijsonder van den geenen die hen tot oeffeninge van het ingenieurschap sullen willen begeven." Molhuysen, *Bronnen*, 1:122. The program of the new school, later renamed as the *Nederduytsche Mathematique*, had been designed by Simon Stevin at the instigation of Stadtholder Count Maurice of Nassau. See Ibid., 1:389*. See also Westra, "Nederlandse Ingenieurs," 82. The first directors of the so-called *Nederduytsche Mathematique* in Leiden were Ludolph van Ceulen (1540–1610) and Symon Fransz van Merwen, who were both closely involved in the examination of privilege applications.

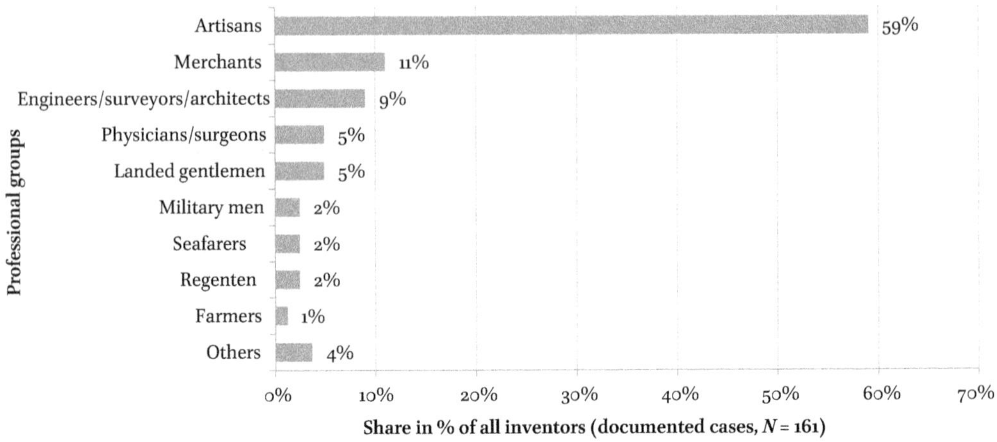

FIGURE 38 The professional status of inventors, 1589–1621

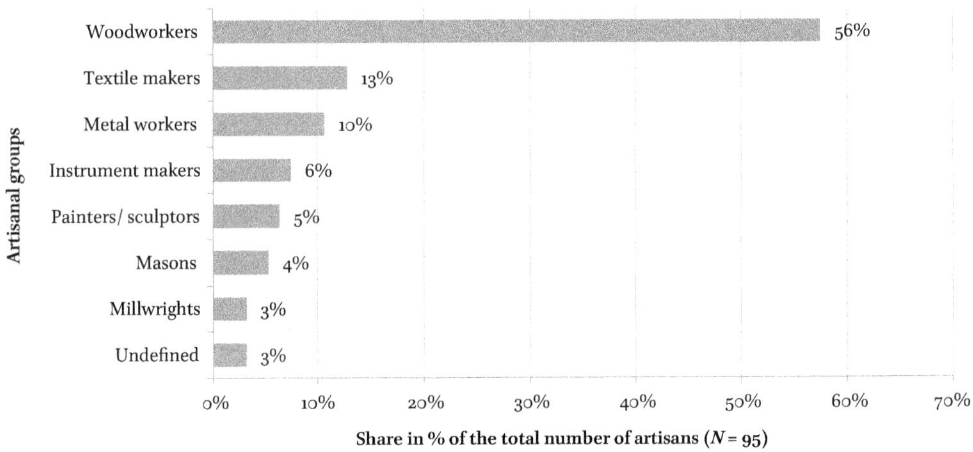

FIGURE 39 Breakdown of artisans according to material used, 1589–1621

the daily operations of the school. However, we do know (based on an attendance list from 1611) that the program was attended by carpenters, architects, and masons.[35] New students were expected to learn some trigonometry, some basic Euclidian geometry, how to calculate surfaces, how to make drawings of cities, etc. It was perhaps no coincidence that these were exactly the qualities needed for the proper presentation of an invention. Along similar lines, the dominance of the number of carpenters in Figure 39 could be explained by the fact that they worked with the material needed to make a proper presentation

35 Molhuysen, *Bronnen*, 1:392*.

of invention: perhaps carpenters had more experience with modeling because of the nature of their profession.

Still, it should not be forgotten that the title 'carpenter' merely indicated that someone had a general familiarity of working with wood; it did not necessarily mean that someone was working on an individual basis in a woodshop. And indeed, on closer inspection, it appears that many of the carpenters were so-called Town carpenters (*stadstimmermannen*) who operated in the service of local authorities for the Department of Public Works (*fabriekambt*). In the course of the fifteenth century, urban authorities in the Low Countries had delegated the work on local infrastructure and municipal buildings to a Town master builder (*stadsfabriekmeester*), who was assisted by a number of workers, including a Town mason (*stadsmetselaar*) and a Town carpenter (*stadstimmerman*). As time went on, the Department of Public Works expanded, and the tasks of the Town carpenter became ever more administrative and supervisory in nature.

The operatives of the Department of Public Works played a central role in the privilege business. Not only were they the inventors of new technologies but they examined inventions too. For instance, Cornelis Muys, the town carpenter of Delft, not only obtained a privilege for his newly invented dredging machine but was also called in to judge the invention of a corn mill built for the army (p. 146). The treasurer of the city of Leiden, Simon Fransz van Merwen (1548–1610), on the other hand, was called in to give advice on the invention of brick-making, besides also obtaining a privilege in his name for a new water mill. The text of this privilege clearly spelled out his role in the Public Works' department:

> The Court of the City of Leiden has showed the aforementioned States [of Holland] a certain new invention of a water wheel, equipped with a screw on the inside by Mr. Simon Fransz van Merwen, Treasurer extraordinary of the mentioned city and commissioned to supervise the Public Works department there, which serves [i.e. the invention] to bring up water to great heights, having on the bottom and on the top its center or the disposal point, erected on his own costs in the Leprosy House in Leiden for testing in the ditch there, and operated by three men, the water was drained in abundance to the height of 5 or 6½ feet.[36]

36 "Alsoo die van den Gerechte der Stadt Leyden de Staten voorn, hebben doen verthoonen seeckere nieuwe inventie ende maniere van een Schep-Radt, van binnen slex gewys gemaeckt by Mr. Simon Fransz van Merwen, Tresorier extraordinaris der voorsz. Stede ende Gecomm. tot d'opsicht van des Stadts Getimmerte aldaer, dienende om 't water

The invention by Van Merwen was implemented as part of a master plan developed by Leiden's city secretary, Jan van Hout (1542–1609), to take fight the decay of water quality caused by the fullers in Leiden.[37] There was a pattern in the way that the Department of Public Works played a central role in the making and appraisal of new inventions.[38]

But there were other professional groups among the inventors. One thing worth noting is that merchants took out a considerable proportion of the privileges. Not only was there a close liaison between inventors and merchants because of negotiations on a venture capital market, but merchants sometimes teamed up with inventors to strengthen their market position. A fine example is the contract made between two merchants and the master pewterer Auleander, who had obtained a privilege in his name for a fire extinguisher. The notary contract for the exploitation of this invention stated:

> that both the work and the privilege for it, as well as all other water- or mill works that they shall make from now on, either among the three of them or individually, as well as the privileges that shall be obtained, no matter on whose name of the three they may be, shall for always be and stay to the advantage and profit of the three of them and their heirs, as well as everything that shall be made, earned and won, which shall be divided and enjoyed in three equal parts among the three of them [...].[39]

Inventors generally sought a privilege in their own name first, and then later formed companies for the exploitation of their inventions. As time went on,

---

met quantiteyt in de hooghte te brengen, ende onder en boven hebbende syn centrum of het afpunt om uyt te loopen, 't welck den voorn, van Merwen t'synen koste in het Leproos-Huys tot Leyden voorn, tot een preuve hadde doen opstellen in de Sloote aldaer, ende door werckinge van drie Persoonen 't Water in groote menichte ende overvloedigh uyt syn centre doen lossen ter hooghte van 5 of 6½ voet." NL-HaNA, States of Holland, 3.01.04.01, inv. no. 345, fol. 369 [31 May 1589].

37 For a detailed study of the implementation of this mill, see Buning "Stench and the City."

38 Compare also the remarks in the excellent article Hurx, "The Most Expert in Europe," 8–9.

39 "[...] dat soowel tselve werck ende het octroy, daerop vercregen, als oock alle andere water- oft molenwercken, die onder hun dryën ofte by eenich van hun naermaels noch sullen mogen gepractiseert ende gemaect werden, mitsgaders die octroyen, die daervan noch sullen mogen werden vercregen, tsy op wiens naeme van hun dryën tselve soude mogen wesen, altijts sal sijn ende blyven ten oirbore, nut ende proffite van hun dryën ende hunnen erffven, gelijck mede alle tgene, daermede gemaect, verdient ende gewonnen sal mogen werden, by hun dryën ende hunnen erffven in gelycke derdeparten gedeelt ende genooten sal werden [...]." Van Dillen, *Bronnen*, 2:257. [no. 412, 2 November 1617]. For the privilege, see NL-HaNA, States General, 1.01.02, inv. no. 12301, fol. 46v [G132, 11 June 1614].

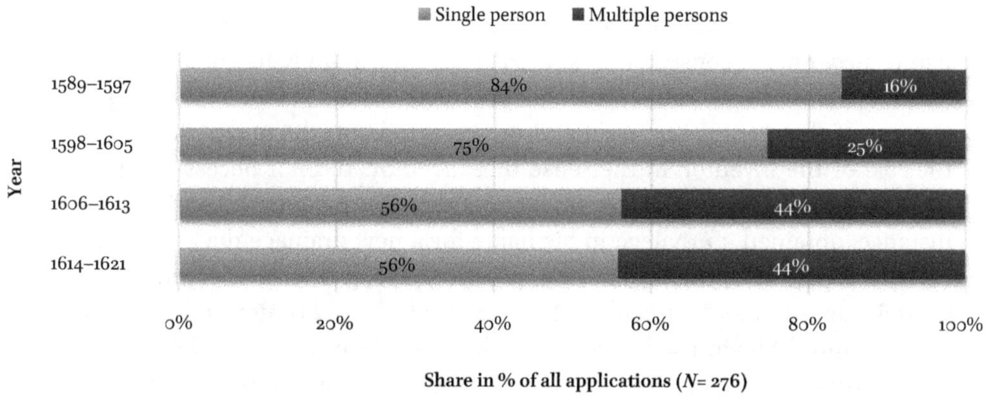

FIGURE 40 The percentage of inventor consortiums, 1589–1621

however, privileges were more frequently made out beforehand to the company that exploited the invention (see Figure 40). Sometimes, the members of these inventor consortia were mentioned individually by name. In other cases, the indication "and company" sufficed.

Another professional group remarkably present in the privilege documents was the architects and engineers. Although not quite the same, both professions were closely related: whereas engineers kept themselves busy with military constructions, architects devoted their time to buildings in the civil sphere.[40] One of the main traits of architects and engineers as a professional group was that they were not limited to one particular craft but were expected

40 The professional status of either 'engineer' or 'architect' was not always clearly defined. There was, for instance, no unity in terms of education; engineers gained most of their knowledge in practice and, except for Simon Stevin, there are no examples of engineers who had followed any university education in the sixteenth century. Westra, *Nederlandse ingenieurs*, 83. For later developments regarding the education of engineers, see Van Winter, *Hoger beroepsonderwijs*. As far as architects are concerned, the traditional view in the historiography is that the term 'building master' (*bouwmeester*) was mostly used in the fifteenth century and term *architect* only came into use in the sixteenth century, when the design process and the execution of projects increasingly grew apart. Meischke, "Het architectonisch ontwerp," 202. But others have argued that the use of the term 'Architect' in sixteenth century Dutch was a neologism, taken from Vitruvius, and that it basically meant the same as *overbouwmeester, werckmeester, Timmermeester*, or *Metselmeester*. Hurx, "Architecten en gildendwang," 5–7. For a broader discussion on the fluid professional identity of early modern architects, as well as the kind of knowledge they produced, see also Gerbino, *François Blondel*; Lefèvre, "Architectural Knowledge"; Long, "Multi-Tasking"; Merrill, "The Professione"; Miedema, "Over de waardering"; Nègre, "Les figures de l'architecte"; Pevsner, "The Term 'Architect'."

to make use of different techniques.[41] Yet this did not mean that they got their hands dirty on the construction site: their task was mainly to sketch out the building plans as well as to coordinate their execution. This understanding of the role of the architect as a 'supervisor' was similar to the understanding of the role of the inventor, in the sense that inventors did not necessarily produce their invention with their own hands either. The inventor Johan Thyll, for instance, obtained a privilege in his name for a new drainage device that he had developed "with the help of [the carpenter] Pieter Arensen."[42] A new type of drainage mill was "remodeled after the way of [the inventor and privilege holder Simon] Stevin by the carpenter M. Joost Govertts."[43] In other words, the inventor was the one who contrived the idea, not per se the person who executed the works.[44]

This leads to the question whether there was such a thing as a 'professional inventor' in the early days of the Republic. One way to test this is by looking at the number of multiple applications by a single inventor: although, in theory, it would be possible for a professional inventor to apply only for one privilege, one would expect him to be involved in multiple projects to expand his invention portfolio.[45] As Figure 41 attests, some inventors did indeed reappear at several instances in the privileges acts; yet, the percentage was not very high in comparison. In addition, it is hard to know for certain whether 'professional inventors' depended on the revenues from the privilege business. Although a

41 For that reason, architects often worked outside of guild obligations. Hurx, "Architecten en gildendwang," 12.

42 "[...] met hulpe van Pieter Arentsoon." NL-HaNA, States General, 1.01.02, inv. no. 12300, fol. 348v [G117, 31 March 1612].

43 "[...] by M. Joost Govertsz Timmerman, op de nieuwe maniere vande voorschreve Mr. Simon Stevin vermaekt ende geerigeert." Stevin, *Principal Works*, 5:387.

44 The distinction between the person who invented something and the person who executed the idea was rooted in the time. In the preface to his famous ten books on architecture, for instance, Alberti stated that: "But before I proceed further, it will not be improper to explain what he is that I allow to be an Architect: For it is not a Carpenter or a Joiner that I thus rank with the greatest Masters in other Sciences; the manual Operator being no more than an Instrument to the Architect. [...] Him I call an Architect, who, by sure and wonderful Art and Method, is able, *both with Thought and Invention*, to devise, and, with Execution, to complete all those Works, which, by means of the Movement of great Weights, and the Conjunction and Amassment of Bodies, can, with the greatest Beauty, *be adapted to the Uses of Mankind*: And to be able to do this, he must have a thorough Insight into the noblest and most curious Sciences. Such must be the Architect." Alberti, *Ten Books*, preface. Emphasis added. The role of the inventor was very similar to the role of the architect: he was the person who devised the invention.

45 Cf. Davids, "Patents and Patentees," 272–273 For similar ideas related to a transformation in the eighteenth century, see Dutton, *The Patent System*, 108.

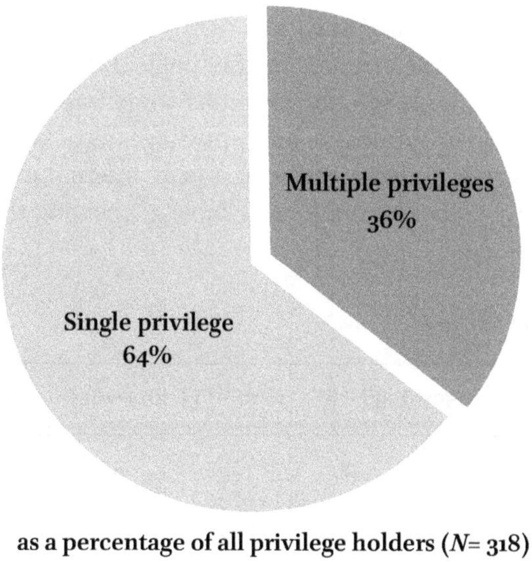

FIGURE 41
Inventors holding a single or more than one privilege, 1589–1621

privilege may have contributed in some cases to the financial well-being of the inventor, most inventors had some additional means of income.[46] It is also striking that inventors repeatedly made inventions in another field than their own. One encounters, for example, a watchmaker who invented a mill to produce shammy,[47] the fortification specialist Samuel Maralois as the inventor of a new type of horse mill,[48] and the famous sculptor Kasper Panten, who obtained a privilege for a drainage machine.[49] While this might be interpreted as a way for inventors to diversify their portfolio in the search for maximum profit, one could also say that was simply still too early in history to confirm that any degree of specialization resulted in a quasi-professional status of any sort.

46 That obtaining a privilege contributed to the financial well-being of the inventor can be derived from the remark of the inventor Pouwels Cornelisz, who petitioned for a privilege as he "was responsible for a house full of children and would like to live of it [i.e. the privilege]" (*die belast is met een Huys vol kinderen ende daer gaerne van leven soude*). NL-HaNA, States of Holland, 3.01.04.01, inv. no. 373, fol. 295 [H32, 17 October 1607].

47 The inventor was Nicolaes de Wagemaker ("Nicolaes the Cartwright"), NL-HaNA, States General, 1.01.02, inv. no. 12298, fol. 137v [G28, 20 February 1596]. In fact, the device served for pressing oil, cloth fulling, preparing of shammy leather and the making of gunpowder.

48 NL-HaNA, States General, 1.01.02, inv. no. 12300, fol. 233v [G107, 8 February 1611]. Samuel Maralois obtained the privilege "following the model" of his invention of a new type of horse mill that would clean and refine all types of melted steel "without the loss of time and with very little cost."

49 Dodt, *Archief*, 7:3,4.

Geographical Origins and Matters of Religion

Privileges had a local validity that was determined by the scope of the sovereign power. Because of this limited legal validity, inventors sometimes travelled from country to country in order to amass as many privileges as possible (see also Chapter 5). The result was the emergence of a truly international system– something that instantly comes to the fore in Table 2, showing the geographical origins of inventors.

TABLE 2 The geographical origins of inventors

Country of origin	1589–1597	1598–1605	1606–1613	1614–1621	Total
Unknown	36	48	55	80	219
France	0	1	4	4	9
Germany	3	1	6	5	15
Southern Netherlands	4	7	1	5	17
England	0	3	11	3	17
Italy	2	1	2	3	8
Probably of foreign extraction	4	2	0	2	8
United Provinces	9	13	10	5	37
Total	**58**	**76**	**89**	**107**	**330**

In most cases geographical indications have been substituted by the modern-day national geographical names (except in the case of the Southern Netherlands). This was done in order to create some broader categories that would be functional when visualizing the data. Thus, for instance, if an inventor originated from the city of 'Andige in Languedocq' he has been included as originating from 'France'.

Although there was some fluctuation over the years, Figure 42 attests that the percentage of foreigners who applied for a privilege initially was more or less in balance to the percentage of Dutch natives. This seems to contradict the assumption that the "principal political and economic objective of [inventor privileges] was that of attracting specialized craftsmen and technicians from other countries."[50] It may well be possible, of course, that the import of knowledge was still the highest ideal aimed for; the fact that many Dutch natives applied for privileges as well could then be interpreted as a mere side effect of this aspiration. However, it is more likely that by the beginning of the seventeenth century the nature of the privilege system had changed in a such way

50 Belfanti, "Between Mercantilism and Market," 322–23.

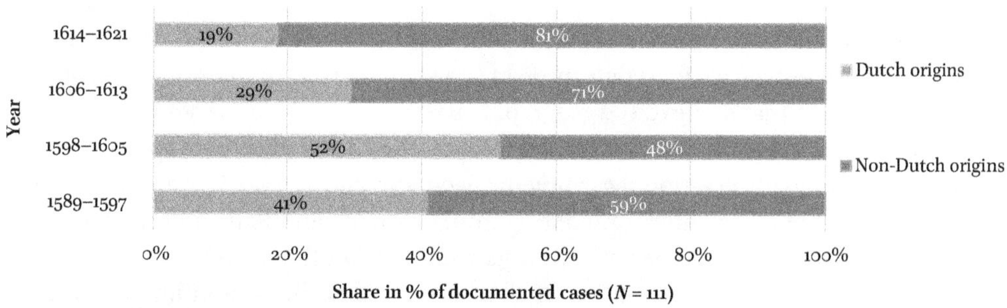

FIGURE 42 Percentage of Dutch vs. foreign inventors

that its primary goal was no longer to import knowledge as such, but rather to reward 'good inventions' in a more general sense.[51]

Perhaps the relatively high percentage of Dutch applicants for privileges could also be explained by a deliberate political choice to supplant artisans that fled the country because of the imminent threat of war. This idea returns hidden between the lines of a pamphlet written by Flemish merchant Willem Usselincx (1567–1647), who delivered a fiery plea for the establishment of a Trade Company under privilege that would monopolize the trade on the West Indies (later to become the Dutch West Indies Company). Usselincx argued that:

> what requires privilege is that one manages to keep business, shipping, merchants and artisans here [in this country] with all proper means, and that, if some were to leave, others may take their place. Now, to achieve this, a means must be found to make the merchants from these united lands trade with more profit; to have artisans earn greater daily wages than elsewhere and to have their handiwork sell easily; to let the seafarer establish exclusive navigation so that he may sail with more profit than in enemy countries; and if we can find this means, the people and the business will be kept without a doubt.[52]

51 For similar observations compare the conclusion of Veluwenkamp, "Buitenlandse octrooien," 554.

52 "[...] uit het gene is boven alreede verhaelt hebbe canmen genoech sien dat het voornaemste daer naest de Religie ende der Landen Privilegie op behoort gheledt te worden, is datmen de Neiringhe/ Zeevaert/ Cooplieden ende Ambachtslieden/ met alle behoorlicke middelen hier sie te behouden/ ende dat/ so eenighe quamen te vertrecken/ andere in hare plaetse mochten crijgen. Nu om hier toe te moghen comen soo moet eenen middel ghevonden worden/ om den Coopman uyt dese vereenichde Landen met meerder voordeel als uyt andere landen te doen handelen/ dat den ambachtsman grooter dachhueren mach verdienen als elders/ ende dat zijn hant-werck wel mach vertiert worden/ dat den Zeeman eenighe vaert hebbe daer hy oock met meer profijts mach vaeren als uyt t'svyants

The appropriate means that Usselincx had in mind consisted of granting exclusive privileges for a determinate time, whether to trade companies or to artisans. In that context, the geographical roots of 'inventors' did not matter so much: the main point of the privilege was to incite local business opportunities.

A second observation regarding the geographical origins of inventors is that the number of inventors originating from Germany and England was more or less equal to the number of inventors coming from the Spanish Netherlands. This observation puts a different slant on the idea that the influx of skilled laborers from the southern provinces was a crucial factor in the occurrence of the economic boom that the Dutch Republic experienced shortly after its independence.[53] Of course, the data does not tell us to what degree southern Netherlanders were involved in the exploitation of profitable inventions. But it tells us that they were not necessarily the primary inventors of privileged industries.

The high percentage of Englishmen and Germans among the inventors should probably be traced back to the strong religious (and hence economic) ties that existed between them and certain religious groups in the Republic. It is tempting, for instance, to assume that the increased influx of German natives was related to the beginnings of the Thirty Years War (1618–1648), when large groups of religious refugees ended up in the Dutch Republic in search of a safe haven. The information that has come to light so far, however, does not confirm or negate this assumption; perhaps a different type of research with a more proposographical approach can in the future provide clearer findings on the issue of religion as a driving power for wandering inventors.

One can still make some observations regarding the importance of religion in the constellation of the Dutch privilege system. First of all, the fact that people in power considered it a matter of minor importance to note religious convictions already tells us a lot: apparently, according to official rules at least, religion was of subordinate importance when it came to obtaining a privilege. Most religious groups had little trouble accessing the privilege system. A selected number of French and Italian inventors, for example, must have been of Catholic origin. But also inventors from anomalous protestant groups frequently appeared as beneficiaries in the privileges acts. By that token inventors such as Leamer, Maralois, Leeghwater, and Drebbel were not in any way

Landen/ ende dat de reyse boven dien niet al te ericulaeus oft lanck en sy/ ende soo wy dit connen vinden/ sullen het volck ende de neiringhe hier sonder twijfel wel behouden." Usselincx, *Vertoogh* (1608).

53 For the immigration from the Southern Netherlands, see Eggen, *De invloed*; Rogier, "Over karakter en omvang," 1938; Rogier, "Over karakter," 1937; Van Schelven, *Omvang en invloed.*

obstructed when applying for a privilege, despite them being Huguenots, Anabaptists or of some other non-Calvinist creed.

That said, it seems that non-Calvinist protestants were increasingly sidelined after the coup by Maurice in 1618 that concluded the so-called Truce Troubles, named after the Twelve-Year Truce between the United Provinces and the Spanish Habsburg Empire, which lasted from 1609 to 1621. During the Truce, a conflict broke out between two parties within the Dutch Reformed Church, the Arminians and the Gomarists, who had come into conflict with each other over the question of free will.[54] What started out as a religious disagreement, around the year 1610, turned into a political conflict of unprecedented scale, bringing the Republic to the edge of a civil war.[55] The urban regents took their position on the side of the Arminians against a Gomarist Stadtholder who increasingly tried to position himself as the supreme ruler of the Republic. The conflict had escalated by the end of 1617, when Maurice set off on a military campaign to disarm his opponents.[56] After a show trial, the political unrest was dramatically concluded with the beheading of Johan van Oldenbarnevelt, a member of the Arminian fraction, at the Inner Court in The Hague on 13 May 1619. The religious bickering was settled with the international Synod of Dort, which had its last meeting on 9 May 1619.[57] The Synod ruled in favor of the Gomarists and shortly after some two hundred Arminian preachers were

54 The names 'Armenians' and 'Gomarists' were derived from the respective leaders, Jacobus Arminius (1560–1609) and Franciscus Gomarus (1563–1641). They had come in conflict with each other over the question of free will: whereas the Gomarists denied man any form of free will in its most extreme form, the Arminians were willing to pretend a certain ignorance in these mattes and to speak of God's 'stationary' or 'permitting' will (*voluntas remissa*). Dekker, *Rijker dan Midas*, 115, 188. The Arminians were strongly influenced by the Spanish scholastics on this point, and notably by Francisco Suárez (1548–1617) and Luis de Molina (1535–1600).

55 Price, *Holland and the Dutch Republic*, 268; Israel, *The Dutch Republic*, 441–42.

56 Slowly but surely, the Stadtholders sided more and more openly with the more radical Gomarist fraction. In 1617, the conflict suddenly escalated when the States General agreed to summon an international church synod despite the objections of Holland. The States of Holland reacted by passing a so-called 'Sharp Resolution' (*Scherpe Resolutie*, 1617), which gave them the liberty to raise a military force (*waardgelders*) of *c.* 4000 men to secure the order. Maurice proclaimed the Sharp Resolution an act of war and as "an affront to the true reformed religion and our person." Smit, 'Prins Maurits en de goede zaak', 60–1, as cited in Israel, *The Dutch Republic*, 441. He set off on a campaign to disarm the *waardgelders* and effectively committed a coup taking over power of the entire country.

57 The Synod confirmed the authority of the Belgic Confession and the Heidelberg Catechism, while it drew up ninety-three new canonical rules. The Synod of Dort also decided upon several other issues for that matter, such as a Bible translation into the vernacular on costs of the States (which resulted in the *Statenbijbel* of 1637).

dismissed from office.[58] From that point onwards, inventors and examiners who were known for supporting the Arminian fraction no longer appear on the privilege rolls.[59]

To conclude, it is worth remarking that one religious group is never mentioned in the privilege rolls, namely the Jews. This is not surprising when one takes into account the existence of a legal rule that excluded heretics from obtaining a privilege.[60] But given the fact that Jews could, for instance, obtain privileges in the Republic of Venice,[61] it is still remarkable – and I have not been able to find any satisfactory explanation for it – that in the 'tolerant' Dutch Republic, Jews were entirely absent from any of the official privilege documents. Once again, this does not exclude the possibility that Jews were involved in the exploitation of certain inventions behind the scenes. But one would have expected the Jewish community to have been more clearly visible in the official documents as well. Perhaps the Jewish community in the Republic was simply still too small at this point in time and the situation would change later in the seventeenth century.

Place of Residence

The applicants usually had their (temporary) place of residence in the Republic at the time of filing a petition for a privilege. As Figure 43 attests, they were

---

58 Nijs and Beukers, *Geschiedenis van Holland*, 2:311. A very small percentage was actually driven to expulsion.

59 This development can adequately be illustrated by the case of Robbert Robbertz Le Canu, who had been involved in the privilege business as an examiner (see pp. 182–184). Robbertz initially had easy access to the highest circles of power. But his choice to side with the Arminians would appear the beginning of his downfall. In 1611, Robbertsz wrote a lament on the death of Arminius. When he requested an augmentation of his payment in 1615, it was denied on the explicit orders of Stadtholder Maurice. Dodt, *Archief*, 6:372 (6 August 1615). Instead of laying low, Robbertz continued to publish material supporting the Arminian fraction. In 1618, the States General decided that it had been enough: Robbertsz's "various yearly salaries that he enjoys on behalf of the country [...] should be cancelled" (*diverse tractementen jaerlixc genietende van wegen het landt* [...] *behooren opzeggen*) because he had published another "scandalous little book" (*schandeleux boecxken*) in favor of the Arminians. Dodt, *Archief*, 7:76 (17 July 1619). The States first wanted to see the "scandalous little book". On 26 August 1619, it decided to carrry on with the descision. Dodt, *Archief*, 7:83. A later request by Robbertz, which was forwarded to Prince Maurice was of no use. Dodt, *Archief*, 7:89 (26 October 1619). On 30 November 1619, Robbertsz salaries were cancelled. Dodt, *Archief*, 7:95. Robbertsz then slowly disappeared from the scene.

60 Merula, *Manier van procederen*, 49.

61 Berveglieri found that about 15% of all the inventors in Venice were Jews. Berveglieri, *Inventori stranieri a Venezia*, 42.

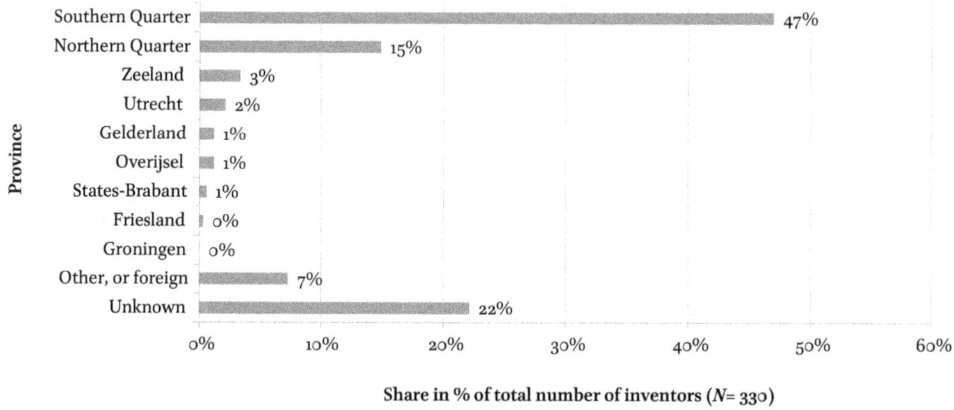

FIGURE 43 Present whereabouts of the applicants, 1589–1621

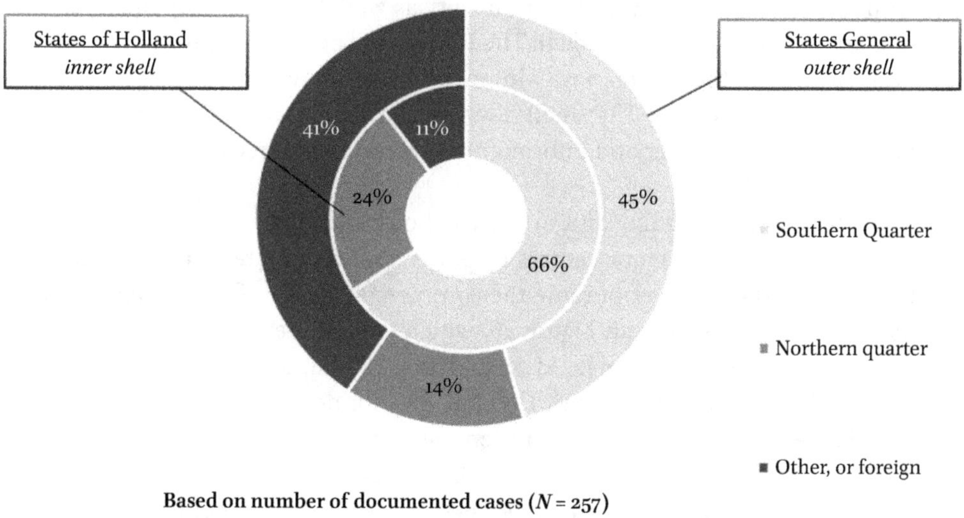

FIGURE 44 Share of applicants from Holland, 1589–1621

spread out all over, even if most of them resided in the Province of Holland (see also Figure 44). This predominance was probably related to the fact that Holland was the richest province with the most cities, and the place where one had the most chance of commercial success. The chance of a successful application was, however, not really dependent on the inventor's place of residence: what mattered was whether the invention was profitable for the local economy, not where the inventor came from.

The actors in the privilege system clearly operated in an urban context. Although it does not seem to have been obligatory to be a burgher (*burger* or *poorter*), it was noted if someone was (see Figure 45). Besides the fact that most

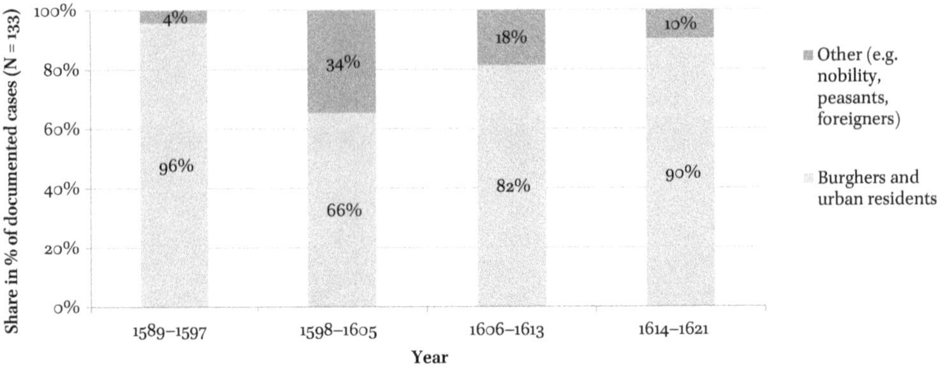

FIGURE 45 Burghers and urban residents

inventors had their residence in the cities, it should not be forgotten that the entire privilege process unrolled in the city as well. Physically, the final decision about a privilege was made in The Hague, and even when inventors came from elsewhere, they were mostly in town to give more detailed information about their invention.[62] The small town of The Hague was thus transformed into the center of power and information, a centripetal force of knowledge so to speak.

In stark contrast to England or France, however, where London and Paris soon developed into major centers of power that dominated the entire hinterland, The Hague never became the commercial center of the United Provinces. This is also shown in Figure 46, which enumerates the cities in which inventors kept their residence. Most inventors, by far, dwelled in South Holland (*Zuiderkwartier*); yet, South Holland was not dominated by the 'capital' The Hague. Even if all information eventually passed through The Hague, the political center of the Dutch Republic did not develop into the major center of power par excellence. Other cities were just as important, not least the city of Amsterdam, which acted as the commercial center of the Republic.[63] As Fernand Braudel remarked, there was a clear division:

62 At least, this can be inferred from the fact that further research, if needed, was mostly concluded within a couple of days. For illustration, during a discussion at the States General about an invention of more economical oven for baking tiles, the inventor "was not at hand, so it is [decided] to postpone until tomorrow to resolve" (*hy was nyet en by de handt, maer is uytgestelt daerop te resolveren tot morgen*). Dodt, *Archief*, 7:38.

63 For an analysis of Amsterdam as a staple market of knowledge, see Cook, "Amsterdam, entrepôt des savoirs"; Lesger, *The Rise*, especially 214–257. For a more gereral account of Dutch commerce and knowledge accumulation, see also Cook, *Matters of Exchange*; Margócsy, *Commercial Visoons*; Roberts, *Centres and Cycles*.

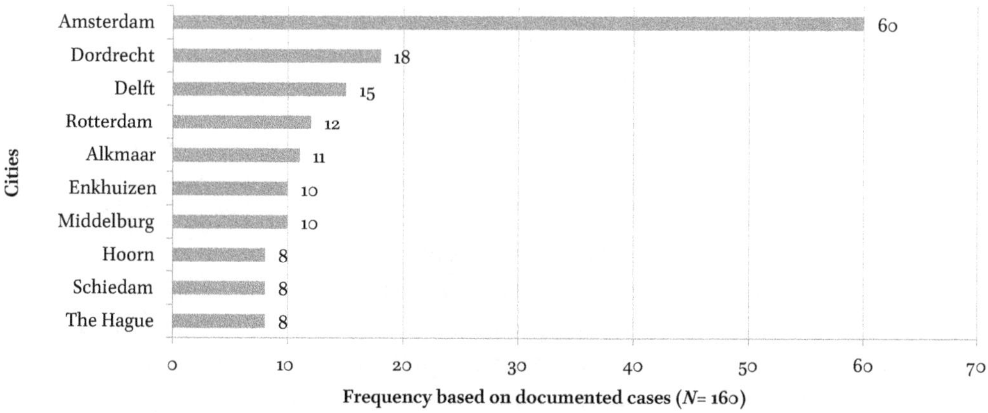

FIGURE 46 The ten biggest cities in which inventors kept their residence

> industry prospered in Leyden, Haarlem and Delft; shipbuilding in Brill and Rotterdam; Dordrecht made a living from the heavy flow of traffic along the Rhine; Enkhuizen and Rotterdam controlled the fisheries of the North Sea; Rotterdam, again, the most important city after Amsterdam, handled the lion's share of trade with France and England; The Hague, the political capital was something like Washington in the United States, both now and in the past.[64]

Because of this division, the diffusion of knowledge in the Republic was more pronounced than in other sovereignties. Whereas on some occasions cities tried to play each other off, at other times they had to form coalitions for survival. During discussions at the Provincial States and the States General new technologies were discussed, and so, because of the privilege system, knowledge about the existence of new technologies spread around the different territories of the Republic.

Customers and Clients

Earlier in this chapter, we saw how inventors could use a privilege to seek investors, who were prepared to finance the implementation of an idea in exchange for a percentage of the profit. Once the invention was reduced to practice, the company that exploited it would try to seal the deal on the private market to make some money. In other cases, urban authorities entered directly into a contract with the inventor, or they handed out assignments for

64 Braudel, *Civilization and Capitalism*, 184.

the development of new inventions. But, next to the lively trade of the local market, it appears that the States General handed out assignments for the production of new techniques as well.

The assignments for the development of specific technologies by the central authorities were not always coupled to a privilege. On 18 November 1602, for instance, the inventor Thomas Odel obtained 40 pounds "to make his instruments for the production of army bridges on a large scale and to communicate them to the States."[65] He never obtained a privilege for this invention. In other cases, the authorities used the privilege as part of a negotiation process to stimulate creative solutions to particular problems. This procedure generally began with the States reimbursing the inventor for the production of a scale model of his invention. Sometimes, the project would be abandoned after paying for the model, whereas in other cases the authorities decided to push the project through. The carpenter Gysbrecht Pietersz, for example, obtained 92 pounds on 8 January 1603, "in his honor for a sort of bridge presented on a small scale."[66] This invention was never realized. On 21 January, however, the same inventor obtained another 30 pounds "for the model of a certain watermill presented by him to the States."[67] He then obtained a six-year privilege for the exclusive production of this mill on 17 February 1603.

The States General usually provided the inventor who worked under commission with all the necessary means to develop a particular idea. These means could range from a sum to produce the scale model to subsidizing the entire production process. A good example is that of Johannes Ketschet, who had obtained a privilege for his invention on 24 January 1615. The privilege contained a six-month exploitation clause and specified that "once the invention were brought into effect, the High and Mighty retained the right to purchase [it] for a reasonable compensation, or shall arbitrate for how much the Suppliant will leave [his invention] to any private person, who would want to make use of it."[68] The inventor filed a second request on 31 July 1615:

---

65 "[...] tot het betalen aan Thomas Odel voors. Staten gepresenteert hebbende sekere inventie tot het maken van bruggen tot oorloge de somme van xl ponden van xl grooten t pont, toegelegt om dezelvde syne instrumenten int groot te maken ende voors. staten te communiceren [...]." NL-HaNA, States General, 1.01.02, 12548.119, fol. 84.

66 "Gysbrecht pietersz timmerman 92 pond to vereeringe toehge voor zekere soorte brugge int cleyn gepresenteert." NL-HaNA, States General, 1.01.02, inv. no. 12504, fol 1.

67 "voor het patroon van sekere watermolen by hem aan de staten gepresenteert." NL-HaNA, States General, 1.01.02, inv. no. 12504, fol. 23.

68 "[...] reserverende haere Ho.Mo. aen haer, de voorsz. inventie ten effecte gebracht synde, deselve aen haer te nemen voor eenen behoorlyken recompensie ofte arbitreren, waervoor de suppl. deselve elck in 't particulier sal hebben over te laten, die deselve sal begeren te gebruiken." Dodt, *Archief*, 6:362.

> Upon the request by Johannes Ketschet, who asks to be assisted with a reasonable amount of money, to be furnished in the form of a loan, to make his invention, to make ships go against the stream, without the use of humans, horses or wind; it has been agreed that the Delegated Councils of the Admiralty of Rotterdam are to assist the before-mentioned supplicant with an amount of three or at most four hundred guilders at once, without further ado, which is destined to pay for the necessary materials and the carpeting of his works, of which they [= the Admiralty of Rotterdam] shall advert the High and Mighty from time to time, [that is] of the manifestation that they will notice in the success of those works.[69]

So, the central authorities closely monitored the development of new technologies and remained in contact with the inventor after the privilege had been granted.[70]

On occasion, the States General also directly purchased the machinery produced under privilege. On 12 January 1602, for example, a merchant from Amsterdam by the name of Gerrit Ryckelsma had obtained a ten-year privilege for a transportable corn mill, which he had "recently invented and [...] had made." Rijkelsma was married to Jannetje Verschuil, who had just invested 600 guilders in the first Dutch East Indies Company shares. The couple also leased some land in Amsterdam, on the former grounds of the Carthusian cloister, where they kept busy with the production of roof tiles.[71] Rijkelsma established a workshop to produce his movable corn mills on the same spot. The mills clearly served a military purpose.[72] They "could grind enough in 24 hours to

69 "Opte requeste van Johannes Ketschet, versouckende geassisteert te worden met eene redelycke somme gelts, by forme van leeninge te verschieten, tot het maecken van syne inventie, omme de schepen tegen de stroom te doen opvaren, zonder trecken van menschen, paerden ofte wint; is geacc., dat men sal schryven aen de gecommiteerde raden ter admiraliteit binnen Rotterdam, dat sy, ten fine voorsz. den voorsz. suppliant assisteren totte somme van drye ofte vuyterlyck vier hondert guldens eens, sonder meer, te weten tot het betaelen vande noodige materialen ende de timmeringe van zyne wercken, dies sullen sy haere Ho.Mo. van tyt tot tyt adverteren vande apparantie, die zy sullen vermercken van het succes van deselve wercken." Dodt, *Archief*, 6:371 [31 July 1615].

70 Another example is that of Casper Benoist (see pp. 199–200). In 1604, Benoist obtained a privilege to start up the silk production in the United Provinces. In 1605, the States General ordered the administrators of the VOC to deliver three bales of raw Chinese silk and to provide security for the payment; in 1606 they allowed that the inventor paid back the money spread over four years. In 1608, the inventor is f4000 behind on the payment; the States General then decide that he can pay-off the debt with f500 per year (to the f16000 in total). Benoist probably did not succeed in doing so.

71 Van Eeghen, "Van karthuizerklooster," 52–53.

72 Actually, these moveable mills were invented by the Italian engineer Pompeo Targone, who had made such mills at least since the 1580s for the Spanish army led by Spinola. It

feed more than ten thousand men [...] and could be erected and dismantled in less than five hours and [were] suited to be transported on carts and boats to all places, and to be used in the battlefield of besieged cities, and otherwise."[73] The benefits of the moveable mills "could be confirmed by His Excellency Stadtholder Maurice" and by several members of the States General. Together with the privilege, the States General initially gave Rijkelsma 500 pounds "for the [future] production of some mills."[74] After inspection by the city carpenter (*meestertimmerman*) of Delft, Cornelis Dirckz Muys, Rijkelsma obtained another 4000 pounds for "the full payment for 5 horse mills, being at the service of the Generality."[75]

Sometimes, the States also managed to lay down conditions upfront to ensure that the novel technology would fall into their hands. The inventor of a new type of artillery, for instance, was "obliged, prior to using this privilege, to make two more [guns] of six pounds iron, and two of three pounds iron, which [...] are to be offered to the country for a reasonable price."[76] Hence, the privilege became part of a negotiation process to push down the price when acquiring new technologies. A clear example was the invention of a new dredging machine, which had been "brought to work in Delfshaven with the consent and at the cost of the States [of Holland]." The States had:

---

was Targone's version of the mill that also made it to Vittoria Zonc'a engineering treatise *Novo Teatro di Machine et Edificii* (Padua 1607). Later, the mill was also featured in the *Collected Diagrams and Explanations of the Wonderful Machines of the Far West* (*Yuanxi Qiqi Tushuo Luzui*, 1627) compiled by the by German Jesuit scholar Johann Schreck (1576–1630) and Ming Dynasty Chinese author Wang Zheng (1571–1644).

73 "Corenmolens, werekende met vier paer Steenen, [... ... ...] Ten welcken aensien (ende dat wy, ende Syne Exellentie oyck metter daet bevonden hadden, dat een van de voorn, moelens, binnen den tydt van vierentwintich uren, soo veel conde gemalen, daermede men dagelycx meer als thien duysent mannen conde spysen, Welcken molen binnen den tydt van vyff uren conde wordden gestelt ende ontstelt, ende oversulex bequaem, om op wagens ende schepen tot alle plaetsen te vervoeren, ende te gebruycken in belegerde plaetsen te velde, ende anderssints [...]." NL-HaNA, States General, 1.01.02, inv. no. 12299, fol. 51v [G65, 12 January 1602].

74 "voor enige molens te maken." NL-HaNA, States General, 1.01.02, inv. no. 12503, fol 363v.

75 Documents relating to the contract (dated 14 March) can be found in NL-HaNA, States General, 1.01.02, inv.nr. 12548.119. See also RSG, 12:324. For the payment to Rijkelsma, see NL-HaNA, States General, 1.01.02, inv.nr. 12503, fol 363v. Rijkelsma also sold his mills to Christiaen IV of Denmark, with whom Dutch had extremely very good connections, and he later worked for the Prince of Hesse as well.

76 "[...] welverstaende nochtans dat den voors. Bartelt Cornelisz Smidt alvoorens dit Octroy te mogen gebruycken gehouden sal syn noch te maecken twee stucken van ses pont yser, ende twee van drie pont yser om de selve tegens andere s'lants stucken geprouft synde voor 't landt op redelycken prys aengenomen te worden." Doorman, *Octrooien*, 192. Privilege made out to Bartelt Cornelis Smidt on 13 May 1633.

> found indeed (*metter daet*) that the instruments that have been used to that end and [those instruments] that will be used in the future are very useful (*dienstigh*) for these Lands to clean and deepen inland waterways, harbors and canals [... and therefore ...] the States have [accepted] the request to be privileged that the art and instruments cannot be copied [...] unless it is of great importance to the aforementioned Lands.[77]

The inventor of the new dredging machine was the city carpenter Cornelis Dirckz Muys, whom we have just encountered as an examiner of the corn mill of Rijkelsma. The States of Holland specified that before the privilege took effect, Muys first had to negotiate with the Delegated States of Holland about the price for which his invention could be put to use. These conditions were laid down in more detail on 5 May 1589, when it was agreed that:

> [The inventor] will enjoy from each of the following cities, Dordrecht, Haarlem, Delft, Leyden, Amsterdam, TerGoude, Rotterdam, Gorckum, Alcmaer, Hoorn and Enkhuizen the sum of two hundred carolus guilders at once, when and in case they desire to use the art during the time of the privilege, proviso that the other cities of Holland are allowed to use the same art [...] for the sum of 150 carolus guilders at once. Provided that, when the States or her delegates, as well as his Excellency [Count Maurice] would desire to use the aforementioned art, only in the general land [of Holland and West-Frisia] and not affecting or concerning the cities, for some time in some bulwarks of forts, this can be done at any time without the aforementioned Cornelis Muys enjoying anything for that.[78]

The dissemination process was clearly marked out here: the bigger cities would pay the full price, whereas the smaller cities got a sort of discount. In exchange for the privilege, however, the invention could be used for military purposes without any extra costs.

Aside from (inter)provincial authorities, the different trading companies in the Republic pushed for the development of particular technologies too.

77 "[...] geinformeert wezende dat Cornelis dircks Muys seeckere nieuwe goede Inventie ende Kunste tot Delffshaven met consent ende tot costen vande Staten van Hollant besocht ende int werck gestelt hebbende metter daet bevonden is dat d'Instrumenten die daertoe gebruyckt zyn ende noch voirder gebruyckt zullen mogen worden seer dienstelyck zyn voor voorsz Landen omme die binnenwateren havenen ende graften te mogen reynigen ende diepen." NL-HaNA, States of Holland, 3.01.04.01, inv.nr. 5022, fol. 185 (9 March 1589).

78 NL-HaNA, States of Holland, 3.01.04.01, inv.nr. 347, fol. 297v; inv.nr. 5022, fol. 336.

Trading companies were, for instance, directly involved in the invention of a reliable method for finding longitude at sea.[79] But they were interested in the production of finished products, as well.[80] The 1618 privilege for English inventor John Osborn "to prepare whalebones in the United Provinces" was produced at the direct request of the Chartered Nordic Company (*Noordsche Compagnie* or *Compagnie van Spitsbergen*), a trade company that had been established in 1614 for whaling around Greenland (see Figure 16).[81] A year later, a certain Abraham Vos obtained a similar privilege for working whalebones. In the case of Vos, it was specified that the inventor had obtained his experience of working with whale material "having made a voyage as surgeon, with the Nordic Company to Greenland, sailing to catch whales."[82]

Aside from the chartered companies, the guilds and the *neringen* were some of the other organizations with an interest in new technologies. In the previous chapter we saw how these organizations were at times involved in testing procedures. It appears, moreover, that several inventors were guild members too.[83] As Karel Davids formulated it, "guild membership in The Netherlands did not preclude efforts to reap gains from technical inventions by securing private property rights in the form of patents granted by public authorities."[84] But although guilds and inventors were both protected in their activities by privileges, these privileges were not of equal standing. In the framework of a discussion about guild membership, four overseers (*overlieden*) of the Amsterdam carpenter guild declared in the presence of a notary that:

---

79 This appears from a resolution of 3 September 1594, which spelled out that the "Merchants of the big voyages and the trade with the Dutch East Indies" (*Koopluyden van de groote Reyse ende Vaert op Indien*) urged the States of Holland to privilege a method that had been proposed by the protestant minister Plancius. See also pp. 178–186.

80 In 1618, for example, the Englishman John Osborn obtained a privilege that "allowed [him] to make and to put to work his new invention to prepare whale bones in the United Provinces." NL-HaNA, States General, 1.01.02, inv. no. 12302, fol. 174v [G187, 3 March 1618]

81 Muller, "John Osborn," 200.

82 "[...] mede hebbende gedaen een voyage voor chirurgyn, mette groenlantsvaerders, vaerende om walvisschen te vangen" and "[...] gehoort hebende de Bewinthebberen van de Noorsche Compaignie." NL-HaNA, States General, 1.01.02, inv. no. 12302, fol. 95v [G173, 15 February 1619].

83 Davids, *The Rise and Decline*, 1:422.

84 Davids, "Guilds, Guildsmen," 16. Carlo Belfanti found that also in Northern Italy "guilds and patents formed an institutional pairing, aimed at managing and transmitting new production techniques while minimizing conflict." Belfanti, "Guilds, Patents, and the Circulation of Technical Knowledge," 589.

> no one, whether mill maker or house carpenter, from outside or even from inside the city, has the freedom to produce any houses, mills, or other works, *with the exclusion of those who have a privilege* from the High and Mighty [= States General] or works that have been put up to public tender by the city, unless they are citizens, passed the [guild] examination and paid all their dues.[85]

The example suggests that invention privileges had precedence over guild privileges, even if they had a comparable legal validity. It was up to the authorities to find the right balance between the interests of established trades and the introduction of innovations that would put individuals in a privileged position.[86] Considering the limited amount of resistance to the introduction of new inventions, the authorities seem to have been successful in their task.

What may have helped in this process to keep the diferent parties happy are the clauses that determined the price for which an invention was to be put on the market.[87] These pricing agreements generally obliged inventors to leave their invention for a 'civil price' (*civilen prys*) to those who wanted to put it to use.[88] But in some cases the authorities also fixed the exact price at which

85 On 4 June 1622, the authorities of the St. Joseph guild declared on behalf of several carpenters for the city of Rotterdam "[...] ende dat oversulcx nyemant, tsy molenmaker ofte huystimmerman, van buyten ofte selffs van binnen deser stede en vermach binnen derselve stede noch de vryheyt van dien eenige huysen off molens noch de wercken, daeraen dependerende, te maecken *uytgenomen dengenen, die van de Ho. Mo. octroy hebben, ofte wercken, die vanwegen de stat openbaerlijck besteet werden – off, gemaeckt sijnde, binnen deser stede te brengen ende rechten,* off d' selve moeten alvooren burgers sijn, de proeff van 't gilde gedaen ende de gerechticheyden van dien ten vollen betaelt hebben." Van Dillen, *Bronnen*, 2:452. [No. 780]. Emphasis added.

86 For illustration, research was done to assess the effects of the privilege issued to the English nobleman, mathematician, and globe-maker, Emery Molyneux with reference to his newly invented lightweight artillery "that could shoot further than half an English mile." NL-HaNA, States General, 1.01.02, inv. no. 12298, fol. 226v [G38, 26 January 1598]. The artillery built by Molineux was put into the hands of Merten de Donckere on 6 June 1598, who was to examine whether the new invention would not prejudice "the manufactures of the gun-makers" (*de manifaicturen van de roermakers*). RSG 10:370.

87 Doorman compared these clauses to the English system of 'licenses of right'. Doorman, *Octooien*, 26. However, a major difference is that early modern price clauses were not included at the request of the patent proprietor but set by the state. The Dutch were, for that matter, not the first to include these price clauses. For Venice, see for example Berveglieri, *Inventori stranieri a Venezia*, 29.

88 For instance, NL-HaNA, States General, 1.01.02, inv.no. 12300, fol. 185v [G122, 5 April 1613]. Ideas about the 'just price' clearly resonated at this juncture. See also pp. 154–155. For a proposal for the implementation of this invention, see GA, inv. no. 3003, 0307.

the invention was to be put on the market. The inventor Claes Wolbrantsz, for example, was "held to make [his invention] for those who would desire so, receiving in addition to the material and labor costs, thirty guilders for his invention on top."[89] The authorities included price clauses to protect the population against the possibly damaging effects of the privileged monopoly. There was a clear consciousness that inventors could abuse their monopoly position to drive up prices. Moreover, the authorities seemed aware that the implementation effect of inventions might not have the same value for everybody. An English inventor consortium, for example, obtained the exclusive privilege to produce a new economical heating device on the condition that they were:

> to communicate their invention to private individuals who would want to use it, enjoying [...] twenty-five guilders once-only for their invention, on top of the costs of making it, and fifty guilders from any cooperatives (*corpora*).[90]

Aside from noticing how the act of inventing acquired a vivid monetary value, it is interesting to note how different types of consumers were taken into account in the dissemination process of a new product.

Conclusions

This chapter has highlighted that the crux of the legal revolution that unrolled in the sixteenth century was not that a new product was made eligible for legal protection, but that the production of materials was brought into the context of legal action at a stage where it did not yet have a tangible form. Earlier legal systems had certainly recognized the existence of a relationship between the creator and his product, regulating this relationship, for instance, through

89 "Welverstaende die voors. Claes Wolbrantsz gehouden sal syn t'voors. werck te maecken voor den geenen die dat van hem sullen begeren ontfangende van de selve boven de costen daeraen te doen aen materialen ende aerbeyts loon derttich Guldens eens voor syne Inventie." NL-HaNA, States General, 1.01.02, inv. no. 12301, fol. 68 [G133, 13 August 1614].

90 "Welverstaende dat de supplianten gehouden sullen syn hare Inventie te communiceren aen particuliere persoonen die deselve sullen begeren te gebruycken, genietende van de selve voor hare Inventie vyffentwintich guldens eens, boven de costen van het maken van de selve, ende aen de corpora voor vyftich guldens eens [...]." NL-HaNA, States General, 1.01.02, inv. no. 12300, fol. 282v [G111].

trademark law or by means of guild regulations.[91] The difference with the system of invention privileges was, however, that earlier forms of 'intellectual property' had never incorporated the notion of time; they had only dealt with manufactures, never with projects.

The possibility to hedge against the possible future success of projects is what made the privilege an interesting investment opportunity for merchants and projectors. Whilst this might not immediately suggest an early manifestation of a 'risk society', it is remarkable that the insurance represented by the privilege was arranged for by the state for a self-induced risk that was in this way considered to be controllable.[92] Just as the Dutch East Indies Company (the VOC) wanted to hedge the risks arising from its activities using a privilege, which in turn led to the world's first modern securities market,[93] the system of invention privileges fitted in with a new way of thinking about economy, state, and society. Only those inventions considered useful by the state could count on support; the state thus intervened silently, using the law as a technology of control to shape the type of knowledge that was produced. Yet it was mostly the urban elite, and mostly that in Holland, which profited from the system; the occurrence of artisans in the privilege rolls did not mean that the common man could progress independently by dint of the privilege system, as he was still dependent on investors who could provide the start-up capital to implement inventions.

Taking a giant leap forward, one could compare the dynamics of the early modern privilege regime with what the legal scholar Edmund Kitch has also identified as 'prospect patents'. In his 1977 article *The Nature and Function of the Patent System*, Kitch observed that patents not only reward inventors for their efforts but also encourage investment in a technological prospect *after* the property right has been granted. Patents, according to Kitch, therefore had a "scope that reaches well beyond what the reward function would require."[94] Kitch then proposed to use this insight to more effectively organize the market.

---

91 For the relationship between craft secrecy and proprietary attitudes toward craft knowledge, also see Long, *Openness, Secrecy, Authorship*, 13–14, 245.

92 Beck, *Risikogesellschaft*; Giddens, *Modernity and Self-Identity*. One difference is that invention privileges covered the risk of individuals, yet did not deal with risk in society at large.

93 Petram, *The World's First Stock Exchange*.

94 Kitch, "The Nature and Function of the Patent System," 267. "The conventional view of the patent system as a device that enables an inventor to capture the returns from his investment in the invention will be called the reward theory. The reward theory is not questioned on its own terms. Rather, it is argued that the reward theory offers an incomplete view of the functions of the patent system." Kitch, 266.

He pleaded, using US mining privileges as an example, for "awarding exclusive and publicly recorded ownership of a [technological] prospect shortly after its discovery."[95] As a result prospect patents would put their owner "in a position to coordinate the search for technological and market enhancement of the patent's value."[96] This is exactly what happened in the Dutch Republic, where authorities gave out broad 'patents' in the very early stages of technical development to encourage investment. Merchants then jumped on that development and what emerged subsequently was a true market of inventions.[97] The next chapter will examine how this market was legitimized in the legal theory of the time, and how the legal system of privilege examinations worked in practice

95 Kitch, "The Nature and Function of the Patent System," 266.

96 Ibid. Compare Duffy, "Rethinking the Prospect Theory of Patents," 441.

97 The expression 'market of inventions' is borrowed from Molà, "Il mercato delle innovazioni nell'Italia del Rinascimento."

CHAPTER 4

Legitimate Monopolies

Despite the fact that invention privileges belonged to a legal language that was widespread throughout early modern Europe, they received strikingly sparse attention in the learned treatises of contemporary legal scholars.[1] Whereas other types of legal privileges, such as those attributed to creditors, or matrimonial privileges, were extensively dealt with, privileges that specifically mentioned inventions and inventors were systematically neglected. References remained restricted to a vague allusion or a couple of sentences, which is remarkable when one considers the revolutionary aspects of the possibility to claim individual ownership over technological innovation. The first dim glimpses of any scholarly attention for invention privileges can only be observed in the course of the sixteenth century, i.e. about a century after the practice of granting privileges to inventors had found widespread acceptance. This chapter collects some of these scattered references and shows how temporary monopolies were theoretically underpinned in the early days of the Dutch Republic. After examining the legal characteristics of the early modern privilege, as well as with different forms of evidence used within the Dutch privilege system, the second part of this chapter will scrutinize how legal theory was applied in practice.

Historical Backgrounds

Roman law had never mentioned the use of privileges to favor inventors.[2] One explanation for that neglect could be that inventing entailed a form of manual labor, which was generally performed by slaves in ancient Rome and had

---

1 Portions of this chapter have been previously published in Buning, Marius. "Inventing Scientific Method: The Privilege System as a Model for Scientific Knowledge-Production." *Intellectual History Review* 24, no. 1 (2014): 59–70. Copyright © International Society for Intellectual History, reprinted by permission of Taylor & Francis Ltd, www.tandfonline.com on behalf of International Society for Intellectual History.

2 For an overview of existing opinions about intellectual property law under Roman and Greek civilizations, see Vukmir, "The Roots." Although trademark laws have been enacted in Rome, there is no proof that either the Greeks or the Romans proceeded to implement something that could be compared to a patent or a copyright. Cf. Greenberg, "The Ancient Lineage of Trade-Marks."

 | DOI:10.1163/9789004320420_006

therefore enjoyed little respect.[3] A more likely explanation is, however, that the Romans simply considered an invention privilege a crude example of an economic monopoly, for which they had always felt an exceptionally strong sense of natural contempt. This negative connotation of monopolies gained ever more sway as Roman law permeated throughout early modern societies during the so-called 'period of reception'.[4] With immense esteem for their Roman predecessors, rulers repeatedly re-enacted a ban once promulgated by the Byzantine Emperor Zeno (425–491 A.D.) to ensure that "no one shall exercise a monopoly over any […] material, whatever by his own authority or under that of an imperial prescript heretofore or hereafter promulgated."[5] With special regard to the Low Countries, Charles v enacted a law in 1540, declaring everything "that tastes like monopoly" to be forbidden.[6]

In addition, canon law strongly disapproved of any monopolistic control of the market too.[7] The religious abhorrence of monopolies ran parallel to discussions about the 'just price' that were based on the work of Thomas Aquinas.[8] The 'just price' was not simply the price that any individual would pay for a commodity, nor was it the price set by the State. It was the price "which at a given time, can be gotten from the buyers, assuming common knowledge and

3 Doorman, *Octrooien*, 11. For an exposition of the low esteem in Ancient Rome for the arts, see Rossi, *Philosophy*, 13–15. For a contesting opinion, see Long, "Invention," 848–49.

4 The 'period of reception' was a process that lasted from the eleventh to the sixteenth century. During that period, Roman and canon law slowly integrated into other systems of law, such as such as merchant law, feudal law, imperial law, etc. See also Monté Verloren, *Hoofdlijnen*, 270ff.; Wieacker, *A History of Private Law*, 124ff.

5 C.4.59, as cited in Prager, "The Early Growth," 115. See also Berger, *Encyclopedic Dictionary of Roman Law*, 576. For an overview of other places where the ban by Zeno was enacted, see Prager, "The Early Growth," 122.

6 "Monopolien, en onbehoorlijke contracten, by kooplieden en ambagtslieden, of steden, gemeenten en collegien, gemaekt ten agterdeel van andere kooplieden etc […] hetgeen smaekt na monopolie word verboden Plac.Kar.5.4.Oct.1540 art.7." Van der Schelling and Zurck, *Codex Batavus*, 751–52. The literature on early modern monopolies is abundant, but I found the following brief text especially useful: Höffner, *Wirtschaftsethik*. Often, a special role in sixteenth-century debates about monopolies is reserved for Konrad Peutinger (1465–1547), who advised Charles v in his *Consilium* of 1530 to allow for monopolies on luxury goods, such as spices or silks, but not on necessities of life (*res viviles*) such as grain or wine. Bauer, "Conrad Peutinger"; Roover, "Monopoly Theory"; Mertens, *Im Kampf gegen die Monopole*. For an overview and evaluation of the German literature on the role of Peutinger, see Nehlsen- von Stryk, "Die Monopolgutachten."

7 *Corpus juris canonici*, Decr. 11, c. xiv, qu. 4, c. 9. Höffner, *Wirtschaftsethik*, 135. For the general attitude towards to monopolies within the circles of Reformed Amsterdam, see Evenhuis, *Ook dat was Amsterdam*, 2:157–60.

8 Roover, "Monopoly Theory," 495–96.

in the absence of all fraud and coercion."[9] Based on biblical arguments, it was believed that only free negotiations on an individual level would guarantee that the margin of profit would remain reasonable: any attempt to intervene in the economy, either by the State or by a group of merchants, was therefore considered unlawful.[10]

In the course of the sixteenth century, however, some scholars began to show a more liberal stance towards the use of monopolies. They argued that monopolies could be justified in situations where they clearly served the best interest of the common good.[11] This was mostly the case when the exploitation of a certain commodity was so costly or so complicated, that no-one would put himself through the trouble without having a form of security to recover at least a part of the starting capital. Such ideas regarding the permissibility of monopolies were not only to be found among the Spanish Scholastics, but also in the work of the most famous legal scholar of the Southern Netherlands, Joos de Damhoudere (1507–1581), who was an influential member of the Burgundian Council of Finance.[12] De Damhoudere argued that not every form of cartel formation was harmful and that a monopoly might in some cases help to offset costs "so that the common good was helped and nobody would die of hunger or poverty."[13]

---

9 Cajetan, *Comments on Summa theologica*, II, ii, qu. 77, art. I (Leonine edition, VI, I49). As cited in Roover, "The Concept of the Just Price," 423. Medieval theorists thus recognized that the just price would always keep a certain span. Monroe, *Early Economic Thought*, 56. A complete survey of the concept of 'just price' falls beyond the scope of this book, but for an excellent overview, see Höffner, *Wirtschaftsethik*, 113–25. Other relevant articles and book chapters on the topic of just price include: De Roover, "Scholastic Economics"; Roover, "The Concept of the Just Price"; Rothbard, *Economic Thought Before Adam Smith*; Gordon, *Economic Analysis Before Adam Smith*; Monroe, *Early Economic Thought*. For the textual verification of De Roover's observations, see: Langholm, "Monopoly and Market Irregularities."

10 Höffner, *Wirtschaftsethik*, 132–35.

11 Höffner, 107–8.

12 Particularly, disputation 345 by Luis de Molina seems to have lingered in the collective memory; De Molina, *Opera Omnia*, 2:224–26. The work by the Spanish Jesuit scholastic was widely read and cited in the Dutch Republic, as was the work by other authors of the School of Salamanca, such as Francisco Suárez (1548–1617), Domingo de Soto (1494–1560), and Juan de Medina (1490–1547). De Medina particularly pointed to the public necessity (*publica necessitas*) that led the ruler to allow monopolies. See also Strieder, *Studien*, 91. Often with reference to D.9.2.51.2 ("It can proved by innumerable instances that many rules have been introduced by the ius civile in the public interest against the principles of reasoning.") and D 1.4.2 ("When new rules are introduced, their utility must be evident as to whether a law which has been considered just for a long time is to be changed"), the central point was that a (privileged) monopoly *had to be useful* to be permitted.

13 "ut hoc modo reipublicae consulat, ne fame aur rerum penuria aliqui moriantur [...]." De Damhoudere, *Praxis Rerum Criminalium Iconibus Illustrata*, cap. 132, pp. 383–386, on p. 384.

Yet merchants were not to take the law into their own hands: only a strong State could bear the responsibility of legitimizing monopolies, exceptions to the *ius commune*, by using the legal instrument of a privilege.

Privileges as a Legal Instrument

By the turn of the seventeenth century, privileges had enjoyed a time-honored history. Both Roman law and canon law had dealt extensively with the topic, and from there privileges had slowly found their way into the more local legal systems.[14] Roman law regarded the privilege (*privilegium*) as an exception to the *ius publicum*, which was the law concerned with the existence, organization and functioning of the State.[15] Likewise, a privilege became identified in the course of time as an *ius singulare*, which was a special law issued to the advantage of an individual or a class of persons, such as minors or soldiers.[16] But since it was generally believed that it was against the interests of the State to generate legal exceptions, privileges were used with caution.[17] Indeed, they even stood in bad repute.

In the course of the fifteenth and sixteenth centuries, however, the opposition between private and public law gradually appeared in a different light. Gradually, individual interests were equated with the interests of the State and it was now no longer taken for granted that a special law was by definition to the disadvantage of the general public. The underlying thought was that the State would guarantee the well-being of each and every individual, in which case monopolies that augmented the well-being of the State could simply not

14 There was no uniform system of law in early modern Europe. Next to Roman and canon law, which both served as umbrella systems, there was a wide range of other legal systems that claimed legal authority, such as merchant law, feudal law, imperial law, etc. Roman and canon law nonetheless set the standard: it is, for instance, impossible to distinguish mercantile law as an independent discipline in the early modern period, considering how firmly it was embedded in European legal language of Roman law. De Ruysscher, "Naer het Romeinsch recht." For similar conclusions, see also Fortunati, "La lex mercatoria."

15 Berger, *Encyclopedic Dictionary of Roman Law*, 532.

16 Franceschelli, "Lo origini," 160–61. Compare Beerninck, *Dissertatio juridica*, 5. Particularly, the ideas of Henricus de Segusio (*c.* 1200–1272) and Bartolus de Saxoferrato (1313–1357) seem to have had a lasting influence in this respect; see also Scattola, "Das Privileg des Gesetzes," 360, note 70.

17 The Twelve Tables, which were the oldest remains of Roman law that had been posted in the Roman Forum in order for all Roman citizens to be knowledgeable of the law, had ordered that "Privileges should not be imposed" (*privilegia ne inroganto- Tab.9.2*). In practically all of the work by early modern scholars that dealt with privileges reference is made to the Twelve Tables.

be contrary to the interest of any of its subjects. It was exactly along these lines that monopolies on sections of the market were justified, including monopolies for the exploitation of technological inventions.

Nevertheless, tracts that exposed the logic behind invention privileges were rare, also in the Dutch Republic. The most elaborated commentary I have encountered was by the hand of the Leiden University scholar Paulus Merula (1558–1607).[18] Together with Hugo Grotius and others, Merula belonged to an exclusive group of scholars that gathered around the famous humanist Joseph Justus Scaliger (1540–1609) with the idea of developing a new legal framework that would serve as the basis for the administration of justice in the young Dutch Republic. To that end, Merula wrote an incomplete legal textbook in 1592, later quoted extensively in other Dutch legal treatises, which among other things gives some valuable insight into the question how invention privileges fitted in with the overall legal structure (see Figure 47; notice the view on the Inner Court).

Following a typical Ramist way of proceeding,[19] Merula began with a very general definition of privileges as:

> Acts of grace or benefice granted to someone by the Highest Authority or the Prince of the Land, because of provided services: slightly at

18 Paulus Merula was a jurist who had studied in Rostock and Orléans before becoming a professor of history at the University of Leiden. In that capacity, he not only wrote a history of the world from its beginning to the present times – it was later finished by his son – and a Cosmography, but also a teaching book on the Digest under the title *Manier van procederen, in de provintien van Holland, Zeeland ende West-Vriesland, belangende civile zaaken : getrouwelijk ende met neerstigheid by een vergaderd uit alle ordonnantien, privilegien, instructien, missiven, acten, apostillen, &c. tot dien einde in de honderd jaaren herwaards by de Hooge Overigheid, of van wegen de zelve uitgegeven: alles met civile en canonijke regten, nevens der regtgeleerden, zo nieuwe als oude, meeningen bevestigd.* The book was first printed in 1598; I use the reprint from 1705 by Adrian Beeman, in which the pages 39 to 55 deal with the notion of privileges. For a biography of Merula, see Haak, *Paullus Merula*. For the use of Merula's book at the University of Leiden, see Ahsmann, *Collegia en colleges*, 21–23.

19 Peter Ramus (1515–1572) was a French humanist, whose educational reforms had a lasting influence on the development of knowledge in the sixteenth and seventeenth centuries. The standard work on Ramus and Ramism is Ong, *Ramus*. But also see Feingold, Freedman, and Rother, *The Influence of Petrus Ramus*. Merula structured his text in a way that was typical for Ramists; always arguing from the most general to the more specific and with the extensive use of so-called *Technometria* (≈ definition schemes). On the use of these *Technometria*, see Blair, *The Theater of Nature*, 161–64; Evans, "The Geometry of the Mind"; Ferguson, "System and Schema." Blair duly warns, however, against overstressing the importance of Ramus in these matters. Blair, *The Theater of Nature*, 84–85.

disposition to the common laws and ordinances of the country, without doing considerable harm to third parties.[20]

SYNOPSIS
PRAXEOS CIVILIS,
Maniere van Procederen
in dese Provintien/ Hollandt/ Zeelandt
ende West-Vrieslandt/ belanghende
Civile saken.
Ghetrouwelick ende met neerstichept by een vergadert upt alle Ordinancien/ Privilegien/ Instructien/ Missiven/ Acten/ Apostillen/ etc. tot dien eynde in de hondert Jaren herwaerts/ by de Hooghe Overichept/ off van wegen der selver uptghegeven: alles met de Civile en Canonijcke Rechten midtsgaders der Rechts-gheleerden so nieuwe als oude meeningen bevestiget.
Auctore Paullo G. F. P. N. Merula I C.

AEMSTELDAM.
By Cornelis Claesz. opt Water/ int Schrijf-Boeck.
Met Privilegie voor ses Iaren. A°. CIƆ IƆ XCII.

FIGURE 47 Title page of Paulus Merula's '*Maniere van Procederen*'

20 Merula, *Manier van procederen*, 40–41. There were, of course, many other definitions in circulation. One of those was by Hugo Grotius, who spoke of a privilege as an "award, or gift of the state authorities" (*Gunningh, gunst van 'sLands Overheyd*). De Groot, *Inleydinge*, CCC4., coming from the wordlist of "bastard German words, which are translated in this book with proper German words" (*Bastaert Duytsche woorden, die in dit Boeck met goede Duytsche woorden vertaeld staen*). Grotius did not deal with the definition of temporary privileges in any more depth.

Then he subdivided privileges into two categories: a category of time and a category of persons. Dealing first with the category of time, Merula furthermore distinguished between temporal and eternal privileges. Eternal privileges could be recognized by clauses, such as "from now on" (*Van nu voort aan*) or "for us, our inheritors and their offspring" (*voor Ons, onsen Erven ende Nakomelingen*).[21] Invention privileges, however, belonged to the category of 'temporal privileges'; they were privileges limited in terms of time.[22]

Privileges were limited in terms of time, either because the cause for which they were granted ceased to exist, or because "the time will cease to exist."[23] Privileges granted to soldiers, for example, or to the elderly, or because of infirmity, all ended in the course of time. In many of these cases, moreover, time was silently assumed. For instance, a privilege with bearing on a minor was only valid until that person reached twenty-five years of age, without the duration of the privilege having to be explicitly mentioned.[24] In other cases, however, the duration of the privilege was explicitly mentioned. This was particularly the case when a privilege was granted as a form of 'extraordinary help'. These privileges could be recognized by the clause "granted for the time of [*xx*] years."

A decisive characteristic of the privilege was that its cause had to be proven unless it was generally known. This was not the case with other laws, such as by-laws, decrees, or ordinances.[25] Speaking of the presentation of evidence that related to a privilege, Merula argued that it should be equal to all other 'instruments' used to make a claim in court.[26] This meant that the evidence

21 Merula, *Manier van procederen*, 42.

22 The exact interpretation of the eternal privileges was an issue that led to much dispute in the Republic. I shall not deal with these debates, as invention privileges belonged without exception to the category of temporal privileges. For the importance of eternal privileges in the Dutch Republic, see for instance Woltjer, "Dutch Privileges, Real and Imaginary." Van Gelderen, *The Political Thought*, 121.

23 Merula, *Manier van procederen*, 43. Merula speaks of the "ophoudende oorzaak" and "omkomenden tijd."

24 Ibid. Nowadays the age of majority which legally demarcates childhood from adulthood is often set at 18 years. Another example of a privilege that ended with time was that of a magistrate: when he stopped working, his privileges automatically came to an end.

25 For some of the particularities of *keuren, placaten, statuten,* or *edicten*, see Monté Verloren, *Hoofdlijnen*, 224–228.

26 "[...] moet daarvan doen exhibitie, gelijk van andere Instrumenten." Merula, *Manier van procederen*, 510. A common distinction made in the legal literature of the time was that between a 'private instrument' and a 'public instrument'. A public instrument was a letter that was signed by a public official, a prince, or a notary, for example. It demanded no further proof but was considered "stronger than a witness." A private instrument, in contrast, was a letter that was signed by a private person. Within the category of private

had to be presented to the court in writing. Moreover, the exhibition of this evidence was "necessary in order for the whole world to be informed of the privilege."[27] Although used in a different context here, this juridical review would later form the basis for what became known as a 'patent specifications' (see also pp. 76–87).

When it came to the revocation of a privilege, Merula was very reluctant. This uncertainty should be seen in the light of (a general portrayal of) the Dutch Revolt as a movement against the revocation of eternal privileges.[28] Against this backdrop, Merula distinguished between privileges granted to foreigners and privileges granted to subjects. When a privilege had been granted to "a foreigner, a stranger and not a subject," Merula argued, the privilege should be considered as "a contract" to which the Prince had committed himself, even if he was to deal with "the lowest and most evil of the world."[29] In this case, the Prince had acted in the capacity of a "private person" (*particulier mensch*) and so he was not allowed to revoke the privilege.[30] In much more uncertain terms, Merula argued that the Prince had also concluded a contract, when he granted a privilege to one of his own subjects. In any case, the Prince was simply not "supposed to" revoke a privilege as long as it was founded on good reason and if the proper amount of money had been paid.[31]

A privilege could be invalidated, however, if its owner committed a crime; if he acted contrary to, or misused the privilege; if the privilege was harmful; or if the privilege had not been "worked" (*niet en gebruykt*) within the allocated period of time.[32] In the next section of this chapter we will see how the States General did not hesitate to invalidate privileges, especially when inventors failed to deliver on what they had promised. On the other hand, the States General sometimes decided to extend the period for which the privilege was valid. According to Merula, this prolongation was called a 'special

instruments, there was a distinction between different levels of reliability. Thus account books, for instance, were considered as 'half proof'. See Ibid., 511–512.

27 "Dese Exhibitie is gantsch noodig, ten ware alle de wereld van soodanigen Privilegie kenisse hadden." Merula, 51.

28 See note 22.

29 "Want Privilegie by den Prince verleend aan een Uytheemsch, Vreemdeling ende geen Onderdaan zijnde, en moet niet gehouden werden ende bekend als een Privilegie, maar als een Contract. […] de minste en slechste vande Wereld […]." Merula, *Manier van procederen*, 51.

30 Merula, *Manier van procederen*, 51.

31 "[…] Extraordinarise Beden ende Subventien […] behoord […] niet te weder-roepen of breken." Merula, *Manier van procederen*, 52.

32 Merula, *Manier van procederen*, 53–54.

confirmation.'[33] It was used, for instance, when an inventor needed more time to reduce his invention to practice. The inventor Macus Duyvoet, for example, had "obtained a privilege [...] on 2 September ninety-five, on the condition that he would put [his mill] to work within a year." He then obtained an extension of six months on 23 January 1597, in order to "effectuate the trail of his invention [...] notwithstanding that the anticipated time that he was given in the privilege had expired."[34] The second form of 'special confirmation' occurred when the States decided to extend the duration of a privilege that had already been put to practice.[35] Sometimes, the conditions of the privilege, mainly the duration, were adjusted in the process, but in most cases the privileges were prolonged "on the same terms as the previous concessions of the Principal Privilege."[36] It appears that this form of confirmation was not based on any systematic logic; the States simply prolonged a privilege if it believed that the privilege had a positive effect on the local economy. This is further evidence that privileges originally granted for new products or methods could later assume a different function, without it being possible to draw a clear line of distinction between the various categories.

Besides using the category of time, Merula categorized privileges on the basis of the persons to whom they applied. This made privileges either 'general' or 'special'. General privileges were privileges that concerned the whole country, such as the rule not to admit any foreigners to the rank of officer, or

33 The confirmation of privileges was not a new gift, but merely a restatement of what had already been given. The confirmation of privileges could again be subdivided in a general and a special form. In its general form, especially when a prince died, the confirmation by the new prince simply reaffirmed all the privileges that had been granted by the predecessor. This happened usually during the ceremonial first visit of a prince to his country, the so-called *Joyeuse Entrée*. The special confirmation, also called *renovatie*, on the other hand, only concerned a specific privilege without regard of other privileges. Only the method of special confirmation was used for invention privileges. Merula, *Manier van procederen*, 49–50.

34 "[...] (daer van hy den IIen Septembris vyffent-negentich van ons Octroy hadde vercregen, mits dat hy tselve binnen tsiaers daernae in treyn soude brengen) by soo verre ons gelieffde hem daertoe octroy te verlenen [...] niettegenstaende den tyt hem daertoe by het Octroy geprefigeert, was overstreken. [...] ende dat die Suppliant die proeve daervan gehouden wort te doen binnen den tyt van zesse maenden naestcommende [...]." NL-HaNA, States General, 1.01.02, inv.nr. 12298, fol. 180 [G33, 23 January 1597]. Six months seems to have been the common amount of time to extend the trail period of a privilege.

35 See, for instance, NL-HaNA, States General, 1.01.02, inv.nr. 12302, fol. 41 [G164, 5 October 1617].

36 "[...] opten selven voet van de voorgaende concessie van het Principael Octroy." NL-HaNA, States General, 1.01.02, inv.nr. 12298, fol. 356 [G55, 26 May 1601]

not to exempt foreigners from paying tolls.[37] Special privileges, on the other hand, concerned the individual districts (*quartieren van den Lande*), the provincial courts and the States, particular cities or villages, and finally private persons.

With regard to privileges granted to private persons, Merula again made a distinction between privileges that concerned one person only, and privileges that were equally valid for a number of persons. This second category related to individuals who were "many of one kind" (*unius generis personis*), such as widows, orphans, minors, the poor, the insane, students, etc. The first category, that of privileges granted to one person only, related amongst others to privileges for inventors. On this point, Merula explicitly quoted several examples from the books of the Court of Holland that related to the exclusive rights for the exploitation of innovative methods.[38] But whereas all his other examples were backed by an impressive number of references to other jurists and Roman lawyers, Merula had to admit that Roman law had not dealt with the topic of individual privileges at all. He concluded that Roman lawyers had probably not considered it necessary to deal with privileges that "die out" with the individual person.[39]

Finally, this brings Merula to speak of the inheritability of privileges. He claimed that privileges granted to individuals, and particularly privileges that had been granted by the Prince as an explicit favor, always lost their legal validity on death.[40] This seems remarkable, considering that on several instances the heirs of an inventor made legal claims based on an earlier privilege. It appears, however, that in those cases the privileges had been explicitly granted "for the Supplicant and his heirs."[41] The possibility of inheriting a privilege thus had to be explicitly mentioned in the document issued by the authorities.

Nonetheless, a privilege could be transferred to another person. This is apparent not only from the countless acts and contracts in which the invention privilege was transferred and actively negotiated, but also from a legal advice that was given in 1632 by the Dutch jurist Hugo Grotius. The case in question was a sawmill, for which the privilege was sold by the original inventors, in the

37 Merula, *Manier van procederen*, 45.

38 In fact, the word 'innovative' is not used here; what we call 'patents' were simply presented as examples of privileges granted to individuals. Merula refers to the *memoriael boeken* of Bertolus Ernst; I have not been able to trace these references.

39 "[…] die met de Persoone haare kragt verliesen." Merula, *Manier van procederen*, 47.

40 Merula, *Manier van procederen*, 48.

41 The standard clause was "zijn suppliant, ende zijne nakoecomelingen te vergunnen octroy/ ende privilege […]." NL-HaNA, States General, 1.01.02, inv. no. 12299, fol. 102v [G76, 25 February 1603].

presence of a notary, to a company of timber merchants. Apparently, a few copycats had set up knockoff mills, defending themselves against accusations of infringement by raising the argument that the authorities had never granted any privilege to the timber-merchants. Yet Grotius argued that:

> the Privilege has been granted not with respect to the person/ but with respect to the invention/ and that therefore the right/ thus obtained/ has been lawfully and properly transposed to other persons/ who will put the invention to work: And that the mentioned timber merchants/ having obtained the said right/ having properly summoned the infringers upon the aforesaid Privilege/ will be allowed to address themselves to the local prosecutor/ in case [the violators] will continue the contravention: The prosecutor/ in compliance with the aforesaid Privilege/ and with the general imposition it carries on all prosecutors/ will be allowed to immediately cassate [the contravention]/ by taking away the instruments/ as specified in the Privilege/ or by any other appropriate means. In addition, the prosecutor will be allowed to start proceedings on behalf of the timber merchants against the infringers and plead for the confiscation of counterfeit products/ as well as a fine/ as specified in the Privilege.[42]

The advice of Grotius would resonate through the ages, as illustrated by a case from 1719, which also happens to be a nice example of how the legislation pertaining to inheritability of a privilege was later amended. The subject was a certain Servaas van der Wielen, the inventor of a dredging machine, who was succeeded upon his death by two sons. When one of the sons died, the Court of Holland had to decide whether the inventor's grandchildren could make a claim to the privilege rights, or whether the (still valid) privilege should pass

42 "Dunkt (onder correctie) dat alsoo 't geseide Octroy is verleend niet ten aanzien van den persoon/ maar ten aansien van de inventie/ dat daarom het recht/ daar by verkregen/ aan andere personen/ die deselve inventie in 't werk sullen stellen/ wel ende rechtelijk is getranspoztert: Ende dat de voornoemde Houtkopers/ als het zelfde recht bekomen hebbende/ aan de contravenenteurs van het zelve Octrooi gedaan hebbende behoorlijke insinuatie/ zoo de selve noch blijven continuerende in de selve contraventie/ haar zullen mogen adresseren aan den Officier van de plaatse: Welken Officier/ tot nakominge van het gemelde Octrooi/ ende generale last daarby aan alle Officieren/ deselve contraventie/ door afhalinge van de Instrumenten/ gespecifieert in het zelfde Octroy/ ofte andere bequaamste middelen/ datelijk dat mogen doen cesseren. Ende dat voorts den selven Officier/ nevens de voorsz Houtkoopers de contraventeurs soude mogen trekken in rechten/ ende aldaar contenderen tot de verbeurte van de nagemaakte wercken/ ende voorts tot de peine/ in den voornoemde Octroije gespecificeert. Alsus geadviseert den 6 April 1632." *Consultatien*, 509 (Advise 187, H. de Goot).

in its entirety to the surviving son.[43] The jurists who issued the advice in this case, stated, with explicit reference to Grotius, that "this patent is not personal, or founded on any graceful personal qualities that would reside in the person of the petitioner."[44] Instead "the underpinning of the patent is the invention of the patented instruments, and the compensation of time and cost spent on doing so."[45] Their advice was therefore to consider the privilege as *jura realia* and inheritable property.

This brings us slowly but surely to the end of the theory on privileges in the early modern Dutch Republic. In summary, privileges differed in three important respects from other laws issued in the early modern period. The first difference was that privileges were paid for, even if the official reasons for awarding a privilege were embedded in a language that revolved around merit and service. The second difference was that privileges were always in slight contradiction to existing laws (*contra lex*). This aspect of the privilege notion in particular played on the minds of contemporary legal scholars, as it seemed contradictory that a privilege was 'a law against law'.[46] However, the point was not that a privilege freed a legal subject from the law, but rather that it changed the legal status of the person to whom it applied. Hence, even if a privilege created new law, it still had to comply with existing legal structures.[47] The last, and maybe most striking, difference between privileges and other

43 *Vervolg op de Hollandsche Consultatien*, 381–86.

44 "dat dit Octroy niet is personeel, of gefundeert op eenige personeele gratieuse qualiteiten, die in den person van den Verzoeker zoude resideeren." Ibid., 383.

45 "[...] het Octroy [...] niet en is een Privilegie uit eenige personeele consideratie voor Servaas van der Wielen particulier, en alleen *bene placito* van de Souverain verleent, nemaar, dat hetzelve Octroy en de gifte van dien, tot zyn fundament heeft de inventie van de geoctroyeerde Instrumenten, en het dedomagement van de tyd, en de kosten daar aan gespendeert [...]." Ibid., 382.

46 As we saw, Merula defined privileges as: "Acts of grace or benefice granted to someone by the Highest Authority or the Prince of the Land, because of provided services: slightly at disposition to the general law and ordinances of the country, without doing considerable harm to third parties." Merula pointed out that a privilege should not be entirely (*ex diametro*) in conflict with the common law (*gemeene rechten*) but that "the word Slightly [...] softens the contradiction" (*Het woord* Eenigsints [...] *versacht de Contrarieteyt*). Merula, *Manier van procederen*, 41. See also Franceschelli, "Lo origini," 163.

47 See also Quaritsch, *Staat und Souveränität*, 129. Quaritsch makes the interesting observation that, as long as the privilege (an *Einselfallgesetz*) only transformed the legal status of the person to whom it applied (namely, from general to particular), common law and privilege were pretty well matched. Only when general laws came to set aside privileges was the way freed for a absolutist understanding of legal power that turned the king into the *origio iuris*. Ibid., 129–136. Quaritsch traces the roots of this transformation to events in thirteenth-century Germany.

legal remedies was that the cause for a privilege had to be proven.[48] This was in stark contrast to other laws, which demanded no proof whatsoever.[49]

Indeed, it was because of the obligation to put the grounds for a privilege to proof that standard procedures concerning the law of evidence started to play a role in the appraisal of technological advancement. Juridical evidence in this regard came in various forms and degrees. It ranged from 'full proof' to 'half proof', and even to 'quarter proof', which were all qualifications based on the degree of certainty attained by means of the truth claims. Full proof was, for instance, just another way of saying that the matter was clear, whereas half proof was "a claim that informs the judge to some degree, yet not completely, or not to the degree that it might serve as the basis for a verdict."[50]

Another important distinction regarding the rules on gathering evidence was whether the topic of dispute was tangible or not.[51] If the topic was intangible, but instead concerned a trial of legal rules, the case was supposed to proceed 'according to laws' (*in regten*). This meant that the admissible evidence could only be presented by means of so-called *memoriën*, which were short written statements describing the matter at hand as well as the current state of affairs.[52] If, instead, a justiciable issue bore upon something tangible, the case was supposed to proceed 'according to facts' (*in feyten*). In these cases, admissible evidence was primarily presented in the form of testimonies and by 'ocular inspection' (*inspectie oculaire*).[53]

Ocular inspection was the legal term for a procedure in which the court sent out commissioners to inspect a certain issue. These inspections provided full proof in matters of land division and easement on a property, but they were also used "to determine whether the wound of someone who was struck down had been deadly, and [to judge] about divorce for the reason that someone is unable to reproduce, and to determine whether a woman is pregnant,

48 Merula, *Manier van procederen*, 41.

49 Ibid., 52.

50 "Halve preuve, is sodanige bewering, daar by den Regter wel enige kennis van de saak krygd, maar niet volkomen, of sodanig, dat daar uit de saak by vonnis mag werden gewesen, of voor Regt kan werden uitgesproken." Van Leeuwen, *Rooms-Hollands-regt*, 618. Cf. Merula, *Manier van procederen*, 527. The focus is here on the Dutch Republic, but similar standards of proof were used elsewhere in Europe too.

51 Van Leeuwen, *Rooms-Hollands-Regt*, 103.

52 The claimant would provide a brief statement with the arguments why he thought he was right, whereupon the defendant would have the possibility to make a statement to the contrary. It was then up to the judge to further decide upon the matter. Van Leeuwen, *Rooms-Hollands-regt*, 695.

53 Ibid., 616. Other methods to furnish evidence, such as presumption, rumors, and hearsay, were never conclusive in the court of law. Ibid., 619.

and suchlike […]."[54] It was particularly in this context that non-legal experts, such as medical practitioners or land surveyors, began to play an increasingly important role in fact-finding procedures throughout the sixteenth and seventeenth centuries.[55]

Experts could also be called in as witnesses to provide testimony, which was the other way to argue a case proceeding 'according to facts'. Generally speaking, one needed two or three people unanimously to pledge upon a 'thing' (*het ding*) in order to give a testimony the value of a public instrument. But there were other cases in which larger groups were required, and in some cases one witness would suffice.[56] In addition, it was commonly agreed that one ought to attach more credibility to the testimony of "those in Dignity than [to the testimony of] normal People" and more to the testimony of "the Rich than [to that] of the Poor, for fear of Corruption."[57] However, this did not mean that all the witnesses belonged to the higher strata of society, since the main concern was still to obtain testimonials from those who were knowledgeable in the topic.[58]

Experimental Practices

Now that we have dealt with the legal theory that formed the basis for the justification of invention privileges, it is time to examine how different forms of evidence-gathering were carried out in practice. I have decided to do so by

54 "By oculaire inspectie, dat is, vertoog en aanwysing op en omtrent de ding-pligtige saak, werden ten vollen bewesen alle grond-sscheidingen, of Erf of Land-diensstbaarheden. […] De welk ook in veel andre gelegentheden nodig is: als om te oordelen van een wonde van een nedergeslagen, of deselve doodlyk is geweest, en over Egt-scheiding om oorsaak dat yemand tot voortteling onbequaam is, en om te oordelen of een Vrouwe swanger gaat, en diergelyke, werden defelve personen aan de ervare Medicyns en Vroed-wyven ten onderfoek voorgefteld, de welk daar van na genomen inspectie en klaar betoog, oordelen wat van de saak is." Van Leeuwen, *Rooms-Hollands-regt*, 617–618. Cf. Merula, *Manier van procederen*, 485.

55 Crawford, "Legalizing Medicine"; Pastore, *Il medico*; Di Renzi, "Medical Expertise". For later trajectories of expert witnessing, see also Golan, *Laws of Men*; Jasanoff, *Science at the Bar*.

56 Merula, *Manier van procederen*, 502–503.

57 "[…] meer den geenen die in Digniteytt zijn, dan andere gemeende Luyden […] Meer den Rijcken dan den Armen, om de vreese van Corruptie […] meer den Armen van goeden leven ende opregt in haaren handel; dan den Rijcken quaad van Leven ende gediffameerd." Ibid., 500.

58 Ibid., 501. As Merula formulated it, the most reliable testimony was that of "each in his art" (*Maar voor al een eyder in zyne konste*). For similar observations on the reliability of witness statements in the English context, see Shapiro, *A Culture of Fact*, 75–76.

providing a detailed reconstruction of three cases in which the efficacy and utility of new inventions was thoroughly reviewed and assessed. Making a comparison with contemporaneous scientific developments, I will deal with the privilege for a drainage mill and with the importance of mathematical expertise for the assessment of a new method for longitude determination. But I begin with a case concerning a new method of processing alloyed silver thread in various embroideries.

In a letter to the States General, dated 15 July 1615, the merchants Ian de Carpentier and Servaes Hellings wrote:

> how finally, after long and expensive research, they had obtained the knowledge (*wetenschap*) of making and using alloyed silver in all kinds of golden and silver embroidery, which all experts, based on the touch and other proofs, considered just as good and beautiful as embroidery that hitherto had been made with finest silver and gold.

The inventors claimed that the utility of their invention could be "proven by the attestations that have been attached [to the request]" and "that no-one had made [similar work] before in these United Provinces." Similar embroidery, moreover, was "costly in these lands, because it is sent in abundance from Italy and other places." Arguing "that otherwise someone else might copy the method and thus would take away the profit of the art that they have found with so much effort," and expressing their intention "to work with many workmen in producing large quantities of these textiles in these Lands," the inventors therefore requested a privilege for fifteen years to come.[59]

---

59 Ian de Carpentier en Servaes Hellings originated from Luik, which was a centre for trade in precious metals. Klein, *De Trippen in de 17e eeuw*, 88–89. The inventors were both residing in Dordrecht at the time of the application. Information about the case can be found in Coren, *Observationes Rerum* [1633], 159–246. The entire quotation from which different parts are taken in this paragraph is: "Hoe dat syluyden nae lande ende kostelicke recherche, eyntelick gekomen zijn, tot de wetenschap van gealloyeert silver te verarbeyden tot alle stoffen van goude ende silvere borduyrwerckerye, de welcke op de touche, ende andere preuven, by alle kenders, alsoo bequaem, ende schoon werden bevonden, as 't gene men tot deser tijdt toe, van 't fijnste goudt, ende silver heeft gemaeckt, blijckende by de attestatien aen desen annex, Sonder dat oyt by yemant in dese geunieerde Provincien soodanige goude ende silver stoffen, als voorsz. is, van gealloyeert silver is gemaekt. Ende mits den arbeydt hier te lande kostelick valt, voor desen uyt Italien, ende andere plaetsen, in abundantie, in dese provincien de voorsz stoffen zijn gesonden. En alsoo de Supplianten van meeninge zijn met veele Arbeyders, in groote quantiteyt de selve stoffen in dese Landen te maecken, ende te doen maecken, ende beducht zijn, dat yemant anders haer-luyden voorsz Inventie soude mogen naer conterfeyten, ende daer door benemen de

One of the attestations attached to the request was by the hand of a certain Gilles Nagels, a merchant in embroidery, and Isaac Willemsz, an embroiderer, both residing in Amsterdam. They confirmed that they had assessed the work of the inventors and declared that it would serve them well, since it was of better quality than any work that was made up till now "in these Lands" (*hier ten Lande*).[60] A second attestation attached to the request came from two wire drawers, named Lenart Nulck and Athonis Willemz. They, too, declared that the invention was unknown in the United Provinces.[61] On this point, the States General was sufficiently convinced and on 15 July 1619, it granted Ian de Carpentier and Servaes Hellings a nine-year privilege to commercially exploit the new method of processing alloyed silver thread.[62]

The affidavit by Nulck and Willemz had also contained a little box, sealed-closed, in which the attestants had enclosed some of the samples they had used. Apparently, the box "attached to the request [had been] not opened nor examined" until several complaints had come in about the privilege on 16 July 1622.[63] Some of these complaints came from the jewelers Dennis Nootstock and Matthias Schouwerick. They declared on behalf of two gold drawers (*gouttrekkers*) that the high master (*opperdeken*) of the goldsmith hall in London, Mr. Simsom, together with an unnamed London based jeweler, had recently visited Amsterdam to inform them that several boxes containing the work of De Carpentier (with a value of 100 pounds sterling) had been "arrested" in London.[64] The merchandise of De Carpentier had been impounded in the Customhouse, where three more boxes with his embroidery were kept. In addition, Nootstock and Schouwerick declared that De Carpentier was under an obligation to pay a twenty-pound sterling fine to the goldsmith hall in London and that he could no longer export his work to England. He was also obliged to recall all his goods and reimburse anyone who wanted to return any of the purchased embroidery.[65] The quality was poor, and it contained too little silver.

profyten vande kunste die sy met swaren arbeydt hebben gevonden en gepractiseert, soo keeren sy supplianten hen seer reverentelicken aen u, mijne Ho.Mo. Heeren [...]." Coren, *In supremo*, 160.

60 Coren, *In supremo*, 161. The attestation was dated 11 July 1619.

61 Coren, *In supremo*, 161.

62 NL-HaNA, States General, 1.01.02, inv. no. 12302, fol. 130 [G179, 15 July 1619].

63 Coren, *In supremo*, 162.

64 In England, the testing of gold and silver products to determine their ingredients and quality, was a national privilege of the London Goldsmith's Company. Epstein and Prak, "Introduction," 13.

65 Dillen, *Bronnen*, 2:499 [no. 873, 3 August 1623]. Mr. Simpson had furthermore declared that Carperntier's merchandise had passed several times before the Court in England

The States General, however, was reluctant to begin proceedings. Based on personal documents from one of its representatives, it can be inferred that ideas were presented to continue De Carpentier's privilege in the event that the amount of gold were augmented.[66] But the ideas led nowhere, and complaints kept coming in.[67] Finally then, on 14 July 1624, the States General summoned the gentlemen Scotten, Van Vosbergen and Coren from the High Court of Holland, together with Dr. Couwenburch, Wijngaerden, Loenius, and Van der Meer from the Court of Holland, to hear the different complainants against the filigrees as well as De Carpentier and Hellings in their defense.[68] They were to investigate the three following points:

1. On the difference between silver and gold wire from Milan and wire from Hellings;
2. Whether the privilege of Hellings was falsely obtained;
3. Whether the privilege was misused.

The documents present in the registrar's office were handed over to the commissioners, and all parties were encouraged to turn in as much evidence as possible.[69] In addition, the commission ordered that material should be taken from a shop in The Hague that sold material from De Carpentier in order "to fully ensure" that the material in the box was really the same as that sold in the streets.[70] Then, the examination could proceed.

First, the commissioners determined the difference between silver and gold wire from Milan and that from De Carpentier and Hellings (see Figure 48). The outcome of this research confirmed that the wire from Dordrecht contained

---

(*Raedt in Englant*), who had assayed the quality of the material and considered it "very bad and harmful" (*heel slecht ende onlydelijck*).

66 The amount of gold should be augmented to match up to the official standard set by Charles V, and which had been confirmed by the High and Provincial Court. Veenendaal-Barth, Smit, and Vree, *Particuliere notulen*, 1:273, 274. Note made by the deputy for the City of Medemblik (Province of Holland).

67 In another notarial archive we find a declaration by two silk peddlers, who declare on behalf of the governors of the embroidery guild in Amsterdam (*overlieden van het borduurwerkersgilde*) that they had bought merchandise made by De Carpentier for the price normally paid for Milanese wire. Dillen, *Bronnen*, 2:524 [no. 922, 6 March 1624].

68 Dr. Caspar van Vosbergen (*c.* 1575–1649) was a member of the High Court on behalf of Zeeland, whereas dr. Jacob Gerritsz Coren (?-1631) was sitting on the bench for Holland. Scotten is probably dr. Apollonius Scotte (*c.* 1575–1639), member for the Court on behalf of Zeeland. The bibliographical data for the members of the Court of Holland is as follows: dr. Pieter Couwenburch van Beloys (?-1644); dr. Johan Loenius (?- 1641); dr. Abraham Pietersz. van der Meer (1584–1638); dr. Johan Oem van Wijngaerden, Lord of Werkendam (*c.* 1583-?).

69 Coren, *In supremo*, 167.

70 "tot volkomener onderrichtinge...". Coren, *In supremo*, 169.

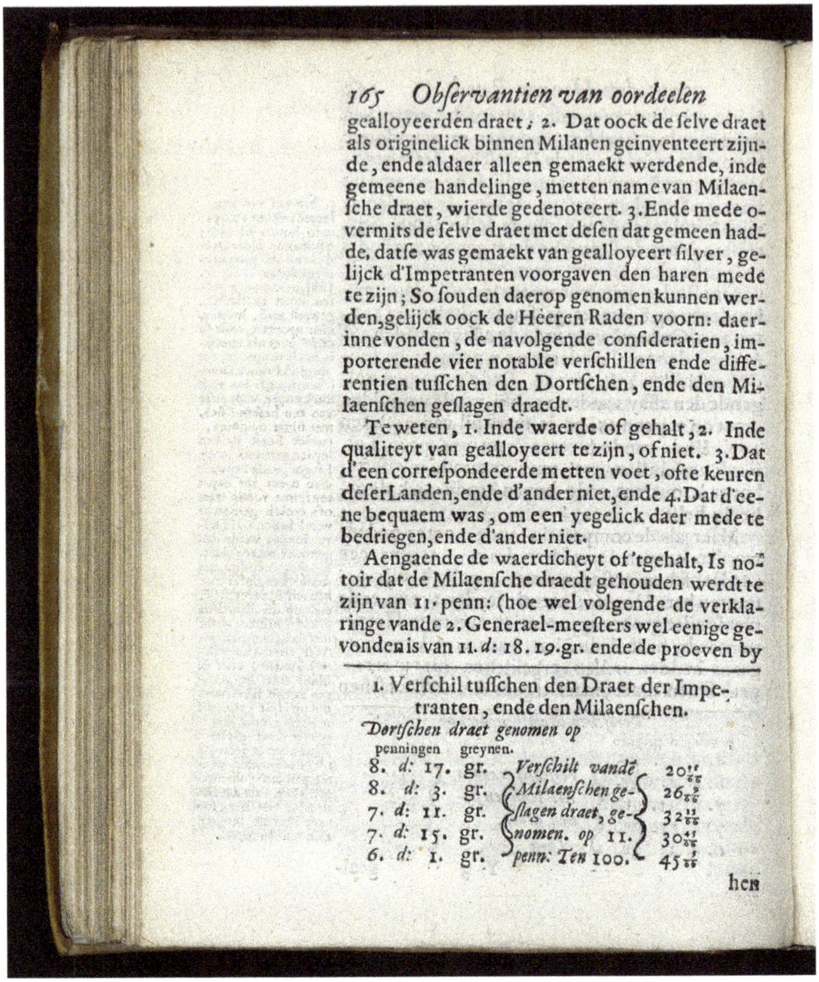

165 *Obſervantien van oordeelen*

gealloyeerden draet; 2. Dat oock de ſelve draet als originelick binnen Milanen geinventeert zijnde, ende aldaer alleen gemaekt werdende, inde gemeene handelinge, metten name van Milaenſche draet, wierde gedenoteert. 3. Ende mede overmits de ſelve draet met deſen dat gemeen hadde, datſe was gemaekt van gealloyeert ſilver, gelijck d'Impetranten voorgaven den haren mede te zijn; So ſouden daerop genomen kunnen werden, gelijck oock de Heeren Raden voorn: daerinne vonden, de navolgende conſideratien, importerende vier notable verſchillen ende differentien tuſſchen den Dortſchen, ende den Milaenſchen geſlagen draedt.

Te weten, 1. Inde waerde of gehalt, 2. Inde qualiteyt van gealloyeert te zijn, of niet. 3. Dat d'een correſpondeerde metten voet, ofte keuren deſer Landen, ende d'ander niet, ende 4. Dat d'eene bequaem was, om een yegelick daer mede te bedriegen, ende d'ander niet.

Aengaende de waerdicheyt of 'tgehalt, Is notoir dat de Milaenſche draedt gehouden werdt te zijn van 11. penn: (hoe wel volgende de verklaringe vande 2. Generael-meeſters wel eenige gevonden is van 11. *d*: 18. 19. gr. ende de proeven by

1. Verſchil tuſſchen den Draet der Impetranten, ende den Milaenſchen.

Dortſchen draet genomen op

penningen	greynen.		
8. *d*:	17. gr.	*Verſchilt vandē*	$20\frac{15}{66}$
8. *d*:	3. gr.	*Milaenſchen ge-*	$26\frac{9}{66}$
7. *d*:	11. gr.	*ſlagen draet, ge-*	$32\frac{33}{66}$
7. *d*:	15. gr.	*nomen. op* 11.	$30\frac{45}{66}$
6. *d*:	1. gr.	*penn: Ten* 100.	$45\frac{5}{66}$

hen

FIGURE 48 The examination of silver thread

a lower percentage of gold than the usual standard.[71] The main issue now was how to interpret the word 'alloyed'. The privilege request had spoken of "alloyed silver in all kinds of golden and silver embroidery." But was the new type of wire really an alloy (i.e. completely intermingled with one another), or was it just a silver coat on a copper wire that gave the illusion of being alloyed silver? To answer this question, the Court carried out an experiment that is worth quoting in detail:

---

71 Coren, *In supremo*, 184, note 41. The standard had been set by ordinance By Charles V on 13 April 1551; it ensured that no silver works could be made with a lesser alloy than set by a declaration made up in Gent on 24 July 1517.

In the presence of the Court and in the presence of Hellings and others, the aforementioned Van Engelen took some wire and put some sulfate on it, which he set on fire with a candle. After he considered it well-enough burned, he split the wire with a small knife and took of the silver from one side. He also took it from the other side, using the before-mentioned Art, revealing thus the copper and consequently that the copper and the silver could be separated preserving its each its substance. Hellings pointed out that by this way of burning sulfate one could also treat a coin in such a way that it could be split with a knife, and the experiment thereof was done by the Assayer General in the presence of the aforementioned Court and others on a half a *rijksdaalder*, from which he took a part, say a scale thick, with the result that the form, weapons, and inscription were still recognizable and readable. [Hellings added that the experiment would have turned out better if saltpeter instead of sulfate had been used.] The aforementioned contra-experiment [sic!] proposed by Hellings to point out the weakness of the first experiment proved that also composed silver or pennings could be cut. But that had not been the intention of the first experiment. But rather, to be able to separate the metals in their substance, [to cut] the coin, which he was not able to bring about.[72]

72 "D'voorsz. Van Engelen heeft voor de Raden, ende in presentie van Hellings ende andere sekere loveren genomen, ende op de selve geleyt hebbende wat swavels, twelck hy daer na met een kaersse ontstack, nae dat hem dochte genoech gebrandt te zijn, heeft met een mesken de loveren gespouwen, ende 't silver aen d'een syde afgenomen, oock wel aen beyde syden. als hy d'voorsz. Konst aen beyde zijden hadde gebruyckt, verthoonende het koper bloodt, ende consequentelick dattet koper ende silver van malkander konden gesepareert werden salva utriusque substantia; Ten selven tijde is by Hellings mede aengewesen, dat met de manieren van swavel te branden, oock gealloyeerde Munte soo gemoillieert konde werden, dat men de selve met een mesken van een snyden ende spouwen konde, gelijck de proeve daer van by den Assayeur generael mede gedaen wierdt in presentie vande voorn: Rade, ende anderen aen een halve Rijcxdaelder, daer hy een stucxken boven afnam, ter dicte allee, van een schubbe, ende sulx dat de forme, wapenen, ende inscriptie noch kenelick ende leesbaer waren gebleven. Verklaeren ooc daer nae dat assay gemaekt zijnde vande overgeblevene penningem d'selve inde alloy niet en was gemineert, maer 't afgenomen stuck was eenichsins slechter gevonden overmits (so Hellings meynde) 't silver door de cracht des sulfers wat geconsumeert was, maer als 't koper door dien 'tsilver daer so wel niet tegen en konde als het koper, gelijck ter contrarie gebruyken de Salpeter 'tkoper eer als 'tsilver verteert soude werden. D'voorsz. contra-proeve wierdt by Hellings voortghebracht, om tot enervatie ende confusie vande 1. proeve te bethoonen, dat gealloyeerde penn: of silver, mede konde gesneden werden. *sed illa non erat intentio primi experimenti. Sed tantummodo, ut ostenderet separari posse metalla in suo subjecto, quod in nummo suo, alter efficere non potuit.*" Coren, *In supremo*, in the margins of page 181 (note 33). Interpunction in translation added by the author.

Helling's point had been to introduce a positive control; and also to show how the coin was separable into two substances. But the coin did not fall apart. The States General decided then to nullify the privilege, taking into consideration that by continuing the privilege "a door was opened to fill the entire Land with false works."[73]

One of the remarkable aspects of this case was the scientific character of the experiment done in the presence of the court. The procedure began with the formulation of a hypothesis, which was then tested by means of experiment. The method of proceeding was described so precisely that, in theory, they could be repeated by others. The goal was thus to examine the truth-value of statements through factual and personal observation, instead of by reliance on previous postulates. Usually, these are characteristics associated with a new way of studying nature, which has also gone down in history under the much-contested label of a "Scientific Revolution". This Revolution took shape roughly in the period between Nikolaus Copernicus (1473–1543) and Isaac Newton (1642–1726), respectively between their major publications *On the Revolutions of Heavenly Spheres* (1543) and the *Mathematical Principles of Natural Philosophy* (1687). It did not consist of one major transformation within the field of science, but rather of a conglomerate of changes in different fields of knowledge.[74] Particularly distinctive for the new way of doing science was that: 1). experience turned into experiment; 2). technology came to be positively valued and admitted as a means to provide scientific evidence; 3). science came to have a practical edge; and 4). mathematics became a trustworthy language with which to describe reality.[75] A glance at the privilege system demonstrates

73 "[...] een deure gheopent was, om 't geheele Landt met valsche wercken [...] te vervullen." Coren, *In supremo*, 156–157. The privilege was nullified on 15 August 1624. The decision was circulated in the form of a pamphlet: *Sententie Van de Edele Eerentfeste Heeren Gedelegeerde,* Knuttel 3543. The inventors had been obliged to hand in their privilege at the registrar's office, but the problems surrounding the silver wire did not immediately come to an end. Carpentier continued to sell his tampered wire and, after many complaints, another placard was issued on 23 July 1627. Cau and others, eds. *Groot Placaet-boeck,* 1:2734–2737 (colums, not pages). Hellings' reaction to the entire issue was telling of his unscrupulous attitude: "Go to the devil and go where one makes better silver, if it is not good enough for you, this is good enough for me" (*Loopt voor den duivel ende gaet daer men beter silver maect, ist U niet goet geneoch, het is my goet genoech*). As quoted in Klein, *De Trippen in de 17e eeuw*, 88–89.

74 The historiography on -what we used to call- the Scientific Revolution has slowly come apart at the seams, and there is no point in trying to summarize it here. A concise overview of some of the issues at stake is provided in Shapin, *The Scientific Revolution*.

75 Clearly, the presented subdivision is not the only way of formulating the problem and, moreover, many refinements could be made (such as the importance of the use of algebra

that these elements, which were seminal for a new way of thinking about gathering evidence in the natural sciences, were in fact already present in the legal practice of accessing invention privileges. It is therefore worth looking more closely at the comparison between both these great 'revolutions'.

Time and time again, the resemblance between methods of evidence-gathering within the privilege system and the scientific method of proceeding comes to the fore. One example presents itself in the examination of a drainage mill invented by the court mathematician Simon Stevin, who is primarily remembered for his work on decimal fractions, and for the discovery of the hydrostatic paradox.[76] Stevin was the pivot of scientific developments in the Dutch Republic at the turn of the seventeenth century. As the Quartermaster-General of the Dutch armed forces, Stevin not only made his appearance in several commissions charged with the assessment of privilege applications (see pp. 59), but he also obtained several privileges throughout his career that affirmed his importance as one of the leading military engineers in the service of the state.[77] One of these inventions was that of a drainage mill that could pump up large amounts of water more effectively; a testimonial passed before a notary to confirm the utility of the invention gave a detailed description of the way in which the mill was put to the test.

On 26 March 1590, a group of people gathered in the vicinity of the city of Gouda to test the new drainage mill that Stevin had invented.[78] The group was divided into two parties. On one side there were two aldermen from Gouda, who vouched for Stevin, and on the other side there were two men chosen to represent the interests of the bailiff and dike-reeves of Stolwijk (a village close to Gouda). The two groups gathered, as they said, to "be informed of the assessment of what the Stolwijckse mill, which had been remodeled after the way of Stevin by the carpenter M. Joost Govertsz, would do against the Beyersche mill

to express geometrical proportions). Nevertheless, I believe that the main characteristics of early modern science are pretty well covered.

76 Stevin was born in Bruges but moved to Holland at the peak of the Revolt. In 1583, he registered as a student at the faculty of Artes Liberales at Leiden University, where he probably became acquainted with Stadtholder Maurice of Nassau. Kubbinga, "Stevin en Maurits." The standard biography for Stevin is Dijksterhuis, *Simon Stevin*, notably chapter 1.

77 On Stevin's 'patents', see Stevin, *Principal Works*, 5:11–38; Van den Heuvel, *De Huysbou*, 72–77.

78 There were a number of testimonials that confirmed the utility of Stevin's invention. These are given in full by Stevin's son Hendric; Stevin, *Wisconstich filosofisch bedryf*, Book X, 3–10. A transcription can be found in Stevin, *Principal Works*, 5:384–390. An English summary in Ibid., 5:391–392.

standing next to it."[79] By reading a surveyor's rod that had been put into the water, the witnesses were able to testify that the water had decreased because "the Beyersche mill had continuously drained for the duration of three hours, as the Attestants had marked and remembered by means of an hourglass."[80] The witnesses further declared "that the Stolwijckse Mill remodeled after the way of Mr. Simon Stevin had then drained only for one hour in the aforementioned soil, according to the above-mentioned hourglass," upon which they had recorded a similar mark on the surveyor's rod.[81] Their conclusion was that "during the conference, the Stolwijckse Mill had drained as much water out of the soil as the Beyersche mill in three hours."[82]

Simple mathematics, one might say. Yet here, because of the necessity to measure the efficiency of the new invention, one can see the emergence of a form of experimental method *avant la lettre*. Translating the results of the Stolwijck meeting into modern scientific discourse, one could say that the Stolwijckse mill was the 'experimental group' and the Beyersche mill the 'control group', while the surveyor's rod and the hourglass served as objective scales of measurement to visualize the results.[83] Other cases, both in the Republic and abroad, revealed the same dynamic, as examiners attempted to ensure identical conditions when comparing an invention against its competitor.[84]

---

79 "[…] omme kennisse te dragen vande prouf die de Stolwijcksche Molen, by M. Joost Govertsz Timmerman, op de nieuwe maniere vande voorschreve Mr. Simon Stevin vermaekt ende geerigeert, doen soude tegens de Beeyersche Molen, daer beneffens staende […]." Stevin, *Principal Works*, 5:387. The Beyersche mill was located in *Het Beijersche*; a hamlet in the province of Holland, between Gouda and Stolwijk.

80 "[…] voorsz Requirant seeckere peyl neffens twater hebbe gestelt […] Beeyrsche Molen in den bosem, een tijt van drie uyren aen malcanderen eerst heeft gemalen, sulx Sylieden Attestanten, by een Santloper gemerkt ende onthouden hebben […]." Ibid.

81 "[…] dat de voorschreve Stolwijcksche Molen op de maniere vande voorschreve Mr. Simon Stevin als vooren geerigeert, daer nae in de voorschreve bosem, den tijt van een uyre alleenlick heeft gemalen, nae het Santloper teycken voorsz, ende dat Sylieden alsdoen insgelijcx een peylteycken daer van genomen […]." Ibid., 5:388.

82 "[…] by de selve conferentie bevonden hebben, dat de voorschreve Stolwijcksche Molen in een uyre so veel water inden voorschreve bosem heeft gemalen als de Beeyersche Molen in drie uyren […]." Ibid.

83 It was because of practical considerations that, in this case, there was no negative or positive control of the experiment. A negative or positive control would have included a third mill, which would have been a somewhat costly affair. However, the attesters for Stolwijk did take confounding variables into consideration, indicating at the end of their declaration "that the Beyersche mill had had the [advantage of] pre-grinding" (*dat de Beeyersche Molen het voormalen heeft gehadt*). Stevin, *Principal Works*, 5:388.

84 In Habsburg Spain, for instance, the invention by two Venetians of an improved bitumen to counter the damaging effects of woodworms on ships was submitted to extensive testing: two sets of oak planks (one of them scorched against worms and one of them

Now, it must be clearly stated that it was obviously not only in the context of the privilege system that the testing of new products took place. Experiments comparing one thing with the other had formed part of legal procedures before the sixteenth century as well; they had figured prominently, for instance, during the English Trials of the Pyx, when newly minted coins were assayed against an objective standard. In the framework of guild examinations, and in other competitive settings, examinations using accurate testing methods were carried out as well.[85] Nonetheless, in these settings it did not always revolve around the advancement of knowledge—and there are other distinctive aspects that make the comparison and parallel between invention privileges and science particularly interesting.

First, it should be noted that several institutions were extremely important for both science and the privilege system. The French *Académie royale des sciences* (1667), for instance, was not only at the forefront of scientific developments in Europe in the seventeenth and eighteenth centuries, it was also the institution that was empowered to examine privilege applications.[86] An early proponent of the Academy, Samuel de Sorbière (1615–1670), wrote in his plans for the Academy that he envisioned "the perfection of the sciences and the arts, and the search in general for everything which can bring utility or convenience to the human race, and especially in France" (1657).[87] His endeavor was to "build an Arsenal of machines to perform all sorts of experiments" and

coated with the Venetians' new bitumen) were sent to the Indies to be compared with one another. Also in this case, it was because of the necessity to test the efficacy of a new invention that a form of controlled experiment emerged. Goodman, *Power and Penury*, 134–36. One could also refer to the case of a certain Frenchman by the name of Giacomo Obertraut had presented himself in Venice with a new invention relating to artillery; the new invention was tested against other weapons presently in use in order to determine whether it was really an improvement. See Berveglieri, *Inventori stranieri*, 111–115 [no. 48].

85 In some ways, privilege examinations resembled the type of experiments that can be encountered in medicinal and alchemic traditions. Yet, without wanting to downplay the importance of these practices for the development towards what we now call 'laboratory science', privilege examinations and alchemic experiments differed on several points. First of all, recipe books generally provided a prescription (an 'itinerary' so to say) to achieve a desired result, whereas the way of proceeding within the privilege examination was almost 'in reverse'. In the examination process of the copper wire described above, for instance, it was not about making the alloy, but an alloy was presented that was then taken apart. The second difference was that recipe books did not include any experimental validation. This had simply been unnecessary. It was as a natural consequence of testing one thing against the other that experimental validation became an integral part of thinking about evidence.

86 Hilaire-Pérez, *L'invention technique*; Baudry, "Examining Inventions."

87 As cited in Brown, *Scientific Organizations*, 146.

he later insisted that the Academy welcomed experimentalists of all sorts, who worked on the apparatus of lodestones, artillery work, and on a "tube made expressly for the examination of the rising of water above its own level," for instance.[88] The type of topics, as well as the ideological underpinnings of the Academy, remind us of the interests that surfaced in the privilege system. This was also the case in other countries: from the point of view of institution building, there seems to have been a structural correlation between the privilege system and the birth of the academies more generally.[89]

Moreover, one is confronted with a question of shared agency. Virtually every "scientist" still considered relevant today appears to have had some sort of connection to the legal profession, and to the system of invention privileges in particular.[90] This was not only true for scientists oriented towards experimental sciences, such as Sir Francis Bacon,[91] but also for those scientists who

88 Brown, 127.

89 On similar projects in Britain, see Feingold, "Projectors and Learned Projects." Particularly relevant in this connection is also the important work of Vera Keller, who, through the lens of the *desiderata* lists, provided a comprehensive analysis of the effects of new practices of political "reason of state" and interest upon the advancement of knowledge. Keller, *Knowledge and the Public Interest*; "Mining Tacitus".

90 See also Bouwsma, "Lawyers in Early Modern Culture." At this point, it is important to briefly note how the number of law students largely outnumbered the rest of the student population at any of the other faculties. Slowly but steadily, lawyers replaced nobility and clergy in their position as primary advisors of lords and rulers throughout Europe and it is beyond any dispute that the Low Countries witnessed a continuous professionalization of its political apparatus from the fifteenth century onwards. See Dumolyn, *Staatsvorming*, 19, 242; De Schepper, "Beleid en bestuur," 43–45; Zijlstra, *Het geleerde Friesland*, 36.

91 Although Bacon is typically remembered as a scientific ideologist, he served as Attorney General and Lord Chancellor of England in everyday life. In this latter capacity, Bacon was also a member of the Privy Council, which was the official body that among others advised on the distribution of new privileges. Thus, Bacon was certainly well-aware of the procedures entailed in the examination of invention privileges. In 1601, Bacon got involved in a fierce battle to preserve the Royal Prerogative on monopolies. Gomme, *Patents of Invention*, 14. During the parliamentary debates, Bacon argued in favor of the existing use of privileges: "Any man out-of his own Wit, industry or endeavour finds out any thing beneficial for the Common-Wealth, or bring in any new Invention, which every Subject of this Kingdom may use; 'yet in regard of his pains and travel therein, her Majesty perhaps is pleased to grant him a Priviledge to use the same only by himself or his Deputies for a certain time." By that rate, Bacon's admiration for the power of invention is beyond doubt. In the *Novum Organum Scientiarum*, for instance, he argued that: "The introduction of noble inventions seems to hold by far the most excellent place among all human actions." Bacon, *Works*, ed. Montagu, 2:399. Other remarks in Bacon's work that remind us of the privilege business include: "The works pertaining to the persons of learned men (besides the advancement and countenancing of them in general) are two — the reward and designation of readers in sciences already extant and invented;

were rather engaged with the mathematical sciences, such as Gottfried Wilhelm Leibniz (1646–1716) and Galileo.[92] It appears that even those scientists with no legal schooling, or who were not named as inventors in privilege documents, often had a thorough knowledge of the legal epistemology of the privilege system. An interesting case in point in the Republic was Isaac Beeckman (1588–1637), a close ally of Descartes at some point, who has also been presented in scholarly literature as the true originator of the mechanical worldview.[93] Beeckman had been schooled in medicine, theology, literature, and mathematics, whereas his professional career included an interesting switch from the candle-maker business to schoolmaster. At first sight, there seems little reason to connect Beeckman to the legal profession and its associated

---

and the reward and designation of writers and inquirers concerning any parts of learning not sufficiently laboured and prosecuted." Bacon, *Works*, ed. Montagu, 1:38. From *The Advancement of Learning*, book II. In his natural philosophy, Bacon advocated the introduction of experiments that used "instruments and machinery" to probe the secrets of nature. Bacon, *Works*, ed. Montagu, 35. This was nothing more than the daily reality in the privilege system. I have argued elsewhere as well for the understanding of the privilege system as a providing "intellectuals and craftsmen with at least one important 'meeting place' where they could become acquainted with each other, and each other's methods, thus enabling actors from different social backgrounds to develop new modes of shared knowledge." Buning, "Inventing Scientific Method," 64. The argument has more recently, with specific reference to Bacon, been taken up by Pastorino, "The Philosopher and the Craftsman." Contemporaries had noted the similarities between 'Baconian ideals' and daily reality as well. Sir Thomas Bodley, for example, founder of the Bodleian Library and representative of the English Crown to the Dutch Council of State (see pp. 38–40), wrote to Bacon that there were already "infinite [numbers of people] in all parts of the world" trying to acquire "a knowledge [of nature] more excellent than is now among us." He added that hundreds of inventions and discoveries were "daily brought to light by the enforcement of wit or casual events." Thomas Bodley to Francis Bacon, 29 February 1608/[9], in Bacon, *Works*, ed. Montagu, 12:83–90. As cited in Harkness, *The Jewel House*, 250. It seems thus that the "bases of Bacon's reformed science were already the bases of London science," as Deborah Harkness concluded. Harkness, 246. For a very vivid description of the daily reality within the English privilege system, see Ibid., in particular chapter 4: *"Big Science" in Elizabethan London*, 142–180.

92 James Franklin has noted this significant correlation before in his excellent study of early modern probability theories. He found that: "[T]he originators of mathematical probability were all either professional lawyers (Fermat, Huygens, de Witt, Leibniz) or at least the sons of lawyers (Cardano, Pascal) and so will have had some contact with at least the broad concepts of legal thought. Bacon and Copernicus, among the peripheral cast, were also lawyers, Montaigne, a judge; Valla, a notary; Machiavelli and Arnauld, the sons of lawyers; and Petrarch, Rabelais, Luther, Calvin, Donne, and Descartes, former law students." Franklin, *The Science of Conjecture*, 350.

93 For a thorough study of the life and works of Isaac Beeckman, see Van Berkel, *Isaac Beeckman*.

epistemology, but upon closer inspection it appears that Beeckman had a thorough knowledge of the privilege system too. Not only does one find extensive notes on the topic of privileges in his diary,[94] he was also called in as the examiner of an application privilege for a perpetual motion machine[95] and for a method of longitude determination forwarded by Galileo Galilei.[96] Thus, one might argue that the privilege system ensured that scientists with no legal schooling came in contact with a legal way of thinking about evidence as well.[97]

In this respect it is interesting to conclude by looking at a case related to a new kind of navigation technology. The prosperity of the Dutch at the beginning

94 For instance, in November 1612, Beeckman studied how to make glue according to the method of the Englishman John Edrington, who had obtained a privilege from the States General on 21 december 1611. Beeckman also closely followed the case of his uncle, Johan van Rhee, who had (falsely) obtained a privilege for the invention of a new type of for several inventions. Beeckman, *Journal*, 4:36. For details on the conflict between Johan van Rhee and Pieter Mybos and Joseph Scelton, see Ibid., 37, note 1. Other records of privileges in the Journal by Beeckman are: Ibid., 3:89, 341; 2:413 (about Stevin's privilege for water mills).

95 On 15 July 1626, Beeckman was called in by Nicolaes Puyck, one of the burgomasters of Dordrecht, to assay a model of the invention. Beeckman, *Journal*, 2:350ff. Puyck was considering to buy a share in the privileged invention. It is interesting to note that Simon Stevin was invoked as an examiner for perpetual motion machines as well, and that he gave his blessing to at least one of them despite not believing in perpetual motion. Dodt, *Archieven*, 7:90 (5 November 1619). It concerned the invention of Jan Allaertsz Cloppenburch. NL-HaNA, States General, 1.01.02, inv. no. 12302, fol. 156v [G185, 3 February 1620]. This falls in line with the observations made by Alan Gabbey who found that, whereas many considered perpetual motion machines mechanically impossible, they believed that a 'physical' perpetual motion machine (as a property of nature) was in principle a possibility. Gabbey, "Between Ars and Philosophia Naturalis," 143–45. See also Gabbey *The Mechanical Philosophy*, 38–84.

96 Beeckman, *Journal*, 4:241; Van Berkel, *Isaac Beeckman*, 75.

97 One might add the idea that the epistemological language of one field (i.e. law) influenced that of another (i.e. natural science) becomes more plausible, when one considers that knowledge in the early modern period was often thought of as being one single entity. As Celio Calcagnini (1479–1541), a professor of Classical languages at Ferrara University, put it: "[Knowledge] is all one body, which the Greeks call *paideia*, and we very elegantly *humanitas* [...]. Thus the disciplines, that is the parts of the *humanitas* are connected among themselves [...]. No one may, therefore, pursue physics without logic, nor logic without mathematics, nor anything without the support of rhetoric." As cited in Rose, *The Italian Renaissance of Mathematics*, 123. The ideal was to become a master in different fields of knowledge. For this reason, it was not so strange to find professors changing academic fields on a regular basis, as was for instance the case of Hilderich von Varel (Edo Hildericus, b.1533) "who taught mathematics at Jena from 1564–1567, and then held a variety of brief teaching positions teaching oriental languages elsewhere in Germany, before becoming Professor of Theology at Altdorf from 1584 until his death in 1599." Barker and Goldstein, "Realism and Instrumentalism in Sixteenth Century Astronomy," 241.

of the seventeenth century was largely determined by the geographical position of their country. Its location on the coast of north-western Europe was not only favorable for fishing (in particular for herring fishing, which generated huge profits through the application of advanced techniques),[98] it was also advantageous for international trade in foodstuffs and goods. Whereas previously the nature of the trade was determined by inner-European trade routes, encompassing the so-called "Mother trade" (*moedernegotie*) and networks established within the framework of the Hanseatic League, at the beginning of the seventeenth century this transformed into a domination of trade across the oceans. In 1602, the Dutch East Indies Company (VOC) emerged from the various pre-companies (*voorcompagnieën*) that had traded in Asia since 1594; it would go on to exert an unprecedented influence on world trade until the end of the eighteenth century. West of the Tropic of Cancer, from 1621 onwards, the Dutch West Indies Company (WIC) would mainly deal with the sugar and slave trade. Thus, led from the European motherland, an empire was born that made its presence felt on every continent in the world.[99] The development of new techniques and knowledge in the field of navigation was closely intertwined with the making of this new empire. Apart from the development of new types of boats (the *fluyt*) and the mastery of for instance rope-making skills, the refinement of maps, globes, and other visualization techniques was essential for the growth of Dutch commercial shipping.[100] In addition, new schools were established where navigators could train themselves in the mathematics needed to solve new problems,[101] which resulted from the new long-distance maritime trade, particularly in the area of fresh water supply and geopositioning. The privilege system was part of the ongoing effort to address these problems by means of new innovations.

And so it was that, on 28 August 1610, the English merchant Thomas Leamer approached the States General with an offer to disclose his new method for finding longitude on land and offshore.[102] The States General showed its inter-

98 Poulsen, *Dutch Herring*.

99 The literature on the Dutch (colonial) empire is voluminous; for a recent introduction, see Antunes and Gommans, *Exploring the Dutch Empire*.

100 For a detailed description of these new techniques and how to place them within the framework of scientific development, see the still unsurpassed work of Karel Davids, *Zeewezen en wetenschap*.

101 For a broader view on academic traditions and training practices across Europe in relation to navigational skills, see Schotte, *Sailing School*. For a concise overview of the major developments in navigation, see Bennet, *Navigation*. For the use of instruments in this context, see also Bennet, *The Divided Circle*.

102 Request kept in NL-HaNA, States General, 1.01.02, inv. no. 7476.

est in the proposal by granting Leamer a seven-year privilege to publish a book in which he would "bring to light" his new method.[103] They specified, however, that by doing so they did not mean to pass judgment on the functionality of the method. Leamer was dissatisfied with the decision. He petitioned for a personal meeting with some of the States' representatives, as well as for a copy of some earlier state papers relating to determination of longitude at sea.[104] The States then commissioned its members Mr. Van Brienen and Mr. Duyck to negotiate with Leamer.[105] In a meeting on 23 November, they reported that Leamer preferred cash instead of a privilege for the disclosure of his invention. After some discussion, the States General decided to offer 10,000 guilders in cash on the condition that Leamer would first prove the functionality of his method within four months and at his own personal expense.[106]

Nowadays, at a time when GPS signals guide us through the world with remarkable ease and precision, it is hard to imagine how it must have been to live in a world where it was impossible to know one's location on earth with any degree of mathematical certainty. Yet it was not until the second half of the eighteenth century that a reliable method of orientation was finally discovered. The main difficulty had not been the determination of latitude – this was done, for instance, by reading off the sun's shadow – but rather the determination of longitude.[107] The issue was that to determine longitude, one needed a fixed point of reference to mark out the current position: at sea, in particular, finding such a point of reference was problematic and not knowing one's exact location could lead to hazardous situations.

The problem was relatively new. Up until the sixteenth century, nautical lanes for the most part ran close to the coast and so one could navigate quite easily by using the coastline as a means of orientation. But with the discovery

103 Leamer, *Een klaer vertoninge*, folio G. Dodt, *Archief*, 5:22. Information regarding Leamer's case can be found in Moes and Burger, *De Amsterdamsche boekdrukkers*, 74, 144–145; NNBW, 790; Van Nierop, *Des aertrycks beweging* [Amsterdam, 1661]: 84–103; Dodt, *Archief*, 5: 22, 23, 24, 240, 244, 247, 248, 249, 253, 254, 262, 263, 266.

104 Request dated 14 November 1610. NL-HaNA, States General, 1.01.02, inv. no. 7476.

105 Hendrik van Brienen was the treasurer (*thresaurier-generaal*) of Guelders, and a member of the delegation to the States General. Dr. Anthony Duyck (1560–1629) was at that point of his career the registrar of the Court of Holland, Zeeland en West-Friesland. *Repertorium van ambtsdragers en ambtenaren 1428–1861*.

106 Dodt, *Archief*, 5:23–24. The States guaranteed that during that period no other application would be "tried or used" (*te proberen ende gebruyken*).

107 The latitude of a location on the Earth is the angular distance of that location south or north of the equator. Longitude is a geographic coordinate that specifies the east-west position of a point on the Earth's surface. For a general overview of the various attempts to solve the problem, see Andrewes, *The Quest for Longitude*.

of the New World in 1492, and with the emergence of a new trade culture that followed, the situation changed significantly.[108] Now that seafarers were embarking on unknown territory, travelling for months in a row without the slightest sign of land, former orientation techniques were no longer adequate. Not knowing one's longitude meant the loss of ships, crew, and material, and States and merchants swiftly became aware that finding a reliable method of longitude determination was essential for the further development of their political and commercial activities.[109]

It was for this reason that, in 1600, the States General decided to hold a competition offering 5,000 guilders plus a yearly pension to the first inventor who managed to disclose a reliable method for longitude determination.[110] Similar rewards had been offered by other rulers throughout Europe, and in particular by Philip II of Spain (1567) and his successor Philip III (1598).[111] The large gains on offer as different states tried to outbid each other attracted a wide range of fortune-hunters who each offered their own solutions. One of these fortune-hunters was the English merchant residing in Amsterdam, Thomas Leamer.[112]

---

108 Because of long distance travelling, sailing became more and more map-oriented. This is why the Portuguese were initially in an advantage with Prince Henry the Navigator (1394–1460) and his so-called School of Sagres. De Albuquerque, "Portuguese Navigation." It was only when Frederik de Houtman (a former student of Robbert Robbertsz) was send out as a spy by the Amsterdam based minister and cartographer Petrus Plancius to secretly obtain Portuguese shipping information that the Dutch could take over. Gaastra, *De Geschiedenis van de VOC*, 16. Petrus Plancius (1555–1622) was a fanatic Calvinist preacher of the Gomarist party, who also ran a navigational school. There, he trained navigators of the VOC; a company in which he had invested a large sum of money. Plancius had proposed his own method of finding longitude as well. For a (somewhat outdated) biography of Plancius, see Keuning, *Petrus Plancius*.

109 From a political point of view, the determination of longitude was important, for instance, to mark the demarcation line that divided the colonial world between the Portuguese and the Spanish; the Treaty of Tordesillas (1494) confirmed the decision that Spain would obtain all (undiscovered) land west of the meridian that run 370 leagues west of the Cape Verde islands, and all the land to the east would belong to Portugal.

110 The yearly stipend was set at 1000 pounds of 20 stuivers a piece. The States of Holland followed the example and had offered their own reward in 1601. The premium of the States of Holland was set at 3000 pounds at once and a yearly stipend of 1000 pounds. Davids, *Zeewezen en wetenschap*, 69.

111 The premium was set at "6000 ducados de renta perpetua más 2000 de de ayuda de costa." García Tapia, *Patentes de Invención Españolas*, 60.

112 Leamer's case was not an isolated one: Between 1600 and 1650 there were eight inventors (among whom Galileo) who claimed to have found a reliable method for longitude determination at sea. All of them failed. For a detailed analysis of the longitude proposals, see Davids, *Zeewezen en wetenschap*, 69–85. See also Jonkers, "Parallel Meridians"; Keuning, *Petrus Plancius*, 120–46.

On 7 July 1611, representatives from the Provinces of Holland, Friesland, and Zeeland, reported to the States General to have been in contact with Leamer again.[113] Leamer had refused to divulge his method and had threatened "to transport his invention elsewhere" if he were not granted a higher premium.[114] The States General, who at that point had not yet seen any proof of the actual existence of any method, decided to raise the reward "considering the high importance of the matter for navigation and the well-being of the country."[115] They promised Leamer 15,000 pounds for the disclosure of his method on condition that he would "give proof and demonstration" in the presence of experts commissioned by the States General.

Leamer then submitted a manuscript in which he exposed the details of his theory. The States General commissioned its delegates Van Brienen, Joachimi, and Van Velsen, complemented with the physician Sebastiaen Egberti, to examine the theory.[116] A week later, the States decided on the wishes of Count Maurice to invite Rudolph Snellius (1546–1613) and Robbert Robbertsz Le Canu (1563–1632) to pore over Leamer's writings, as well.[117] Rudolf Snellius was at the time a professor of mathematics at the University of Leiden.[118] Robbert

---

113 These representatives were Nobel, Joachimi and Van Velsen. Henrick Willemsz Nobel (1568–1649) was at that time representative for the VOC chamber of Rotterdam; he had been burgomaster of that city several times and later would later be a member of the board of the WIC. Blok and others, eds. *Nieuw Nederlandsch biografisch woordenboek*, 1:1383–1384; *Repertorium van ambtsdragers en ambtenaren 1428–1861*. For Joachimi and Van Velsen, see below at note 116.

114 On 9 April 1611 Leamer obtained a small maintenance of 50 guilders "for his maintainance" (*tot zijne teeringen*). Dodt, *Archief*, 5:240.

115 A copy of the "act of agreement" (*acte van verdrag*), dated 9 July 1611, can be found in Leamer, *Een klaer vertoninge* [1612], Folio M-N. Leamer's theory is to be handed in within a month, and should be "understandable to any skipper or navigator in theory and practice" (*dat elck schipper ende zeevarende man die sal kunnen begrypen ende verstaen, ende datelyck practizeren*). Also, wrongly dated at 7 July, in Dodt, *Archief*, 5:245. The price is set two days later.

116 Dodt, *Archief*, 247. Albert Joachimi (*c.* 1560–1654) was a member of the delegates to the States General from Zeeland. Adriaen van Velsen (?-1661) was the delegate of Friesland. Sebastiaen Egberti (*alias* Sebastiaen Egbertsz de Vry, 1563–1621) was a renowned surgeon, immortalized with a number of the surgeon guild's members on a painting (*The Osteology Lesson of Dr Sebastiaen Egbertsz*, 1619) by the hand Nicolaes Eliasz Pickenoy (1588–1650). At the time of his contact with Leamer, De Vry was also a member of Council of Amsterdam. He was a fierce supporter of Oldenbarnevelt. NNWB, 3:1368.

117 Dodt, *Archief*, 5:248. For the literal text of the invitation, see Robbert Robbertsz, *'t Verscheyden antwoordt* [Hoorn, 1612], 32–33. Snellius and Robbertsz received a compensation for their work of 50 guilders. Dodt, *Archief*, 5:249 (1 September 1612).

118 Rudolf Snellius had the first chair of mathematics in the Republic. He was renowned for his Ramism as well as – and mainly – for his triangulation methods. See Van Berkel, "De geschriften van Rudolf Snellius"; Van Berkel, *Isaac Beeckman*, 271–90.

Robbertsz, on the other hand, ran a navigational school at the outskirts of Amsterdam.[119] Moreover, Robbertsz was a prolific religious writer. He had been a member of several different Anabaptist churches, before he was expelled and continued to make his way through life as a religious 'neutralist'.[120]

The theological background of Robbertsz proved a valuable asset in appreciating Leamer's method. Like Robert Robbertsz, Leamer had been a member of a number of Puritan and Anabaptists groups, before he was expelled and created his own belief system that was later labeled 'Arian chiliasm'.[121] Leamer argued that religion and the determination of longitude were intrinsically linked. He argued that the principal reason for not finding a reliable method was that the Dutch worshipped God by the wrong name, which should be Elohim, not God.[122] Without knowing God's true name, so went the argument, one could never know His works.[123] Yet, our further interest lies not with the exact content of the method, but with how it was brought into the limelight.

On 27 August 1611, Robbertsz and Snellius arrived in The Hague for a personal encounter with Leamer during a session of the States General. They posed the inventor a couple questions about his theory in the presence of all the delegates, but Leamer refused to answer on the spot. He claimed that the

119 Most of the important Dutch explorers (such as the brothers De Houtman, Jacob van Heemskerck, and Gerrit de Veer) had passed through this school.

120 Robbert Robbertsz Le Canu had been a member of the Frisian Anabaptist (*friesche doopsgezinden*), before he got expelled and went through life as a religious 'neutralist'. This meant that he despised the habit of avoiding contact with members of the church who happened not to submit to a particular church (i.e. shunning). After having been banned from the Frisian Anabaptist, he started a new sect with Lubbert Gerritsz, called the New Frisians (*nieuwe Vriesen*). He was banned again in 1590. In Amsterdam, he got into trouble with Petrus Plancius (1552–1622) for writing a lament on Arminius's death in 1611. Robbertsz then moved to Hoorn, where we find him at the time of the dispute with Leamer. On Robbertsz Le Canu, see Moes and Burger, *De Amsterdamsche boekdrukkers*, 3:55–185.

121 Leamer had sided earlier with both the Seperatist and Smyth's Anabaptists. Sprunger, *Dutch Puritanism*, 82. On the religious ideas of Leamer, see also Evenhuis, *Ook dat was Amsterdam*, 2:237–38; Lawne, *The Prophane Schisme*, 55–56. Sprunger, *Trumpets from the Tower*, 78–83. On the English Seperatist movement, see also De Jonge, "Franciscus Junius"; Van Schelven, "Engelsch independentisme."

122 *Elohim* is the Hebrew word for God; Leamer claimed in his writings to have learned "that holy language" through self-study as a child.

123 Van Nierop, *Des aertrycks beweging* [1661], 84. In reality, Leamer proposed a method to measure the relative distance between the moon and fixed stars. The method of Leamer was based on knowing the new moon in a meridian. Leamer claimed that he had made some short voyages and had spoken with sailors, who had gone to the East Indies, to inform himself on the problem of longitude. He also claimed to expose the uselessness of certain instruments, notably those that measure the magnetic deviation of the compass needle (i.e. the Plancius' method).

questions were irrelevant, but decided a week later to hand in a reaction in writing as well.[124] The reply was forwarded to Robbertsz, who stood by his earlier conclusions.

Leamer immediately filed complaint. He accused Robbertsz of having infringed upon his rights by informing a colleague of his, a certain Gerrit Gerritsz of Alkmaar, about the details of the method.[125] It is difficult to get at the details of this allegation, but it is true that Gerritsz sought to obtain a privilege for a method for longitude determination from the States of Holland on 1 September 1611.[126] This suggests that Gerritsz indeed had access to the information Leamer had provided. In addition, Leamer claimed that the famous printer and astronomer Willem Jansz Blaeu (1571–1638) had offered him 1000 guilders in cash for his method.[127] I have found no proof to either confirm or shatter this assertion.

Leamer persisted in sending petitions to the States General to question the expertise of Snellius and Robbertsz.[128] His efforts apparently bore fruit and after a couple of months the States decided to let the Admiralty of Amsterdam have another look at his writings under the guidance of Blaeu.[129] What happened next remains unclear. But on 6 February 1612 the Admiralty informed the States General that the conclusions by Robbertsz and Snellius had been correct and that Leamer's method was "totally vain and frivolous" (*gans eydel en frivoel*).[130] The States thereupon decided to return Leamer his writings and not to further engage with the proposal.

But Leamer refused to give up. He continued to send multiple petitions for compensation, all of which were turned down.[131] In the aftermath of his dismissal, he published a book using his earlier privilege, in which he scrupulously described the events related to his application.[132] Robbertsz reacted

124 Dodt, *Archief*, 5:249 (3 September 1611).

125 Leamer, *Een klaer vertoninge*, fol 27v, 28, 28v.

126 NL-HaNA, States of Holland, 3.01.04.01, inv. no. 373, fol. 174. Gerrit Gerritsz was a member of the Anabaptist church in the same region where Robbertsz resided. Van Nierop added that Leamer had accused Robbertz for having tried "to have [Gerritz] get preference, because he was an inlander" (*voor-gangh te hebben/ dewijle dat hij een inlander was*). Nierop, *Des Aertrycks Beweging*, 90. I have not encountered this accusation elsewhere.

127 Leamer, *Een klaer vertoninge*, fol. 30–30v. On Blaeu, see Van Netten, *Koopman in kennis*.

128 A request by Leamer for 200 guilders denied on the 6 December 1611. Another request for reconsideration 21 December 1611. Dodt, *Archief*, 5:254

129 Dodt, *Archief*, 5:254.

130 Dodt, *Archief*, 5:262.

131 Dodt, *Archief*, 5:263, 16 July 1612. Leamer did obtain an allowance of 50 guilders on 13 October 1612. Dodt, *Archief*, 5:266.

132 Leamer, *Een klaer vertoninge*.

by publishing a pamphlet (1612) in which he defended his earlier position.[133] Moreover, one of the most learned ministers of the Republic, Simon Costerus, got involved in the conflict, accusing Leamer of blasphemy.[134] Leamer then slowly disappeared from the scene. The last trace we find of him is when he appeared in person during a meeting of the States General on the 17 October 1614. His demand for compensation was again turned down, because he had failed to give "truth through experience" (*waarheyt by experientie*).[135]

The case of Thomas Leamer is remarkable for several reasons. First of all, it shows that the authorities issued the same type of privilege for the exclusive protection of books and inventions (without making any distinction between methods and products). In this sense, the case was not unique: it merely shows that there was a strange kind of overflow in the early modern period between what we now call 'copyrights' and 'patents'.[136] Second of all, the longitude case shows how the privilege system functioned as a place of encounter where different ways of understanding technological inventions could merge. Inventors, merchants, visionaries, and administrators all had to find a common language to reach an agreement concerning privilege rights. Thus, the privilege system provided intellectuals and craftsmen with an important 'meeting place' where they could become acquainted with each other's methods.[137] Finally,

---

133 Robbertsz Le Canu, *'t Verscheyden antwoordt* (Hoorn, 1612).

134 Coster, *De Grouwelijke Ongehoorde Blasphemien* (Rotterdam, 1612). Costerus attacked Leamer for his suggestion to translate Elohim as "owner" (*eygenaar*). He argued that this suggestion did not have any etymological basis. Moreover, Costerus reproached Leamer to have called Count Maurice "Zabaoth" ("Lord of Lordships") which was a title that appertained only to JHWH, not to man. For bibliographical information about Abraham Coster (1575–1658?), see NNWB, 8:320.

135 Dodt, *Archief*, 6:365 (17 October 1614).

136 The French medicine doctor Johan de Gantelmoes, to give another example, obtained a patent-privilege by the Dutch authorities for the production of a medicine on the basis of rosemary, mentioning the existence of a book that clarified the use and preparation of the invention. Apparently the monopoly rights rested on the product, no matter whether this product was disclosed in book form or by means of the privilege application. Gantelmoes' request was formulated in French, directed to "Son Excellence" and dated 25 August 1611. The Dutch translation was directed to the High and Mighty States General. Both requests are available in NL-HaNA, States General, 1.01.02, inv. no. 7476. Gantelmoes had other privileges *Imperialles*, and got his privilege on recommendation of the Prince Excellence Maurice. The medicine "would cure many diseases [...] as he had explained in more detail in the book he had published" (*naer breeder inhouden van het boekcxken by hem daervan uuytgegeven*). NL-HaNA, States General, 1.01.02, inv. no. 12300, 288v [G110, 3 September 1611].

137 The importance of the interaction between artisans and intellectuals for the development of knowledge in the early modern period has been a constant theme in secondary literature since the work of Edgar Zilsel. It has more recently been put back on the agenda

the longitude case is also noteworthy because it demonstrates that the privilege-granting authorities were open to negotiations with inventors from various backgrounds who sometimes proposed unusual methods, but eventually relied on experts schooled in mathematics to adjudge the theoretical proposals that were beyond their understanding.[138]

Conclusions

This chapter has argued that monopolies held by inventors were considered an exception to the *ius commune*. For that reason, they were legitimized by the means of a privilege. A privilege, in turn, had inherent qualities that had to be respected, one of which was that the cause of a privilege had to be proven in court. This characteristic would not only later give birth to our modern-day notion of 'patent specifications,' it also ensured that standard rules of evidence used in the early modern period became relevant to the appraisal of technological advancement.

In examining the practical implementation of legal rules and administrative obligations, the parallel between legal and scientific method is striking. The method that early-modern 'patent offices' applied to test privilege applications closely resembled the method that was later used to test theories about natural phenomena: it consisted of testing hypotheses by setting up experiments and counter-experiments to arrive at the facts, using the language of mathematics as a trustworthy tool for describing reality. Similar observations on the similarities between legal and scientific methods of reasoning have

---

by the work of Smith, *The Body of the Artisan*; Long, *Artisan/Practitioners*; Roberts, Schaffer, and Dear, *The Mindful Hand*. Following Long, one might say that the privilege system functioned as a "trading zone"; Long, "Trading Zones."

138 These are not the kind of "hybrid experts" that Klein has written about, who personally bridged the worlds of science and of industry. Klein, "Hybrid Experts." But in this case, one can perhaps speak of "expert mediators", defined by Eric Ash "as the intellectual, social and managerial bridge between the central administrators who were his patrons on the one hand, and the various far-flung objects of their control on the other." Ash, *Power*, 8. These experts were especially called upon for projects where mathematics played a major role (longitude, perpetual motion machines, drainage projects). In other cases, it remains unclear to what extent experts in the Dutch patent system "comprehended *how and why things worked*" or that "their expertise was supposed to be more broadly applicable, across a wider variety of situations." Ash, *Power*, 213. For a broader discussion of expertise, see also Ash, "Introduction"; Dear, "Mysteries of State"; Rabier, "Introduction."

been made previously by Barbara Shapiro, Steven Shapin, and others.[139] However, these scholars mainly examined general debates about certainty on the English mainland, drawing a harsh contrast between the legal practices there and on the Continent.[140] One might argue that, in the case of patenting, the method of fact-finding was not that different: it was notably the search for technical solutions to practical problems formulated around the criterion of utility that resulted in a straightforward neglect of first causes.

In effect, the privilege system could in many ways be compared to what Imre Lakatos (1922–1974) has called a 'research programme'.[141] Remarkably enough, however, the privileged research program was not formulated within the walls

---

139 Shapiro, *A Culture of Fact*; Shapiro, *Probability and Certainty*. Earlier references to the interconnections between early modern science and law can be found in Kocher, "Francis Bacon"; Loevinger, "Law and Science as Rival Systems." Of particular importance to the history of science were the comments made on the assessment of witnesses in Shapin and Schaffer, *Leviathan*, 55–60, 326–329. Recent studies have focused more on the emergence of the concept of 'natural law': see e.g. Daston and Stolleis, *Natural Law*. In consequence, less space has been devoted to the particularity of legal practices. Other examples of works in which law and science intersect are, for instance, Ann Blair, *The Theater of Nature* and Toby Huff, *The Rise of Early Modern Science*. Whereas Blair expands more on biographical notions, Huff tends to explain the rise of modern science in terms of institutional (and hence by definition legal) structures. Neither of these scholars worked specifically on the use of legal precepts in the construction of scientific proof, however.

140 Shapiro held a rather teleological view on particularity of 'the English court system' in comparison to legal systems on the continent. See Shapiro, *A Culture of Fact*, 8–32. Along these lines she assumed that "subject to further research, it seems likely that the more formulaic method of legal determination characteristic of Continental courts and the exclusion of laymen [...] impeded the legal carryover into other fields that was so pronounced in England." Shapiro, 215. But when it came to privilege law, the differences between England and the Continent were not as pronounced as often assumed. Thus, we saw for instance that laymen were not in the least excluded from legal practices in the Dutch Republic, but played an important role in fact finding procedures. Steven Shapin, on the other hand, studied scientific practices in the second half of seventeenth-century England and came to the conclusion that the blueprint of scientific etiquette was chiefly based on aristocratic culture. Shapin, *A Social History of Truth*. At this point, he perceived an analogy with legal practices, where the word of a gentleman was supposedly considered trustworthier than that of someone from the lower classes. For that reason, so Shapin argued, gentlemen were considered more reliable as witnesses to scientific experiments than commoners. In fact, further research has shown that Shapin was wrong in assuming that commoners were excluded as reliable witnesses in court. Shapiro, *A Culture of Fact*, 75–76. It should be noted that the influence of legal practices on scientific endeavors was not the primary subject of Shapin's book, but rather the fact that civil behavior conducted the code of honor within the scientific enterprise in a more general sense.

141 Lakatos, "History of Science and Its Rational Reconstructions."

of the universities, but right in the most central organs of the state. The States General, in the Dutch scenario, became the epicenter of new knowledge where all sorts of people gathered to communicate the results of their research and to shape new research questions. They did so, and could only do so, following a set of administrative procedures attendant on the legal framework of the privilege system. The next chapter will further develop how this legal framework related to broader questions regarding the economic well-being of the commonwealth, revealing the type of state that the Dutch wished to build.

CHAPTER 5

Competition and Efficiency

This chapter analyzes the macro-economic framework of the Dutch privilege regime.[1] It begins by unraveling the unstable notions of 'invention' and 'discovery' at the turn of the seventeenth century, before discussing the delicate relationship between 'novelty' and 'imitation' within the system of invention privileges. The different sections of the chapter deal with the debate on how invention privileges operated within a mercantilist way of thinking about economic progress, showing the existence of a multifaceted way of thinking about the economic purpose of inventions. One of the central issues in this matter is the relationship between 'competition' and the circulation of knowledge in early modern Europe. Another aspect is the variegated way of thinking about the economic purpose in terms of efficacy. By tackling the issue of how invention privileges rose together with the emergence of merchant interests in the process of state formation, the final paragraphs will make an attempt to link up the different elements in order to enable a better understanding of how invention privileges operated in an emerging state that quickly set the pace for economic development in Europe.

Inventions and Discoveries

The etymology of the word 'invention' (*inventie* in Middle Dutch) can be traced back to the Latin verb *invenire*, which meant 'to come upon'. Early modern intellectuals thence immediately made a connection with the art of rhetoric, where the *inventio* was the first in a five-step process that served to convince opponents of one's opinion.[2] Invention was about finding the right arguments

---

1 Portions of this chapter have been previously published in Buning, Marius. "Between Imitation and Invention. Inventor Privileges and Technological Progress in the Early Dutch Republic (*c.* 1585–1625)." Intellectual History Review 24, no. 3 (2014): 415–27. Copyright © International Society for Intellectual History, reprinted by permission of Taylor & Francis Ltd, www.tandfonline.com on behalf of International Society for Intellectual History.

2 The *inventio*, etymologically derived from the Greek word *heuriskein*, was the first of the five parts of classic rhetoric, the others being *dispositio*, *elocutio*, *memoria*, and *pronuntiatio*. As Neal Gilbert rightly remarked, the *inventio* stemmed from the Topics of Aristotle (and beyond) and thus was not new, "but the emphasis on its use was." Gilbert, *Renaissance Concepts of Method*, 119. For an excellent analysis (with a focus on in learned circles) of the word

 | DOI:10.1163/9789004320420_007

in debate, or, as Cicero put it, about "the discovery of valid or seemingly valid arguments to render one's cause probable."[3] When it boiled down to the invention of novel machines and methods, however, the goal was not to find arguments but to unveil the unknown. Slowly but surely, the notion of invention was detached from its use within a purely rhetorical framework, and began to play a role in a world that emphasized advancement by means of making new discoveries. As the renowned ideologist of the new sciences Sir Francis Bacon (1561–1626) described it, around 1605: "Invention is of two kinds, much differing; the one, of art and sciences; and the other, of speech and arguments."[4]

The positive appraisal for discovering the unknown was quite a novelty in itself. From the fourteenth century onwards, and especially under the influence of humanist thought, there had been a growing optimism regarding the introduction of arts and techniques. But even though humanists had shown some interest in machinery and mathematics, their interest was for the most part philological.[5] This changed rapidly during the sixteenth century. Although the traditional approach towards ancient texts subsisted, some writers began to describe the way in which things actually worked instead of compiling enumerative lists of contemporary achievements in elaborate neo-Latin.[6]

A modern understanding of the notion of 'invention' was nonetheless still a distant echo from the future. One of the differences between early modern privileges and the current patent system was the rapport between inventions and discoveries. Whereas today discoveries disclose what is believed to exist independently of human perception, inventions come about under the guidance of human intellect: an invention is something man-made and, in this sense, carries a much more active meaning than a discovery.[7] When a 'patent'

'inventor' and 'invention' in the transition from the classical period to into the early modern period, see Atkinson, *Inventing Inventors*, 14–46. On the various trajectories of the invention/discovery binary see also Fleming, *The Invention of Discovery, 1500–1700*; Marr and Keller, "Introduction."

3 Cicero, *De Inventione*, lvii; as cited in Williams, *An Introduction to Classical Rhetoric*, 320.

4 Bacon, *Works*, ed. Montagu, 219. Bacon elaborated the difference by stating that: "The invention of speech or argument is not properly an invention: for to invent is to discover that we know not, and not to recover or resummon that which we already know; and the use of this invention is no other but out of the knowledge whereof our mind is already possessed, to draw forth or call before us that which may be pertinent to the purpose which we take into our consideration. So as, to speak truly, it is no Invention, but a Remembrance or Suggestion, with an application; which is the cause why the schools do place it after judgement, as subsequent and not precedent." Bacon, 222–23.

5 Keller, "A Renaissance Humanist," 345.

6 Keller, 355. For the persistent power of the humanist tradition, see Grafton, *New Worlds, Ancient Texts*.

7 Catherine Atkinson elegantly defined an invention as a "rational process of searching and finding guided by the human intellect or intuition." Atkinson, *Inventing Inventors*, 25. As

was still a 'privilege,' however, the dividing line between invention and discovery was not drawn so sharply. Inventions were more or less synonymous with discoveries and so, for example, the art of writing was considered an 'invention'[8] and mathematics could 'make' inventions too.[9]

This congruence between inventions and discoveries calls for further elaboration. For it was surely no coincidence that invention privileges spread rapidly and widely through all of Europe at a time when the discovery of new continents appeared to be the order of the day. International trade companies operated under privileges, using in practice the same logic built on discovering the unknown. On 13 December 1621, for instance, the Dutch shipmaster Jan Adriaensz Brouwer arrived at the States of Holland with the following message:

> that he had found in Greenland on 69 2/3 degrees a certain new island [...] where no one except for the Suppliant has ever been, and which he considers appropriate for whaling and other affairs, [and] he requests a *privilege with the exclusion of all others* conform the placard issued to that end, dictating that *the first inventor of any new land* shall be permitted to make four voyages.[10]

---

a matter of fact, the difference between invention and discovery is one of the main pillars of modern-day patent law. It is also one of the last reasons why occurrences in the natural world can never become some-one's intellectual property. Admittedly, the distinction between discovery and invention is gradually fading now that plant cells and gene sequences can also be patented; yet, the essence of an invention is still a form of human interference, creating something that nature could not produce by itself. On these issues, see Beauchamp, "Patenting Nature"; Kevles, "Inventions."

8 The idea that writing was invented was based on a tradition that went back to Plinius' *Historia Naturalis*, and especially to Book VII, chapter LVI 'The First Inventors of Things in Life'. Atkinson, *Inventing Inventors*, 161.

9 The mathematician Albert Girard, for illustration, found some "nouveautez en l'Algebre & Geometrie incogneues non seulement des modernes, mais aussi des anciens" which formed the core of his *Invention nouvelle en l'algebre* [Amsterdam: Bleauw, 1629], A2-A3 (Preface, without page numbering). See in this context also Amir R. Alexander, "Exploration Mathematics."

10 "Schipper Jan Adriaensz Brouwer, woonende tot Opperdoes, te kennen gevende dat hij in Groenlant op 69 2/3 graed gevonden heeft seecker nieu eylant, groot ontrent in de langte vijff mijlen, daer noch noyt yemant heeft geweest als hij suppliant, sulcx hij 'tselffde oordelende bequaem omme walvisschen te vangen ende andere negocie te doen, versoeckt *octroy met exclusie van allen anderen*, in conformité van 't placcaet daerop geëmaneert, *dicterende dat den eersten inventeur van eenige nieuwe landen vier voyagie sal mogen doen.* Daerop gedelibereert wesende, hebben de heeren van Amsterdam ende Hoorn versocht copie omme met haere principalen te communiceeren, 'twelck is vergunt." Veenendaal-Barth, Smit, and Vree, *Particuliere notulen*, 1992, 1:253. [no.1271, Monday 13 December 1621]. Emphasis added. The application was put into the hands of the Nordic Company, who had eight days to appeal. Ibid., 302 [no.1510, Monday 7 March 1622].

Reference was made here to a placard issued on 27 March 1614.[11] The aim of the placard had been to encourage investors to take more risks when investing in ventures with an uncertain outcome. This vision fitted within a legal tradition, defended among others by Jean Bodin and Luis de Molina, which sanctioned temporary monopolies for the discovery of new lands because it augmented the prosperity of the state.[12] The logic as well as the terminology used to justify monopolies for the exploitation of new technologies was very similar. Whether it involved the discovery of manufactures or continents, the 'first inventor' (*eersten inventeur*) was given the opportunity to profit exclusively, albeit temporarily, from his new findings on the basis of monopoly rights.

Furthermore, the congruence between inventions and discoveries was visualized in contemporary imagery, as for example in the famous series of engravings entitled *Nova Reperta* (*c.* 1599–1603) designed by Stradanus (Jan van der Straet, 1523–1605). The encyclopedic frontispiece of the second series of engravings (see Figure 49) made reference without any discrepancy to Christopher Columbus as the inventor of the Americas (1. Americe) and Flavius Amalfitanus as the inventor of the compass (2. Lapis polaris). Their inventions were effortlessly grouped with the invention of:

3. Gunpowder (Ignibus Amata Puluis)
4. The printing press (Imprima Volumina)
5. The clock (Rotisq[ue] iugis indita hora ferreis)
6. A treatment for syphilis with tropical wood (Hyacum)
7. A distilling vessel (Ab igne stilla)
8. The cultivation of silkworms (Fila serica)
9. The invention of a saddle with stirrups (Staphaeq[ue]: prisco operta cunta seaculo)

---

11 On 27 March, a couple of merchants requested the exclusive rights to the first six voyages they would make in any passage, land, or island, that they would discover "with God's merciful help." RSG NR, 2:227. The States granted a privilege for four voyages, on the condition that the suppliants would report back after each journey. The States would decide in each case the duration of the timeframe in which the travels should take place. In case different companies would make the same discovery in the same year, they would share in the privilege rights. The *General Charter for Those who Discover Any New Passages, Havens, Countries, or Places* was signed by Oldenbarnevelt; it is reproduced in English in Brodhead, *Documents Relative to the Colonial History*, 5–6. The author gives several examples of the impact of the privilege; Brodhead, 9–19.

12 Jean Bodin (*c.* 1530–1596) and De Molina approved of the monopolies for the discovery of new countries, because they augmented the welfare of the State. Höffner, *Wirtschaftsethik*, 143. For the reception of Bodin and De Molina in the Netherlands, Machielsen, "Bodin in the Netherlands"; Van Gelderen, *The Political Thought*, 62–165.

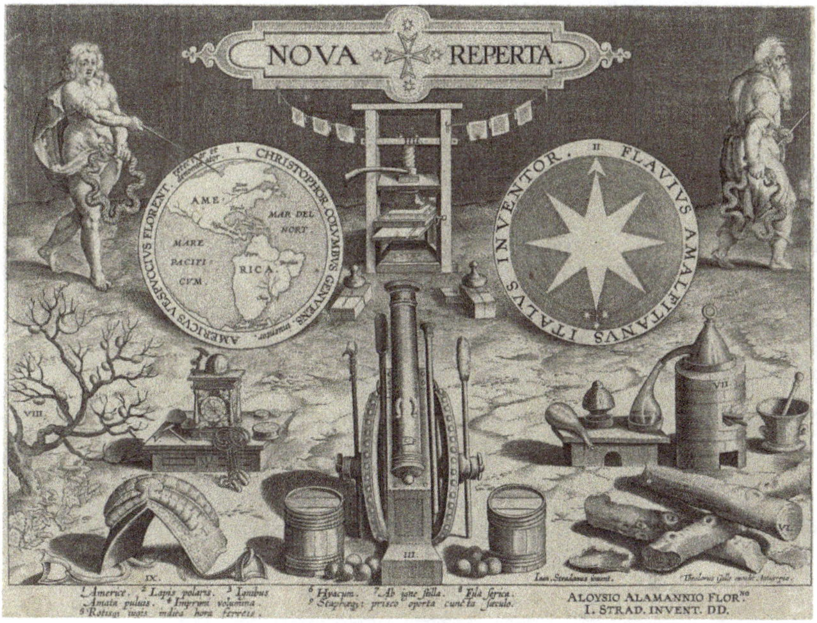

FIGURE 49 Frontispiece to the series *Nova Reperta*

Stradanus's view on the taxonomy of inventions was far from unique; his engraving reminds us that modern notions of 'invention' cannot easily be applied to the past, and that the roots of the veneration of innovation cannot be isolated from the emergence of a mercantile state with territorial ambitions and an active colonization policy.

Novelties and Imitation

Everything within the privilege system revolved around the reinforcement of territorial power. Whereas global expansion was at stake in the case of privileges issued to chartered companies, the authorities were primarily concerned about strengthening the internal market in the case of invention privileges. After all, an invention privilege gave its holder the right "to solely make, have made, and sell [his invention] within the United Provinces," leading effectively to a monopoly on the local market.[13] Whereas, in principle, the injunction to counterfeit privileged inventions had a local character, an impediment was

13 The standard clause was "to solely make, have made, and sell [their invention] within the United Provinces," while making it illegal for anyone else "to imitate or counterfeit [their invention], either in parts or as a whole." For example, "[...] alleene te mogen maken, doen

FIGURE 50 Red glazed ceramic plate, trademarked by DCS (Dierck Claesz Spiegel), 1603

often attached to the privilege that prohibited the import of products counterfeited abroad. That such a clause was not a redundant luxury can be deduced from the case of Dierck Claesz Spiegel, a merchant from Enkhuizen, who had obtained the exclusive privilege for the production of pottery that was "as good as and possibly better" than the pottery then commonly imported from Germany (see Figure 50). In his first privilege, granted in 1602, Spiegel was held to solely use inland soil.[14] Two years later, his privilege was amended because "his assistants counterfeit his pottery from foreign soil."[15] A clause was now

maken, ende verkopen, inhoudende interdictie dat niemant [...] na en maken off conterfeyten, int geheel, ofte deel." NL-HaNA, 1.01.02, inv. no. 12299, fol. 4 [12 January 1602].

14 NL-HaNA, States General, 1.01.02, inv. no. 12299, fol. 77v [G74, 20 July 1602].

15 "[...] Suppliants eygen Dienaers [...] voors aertwerck van uuytheemsche aerde contrefeyten [...]." NL-HaNA, States General, 1.01.02, inv. no. 12299, fol. 173 [G86, 12 March 1604].

included that forbade others from making this type of pottery "with indigenous, foreign or other types of soil" for the duration of the monopoly.[16]

The scope of monopoly was at all times precisely defined. What mattered was the geographical relevance of an invention and, in contrast to modern patent law, privileges were therefore not concerned with 'absolute novelty' or with the general idea behind an invention (the "inventive step" see also Chapter 3, pp. 117–118). In 1615, for illustration, a couple of inventors obtained a three-year privilege:

> to draw gold and silver wire in the Milanese and Parisian style [...] provided it be a new invention [...] and on the understanding that in this injunction [to others to produce wire] does not extend to those who will bring similar style of gold and silver wire from other countries in the United Provinces, and [provided that] the High and Mighty [States General] preserve the knowledge and jurisdiction over question that may arise because of this privilege.[17]

The example shows that invention privileges were all about strengthening the local market. The idea (once popularized by Gerard Doorman and Marcel Silberstein) that the Dutch issued more and more invention privileges instead of trade monopolies after their separation from "absolutist Spain" is therefore difficult to sustain.[18] The true question is what the Dutch authorities considered privileges to be useful for.

In order to answer that question, one must begin by separating the arguments that played a role in the moral justification for creating a monopoly from arguments based on economic considerations. The moral justifications were expressed in general terms that revolved around a discourse of 'reasonability'.

16 "[...] met hierlandtssche, uuytheemsche oft andere diergelycke aerde [...]." Ibid. For the larger context of Spiegel's privilege and Enkhuizer pottery, see Ostkamp and Venhuis, "'tot Soulagemente'." Ostkamp and Venhuis note that the production period of Werra pottery in Enkhuizen mainly took place during the years 1602 to 1613; the privilege was valid from 1602–1617. The "brand" of Dierck Claesz Spiegel (the use of his initials) can be found on the pottery in the period 1603–1606.

17 "omme alleen [...] gout- ende silverdraet te mogen trecken, op het Milaens ende Parysch fatsoen. [...] mits dat het sy eene nieuwe inventie *etc.*, wel verstaende, dat in dit verboth neyt en syn begrepen diegenen, die uyt andre landen gelyck fatsoen van gout- ende silverdraet in de vereen. provicien sullen brengen ende dat haere Ho. Mo. aen haer behouden de kenisse ende judicature vande questien, die op dit octroy ofte ter cause van't sleve in eenigher manieren sullen geraecken te vallen. Die van Seelandt verclaeren hyerinne niet te kunnen consenteren." Dodt, *Archief*, 6:362.

18 Silberstein, *Erfindungsschutz*, 297.

Thus, on the one hand it was considered reasonable that the "first inventor" should benefit from the fruits of his labors, while on the other it was considered reasonable that the inventor be compensated for the time, costs, and effort incurred in the process of inventing. The economic arguments were often much more precise in describing well-defined purposes. In line with a general trend throughout Europe, the arguments used in the Dutch Republic to justify privilege grants can be grouped under three categories. Hence, inventions were useful if they economized on the use of fuels or raw materials; economized on the price of production in a more general sense; or reduced the need to import certain goods.

As Marco Belfanti and others have pointed out, these arguments must be seen in the light of an economic ideology that is known in history as 'mercantilism'.[19] Mercantilism is the (somewhat controversial) shorthand for:

> [A phase in the history of economic policy] based on the premise that the volume of trade was finite, and that each state should adopt protective measures to ensure that its share of trade did not decline, that imports were kept to a minimum and that domestic industries were shielded from competition from foreign imports.[20]

Because the volume of trade was presented as if it were a concrete and finite pie, it was also believed that every state could eventually only have a limited share of wealth. Mercantilist rulers therefore did not shrink from using force to defend their interests. They were in fierce and continuous competition with one another, while trying to streamline their inland economic policy. In contrast to a blind faith in an 'invisible hand' that controlled the economy, mercantilists advocated active interference by the state in economic affairs. As the Dutch are commonly believed to have overseen a comparatively free state, one in which free commerce determined daily reality in tandem with religious freedom and a free press, it is often assumed that mercantilist tendencies were less present in the Dutch Republic than in other European countries.[21]

---

19 Belfanti, "Between Mercantilism and Market"; Silberstein, *Erfindungsschutz*. Both authors suggest, however, that the situation in the Dutch Republic was somehow exceptional.

20 Davis, "The European Economies," 125. The exact definition of mercantilism is a matter of continuous scholarly debate; for purposes of review, I chose to use the working definition of mercantilism given above. Please note that mercantilism was not a system consisting of a number of settled principles, but rather "a description of a trend in economic thought" that cannot be seprated from its political and cultural context. Coleman, "Eli Heckscher and the Idea of Mercantilism," 24.

21 Hekscher, for instance, described the Republic as a "Gegenprobe auf den Merkantilismus." Heckscher, *Mercantilism*, 1:330. More recently, Ormrod still argued that "Debates

But invention privileges tell a different story. Time and time again, they mention the possibility of reducing the import of foreign commodities and increasing export capacities. By way of illustration, one could single out a privilege granted to Philip Brugman in 1617 for an invention to press wool. The new invention was considered worthy of privilege, as many textiles:

> that must be painted in England at present, as the aforementioned invention of pressing does not exist, will from now on be painted and pressed here, [and] because it is commonly known that the paint in this country is much better than the paint in England, Englishmen shall moreover be pushed to bring their white perpetuana in this country, so that the dyeworks and commerce of the before-mentioned perpetuana shall be largely drawn into this country and the staple thereof will be stabilized here in this land, which will be beneficial and augment the licenses [= export tax], dyeing rooms, and other things depending on it.[22]

This is but one of many examples that demonstrate how the Dutch had clear mercantilist objectives in mind when privileging inventions. This observation is in line with the conclusions drawn in the 1960s by Voorthuysen, who argued that mercantilism was less cherished in the Dutch Republic than in its

---

about the efficiency of the decentralised Dutch state remain unsettled, but there can be little doubt that Dutch mercantilism was weak and vulnerable in the face of English and European competition." Ormrod, *The Rise of Commercial Empires*, 26, 27. Debates about mercantilism in the Dutch Republic span over a century: the subject is obviously too comprehensive for incidental treatment in this connexion. For an excellent overview of the older literature, see Voorthuysen, *De Republiek*, 15–27. Very briefly summarized, we have by now moved away from the view of Schmoller, who plainly stated that mercantilism was central to the Dutch Republic, to a more differentiated view of mercantilism. Peter Klein, for instance, argued that the degree of economic liberty differed significantly from sector to sector. Klein, "De Nederlandse handelspolitiek." In their magistral study on the Dutch economy, De Vries and Van der Woude confirmed the conclusion that market forces produced a degree of economic unity despite decentralized economic policy; De Vries and Woude, *The First Modern Economy*, 172–79, 695.

22 "[…] die als nu in Engelandt moeten geverft werden, vermits dat alhyer te lande de voors. invensie van perssinge nyet en is, alhyer voortaen sullen geverft ende geperst werden, omme dat een yeder notoir kennelyck is, dat de verwe van dese landen ongelyck beter is als de verwe in Engelandt, daerover den Engelsman genootsaeckt sal sijn de perpetuanen in dese landen wit te brengen, waerdoor verweryen ende commertie van de voors. perpetuanen ten meesten deele in dese landen sal worden getrocken, ende den stapel van de selve alhyer te lande gestabilieert werden, tot voordeel ende vermeerderinge van de licenten, verweryen ende andere dependentien van dien in dese Landen, ende derselver Onderdanen, daer uuyt sullen genieten." NL-HaNA, States General, 1.01.02, inv. no. 12302, fol. 36v [G163].

neighboring countries because "the goals strived for by the mercantilists (both theoreticians and practitioners) were in many respects already achieved."[23]

Import substitution by dint of imitation did not necessarily lead to slavish copying, but could lead in turn to innovation. This was particularly the case with the imitation of exotic commodities in the textile and ceramics industry. Ever since Marco Polo's travel descriptions, the East in particular had functioned as a powerful "stimulus of desire."[24] A well-known example were the Delftware imitations of Chinese porcelain works. The Portuguese had first introduced porcelain to the European continent around the beginning of the sixteenth century, yet it was only by the beginning of the seventeenth century that porcelain became a more common commodity in the Republic. Notably, this happened after the public sales of cargo taken from two hijacked Portuguese carracks loaded with porcelain in 1602 and 1604, which caused such a sensation on the Dutch market that familiarity with the new product filtered through to all layers of society. Yet it did not end with the trade in imports and stolen goods. A decade after the introduction of Chinese porcelain onto the Dutch market, the stained-glass artist Claes Jansz Wytmans obtained a privilege to produce all sorts of porcelain that was "both in design and the clay used reasonably similar to the porcelain that is coming from far-away strange countries."[25] It would mark the beginning of a branch of industry that was to characterize the Dutch Republic for centuries to come (see Figure 51).[26]

The mimicking of exotic commodities sometimes went on something of a detour. In 1604, for instance, an "inventor of the nature and the workings of Chinese silk" obtained a privilege from the Dutch authorities to exclusively produce "all sorts of silk cloths, which hitherto had to be collected from Spain,

---

23 Voorthuysen, *De Republiek*, 129. My translation. Perhaps it is not superflous to note that the study of invention privileges can only offer a contribution to the general debate about the role of mercantilist ideals played in the Republic; a full-fledged analysis would require the use of different (and more varied) sources.

24 Lemire and Riello, "East & West," 891. On the further impact of the discovery new worlds on the Dutch imagination, see also Schmidt, *Innocence Abroad.*

25 "[...] porceleynen te maecken die van Schilderie, ende aerde (sonder roem) tamelyck gelyckformich souden wesen porceleynen commende uuyt wyde vreemde Landen [...]." NL-HaNA, States General, 1.01.02, inv. no. 12301, fol. 32 [G130, 4 April 1614].

26 The proper heyday of Delftware was actually only in the 1640s, when the Dutch tried to supplant the imports that fell because of the Ming-Manchu dynastic wars. The privilege of Wytmans shows nevertheless that there was a longer history to the genesis of Eastern-styled pottery in the Dutch Republic. Delftware is only thus called because potteries moved into a deserted area in Delft after the explosion of a local gun powder storage magazine. On these issues, see Finlay, *The Pilgrim Art*, 179–83. On the question of how circulating objects on board of Dutch ships transformed knowledge, see also Brendecke, Friedrich, and Ehrenpreis, *Transformations.*

FIGURE 51 Tegeltableau *'In Duijsent Vreesen'*

Italy, and other Lands."[27] The inventor reported in a manuscript that supplemented his privilege application, how he had traveled to Lombardy to study the production of silk: observing how the local population had made use of white mulberry trees, his 'invention' was to transplant these trees to the Dutch Republic (see Figure 52).[28]

27 "[Inventeur] van de natuere ende het wercken van de Chineesche zijde [...]." RSG, 14:625. "[...] alderley soorten van syde lakenen, die men tot nog toe uit Spaegnien, Italie ende andere Landen heeft moeten halen." Dodt, *Archief*, 4:116. Resolution States General dated 22 Aug 1604. For the privilege, see NL-HaNA, 1.01.02, inv. no. 1299, fol. 187 [G88, 23 August 1604]. For more information about the inventor, see Bruinvis, "Caspar Benoist."

28 \ *De gehele conste*. The unpaginated and undated manuscript (35 pages folio, with 3 large drawings in color) can be consulted in the Amsterdam University Library, Ms IC 37. The full title translated into English is *The whole Art, means and Policy to win silk worms, as well as to sow, plant, and raise white mullberry trees to their maintainance, obtained through the investigation of nature and by the experience of many bright minds throughout the times.*

FIGURE 52 The silk production project by Caspar Benoist

Inventions so clearly based on imitation are difficult to place from a modern perspective. Echoing current ideas about intellectual property, we often set imitation and invention against each other.[29] More than that, imitation is not necessarily positively valued: it is somehow contrary to an invention, since imitation repeats what has been done before, and thus does not provide anything new.[30] But in the early-modern period things were not so clear-cut. Imitation was often the first step in a process that in due course could lead to 'real' inventions. One way to illustrate this is by looking at two inventions made by the goldsmith Willem Kick,[31] who obtained an eight-year privilege from the States General in 1609 for his "art to produce lacquer [...] similar to [what is] currently brought from the East Indies."[32] Business being brisk, Kick obtained a second privilege in 1619. This time, the privilege stated that:

> the Suppliant had invented several years ago the art to make all types of gilded woodwork after the Chinese way, and had obtained a privilege in the year 1609 [...] which has expired since a couple of years past [...] [But] recently he had invented with great efforts and large expenses a certain new art of making enameled lacquer ware [...] that even bears the beating of a hammer without flaking off [...] much more black and brown than hitherto made by the Suppliant or others [...] much more fine, plain, and smooth [...] thus that this gold, as well as this lacquer, can

29 I am particularly thinking here about the longlasting influence of Joseph Schumpeter (1883–1950), who divided economic development into three linear progressing stages: invention (the discovery of new things or new ways of doing things, carried out by inventors), innovation (successful commercialization of of a new product or method, carried out by entrepeneurs, combining the new and the old) and imitation (the more general diffusion of new products to various markets). Schumpeter, *The Theory*. For an extensive study of the various conceptual models of innovation in the twentieth century, see Godin, *Models of Innovation*.

30 Admittedly, most intellectual property regimes nowadays recognize something called a 'utility model' or a 'petty patent' too, which has a lower threshold for originality and provides short-term protection against counterfeiting. Nevertheless, as the name 'petty' already indicates, this type of patent is less valued than a 'real patent' and plain imitation would not pass for a petty patent anyway.

31 Willem Kick was pivotal in the network of gold leather technicians in the first half of the seventeenth century. Among others, he was closely involved with the gold-leather factory of his brother-in-law Hans LeMaire from Aken (1586–1641), who also managed to obtain several privileges. See Koldeweij, "Goudleer," 58–60. See also Dilllen, *Bronnen*, 1:585, 691 [No 1036, 1176]. On Kick, see also the excellent article by Baarsen, "Kistjes van Kick?"

32 "[...] de conste van allerhande lackwercken in dese Landen te maecken. In te voegen als alhier vuyt OostIndien wordt gebrocht [...]." NL-HaNA, States General, 1.01.02, inv. no. 12300, fol. 124 [G103, 8 October 1609].

> bear one filling it with hot boiling water, brandy, and even strong water […] and that this gold and lacquer has no smell at all […] it is not sticky […] and cannot be scraped off with a finger nail […].[33]

The new method used by Willem Kick was clearly an improvement: it had passed the threshold of pure imitation and turned into a 'proper' invention. Accordingly, this time Willem Kick obtained a privilege for the duration of twelve years to produce his enameled lacquer-ware (see Figure 15), because it was "much cleaner, more beautiful and firm, than all the works that may come from Japan and China."[34] It is worth noticing that the second privilege had a longer duration than the first privilege. Moreover, the first privilege had specified that although Kick would control the exclusive monopoly for Eastern styled lacquer ware, this was not valid "for the [lacquer] brought from the East Indies by the Dutch East India Company […] or other persons."[35] The second privilege made no mention whatsoever of any restriction regarding imports.

The relative ease with which those involved in the privilege business regarded the problem of technological imitation echoed ideas about imitation in the world of Renaissance arts.[36] In that world, as art historians and literary scholars have extensively shown, imitation was considered part of the learning

33 "[…] dat hy Suppliant eenige Jaeren geleden geinventeert hebbende de conste omme naer de Chinesche maniere alderhande houtwercken te verlacken, ende vergulden, ende daerover In den Jaere 1609 van uwe Hoog Mogende [Heeren Staten Generaal] vercregen hebbende Octroy, […] nu eenige Jaeren geexpireert synde hy onlancx met seer groote moeyten ende costen hadde geinventeert een sekere nieuwe conste van geemailleert Lackwerck te maecken […] ende mogen selffs verdragen dat men der met een hamer op clopt sonder affschelfferen […] veel swarter, ende bruynder als voor desen by den Suppliant ofte andere gemaeckt syn geworden, […] veel effender, fynder, ende gladder […] soo can dit goudt soo wel als dit lack verdragen dat men der heet siedende water in doet, brandewyn, soudt oock selffs sterck water […] heeft dit gout, ende lack gantsch geen reuck […] gantsch nyet cleeffachtich, […] can dit in geenderhande manieren met den nagel affgecrabt, […]." NL-HaNA, 1.01.02, inv. no. 12302, fol. 119 [G176, 31 May 1619].

34 "Willem Kick uit Amsterdam krijgt voor 12 jaar octrooi op het uitoefenen van „zeeckere nyeuwe konste van geëmailleert lackwerck op thin, speaulter, coper, bleck ende diergelijcke substantiën, in dese landen hier te voorens noyt gepractizeert, veel netter, schoonder ende vaster als alle de wercken die van Japan ende China mogen commen." RSG NR, 4:143 [31 mei 1619].

35 "[…] uytgesondert alleene die ghene die uyte OostIndien byde OostIndische Compagnie […] ofte andere persoonen alhier te Lande gebrocgt sullen wordden." NL-HaNA, States General, 1.01.02, inv. no. 12300, fol. 124v.

36 For a fuller version of this argument, with additional references to literature in the field, see Buning, "Between Imitation and Invention."

process; yet, imitation did not mean slavish copying. Authors and artists insisted that imitation (*imitatio*) should eventually lead to emulation (*aemulatio*) and that in the process of copying, something new was created.[37] Similar ideas prevailed in the world of the privilege business, on the understanding that the novelty of a technological invention had to be proven in a legal context.

Local Competition

It was because of the privilege system that novelty became a recurrent subject matter of legal dispute. One of the many instances was when a certain Abraham Lieversz accused his brother Isaac of forgery in procuring a privilege for the production of a red paint dust (*vermilioen*).[38] Abraham contended that the paint dust had been widely known before his brother ever set foot in the Republic and hence his brother's product "cannot be considered as a new invention."[39] In other cases too it appeared that the central issue was related to the question of geography.[40] What mattered was that an inventor was the first to practice a certain craft in the territory of the Dutch Republic; and that he would be the first to reveal the technical details of an invention, thus rendering the invention useful to the local economy.

It was the local authorities that, in discussion with the governing bodies formulated the question concerning what was best for the local economy. A continuous consultation at different levels was supposed to prevent conflicts and ensure a careful balance between the different interest groups and authorities

37 I am especially indebted to the following books and articles on this topic: Achermann, "Unähnliche Gleichungen"; Greene, *The Light in Troy*; Pigman III, "Versions of Imitation in the Renaissance."

38 "[…] *voor geen nieuwe inventie kan gehouden werden.*" For the privilege, see NL-HaNA, 1.01.02, inv. no. 12302 [G158, 5 May 1617]. Isaac had furnished a letter of recommendation by the Amsterdam magistrates (dated 15 January) as well as attestation by several principal merchants. Dodt, *Archief*, 7:8. For the accusation in front of a notary, see Van Dillen, *Bronnen*, 2:238.

39 Van Dillen, *Bronnen*, 2:238.

40 Another example would be the declaration by four damask workers, who testified 1610 on behalf of Lauris Verbeeck, damask worker in the city Alkmaar, that a certain Pachier Lammertijn had falsely obtained a privilege 1601 to produce cinnabar. They claimed that this type of cinnabar had been produced in 'a certain cloister in Gouw', which had sold the pattern about twenty-six years ago to Pieter Ferlijn, who in turn had sold it to Lammertijn. Thus, clearly, Lammertijn "has not been the inventor of the work" (*geen inventor van 't werck en is geweest*). Van Dillen, *Bronnen*, 2:693 [no. 1179, 13 October 1613]. On Lammertijn, see Kaptein, "Passchier Lammertijn."

in the Republic. Existing power constellations shifted to create room for new ideas and technologies, while the structure of the privilege system tightened the grip of the new political rulers. This can also be nicely illustrated with source material by briefly telling the story of the Flemish carpenter Pieter van Kayseel, who had settled around 1600 in Leiden.

On the morning of 1 January 1614, the States General discussed a request in which Kayseel asked to privilege the exclusive production of a newly invented textile polishing machine. As usual, counsel was sought for and the request was forwarded to the States of Holland for further advice.[41] The States of Holland then decided that Kayseel should first address himself to the magistrates of Leiden in order to "reveal his art" and to obtain a "written record of their consent."[42]

Meanwhile, a certain Maritgen Frans Duysten had filed a complaint against Kayseel's application. Maritgen was the widow of Pieter Adriaensz van der Werff, a former burgomaster of Leiden, who at some point during his time in power had obtained the exclusive right to exploit the only calender mill (*klandermolen*) in town.[43] Since textile workers were compelled by law to use this calender mill to polish their fabrics, Maritgen probably feared that the new invention would constitute a threat to her income. She therefore approached the Court of Leiden in person to inform them that she had taken the liberty of visiting the new mill. For the occasion she brought along a piece of textile that had been treated by Kayseel, arguing that Kayseel not only infringed upon the privilege that she had inherited from her late husband, but also that Kayseel's products were of poor quality.[44]

On 9 January, some governors from the *saaineering*, the professional organization of textile manufacturers, were commissioned by the Court of Leiden to examine whether Maritgen Frans Duysten's claim was correct.[45] Two weeks

41 A transcription of the application can be found in Posthumus, *Bronnen*, 4:80 [no. 64, 1 January 1614].

42 "zijne cunste te openbaren ende brengen schriftelick bescheyt van hunnelieder goetduncken, ten einde daernaer, zulx gezien voorts gedaen te worden naer behoren. Gedaen in Den Hage, den 20en January 1614." Posthumus, *Bronnen*, 4:80–81 [no. 64, 20 January 1614, apostille].

43 Posthumus, *Bronnen*, 4:81. Pieter Adriaensz van der Werff (1529–1604) had been one of the burgomasters during the Siege of Leiden (1573), when the Spanish attempted in vain to capture the city of Leiden. The Siege of Leiden was one of the key turning points in the Dutch Revolt. For the twelve-years exclusive right (not an invention privilege) granted to Maritgen Frans Duysten for the exploitation of the calender mill, see Posthumus, *Bronnen*, 3:345–348 [no. 223, 8 March 1610]. The orginal 'contract' (*contract*) between the city of Leiden and Van der Werff dated from 1 August 1601.

44 Posthumus, 3:82–83 [no.65, 6 January 1614].

45 Posthumus, *Bronnen*, 4:82 [no. 65, 20 January 1614, in the margins].

later they reported that the new invention did "not have anything in common with polish-pressing (*klanderen*) but is different from it."[46] The commissioners advised the Court to grant Kayseel the liberty to use his machine for the duration of two to three months "in order to see in the meanwhile what fruit this Kayseel will bring to the business with his privilege, and what harm it might do to the complainant."[47]

The application was now to follow through. On 14 March, the city of Leiden furnished Kayseel with a recommendation letter in which they stated that his invention:

> is useful for our business and will serve to its progress and convenience; being of this opinion, taking into consideration that the supplicant had large expenses and efforts, we consider that he should be given a certain benefice, in which we will willingly go along with the discretion of the States General.[48]

Two months later, the States General issued a privilege to Kayseel, fixing the duration at nine years to come.[49]

The story of Kayseel is interesting for a number of reasons. Firstly, it illustrates the value that was attached to the advice of the examiners that worked for *saainering*. At the end of the day their opinion was decisive in the appraisal of the new privilege.[50] Secondly, the story of Kayseel again demonstrates clearly how the local authorities were subsumed to the central authority of the States General seated in The Hague. After all, the city government was eventually quite happy to go along "with the discretion of the States General." But the story of Kayseel deserves our attention for an entirely different reason as well.

---

46 "Dattet instrument [...niet...] eenige gemeenschap heeft mittet clanderen, maer daervan verscheyden es [...]." Posthumus, *Bronnen*, 4:82–83 [no. 65, in the margins].

47 "[...] om middeler tijt te vernemen wat vrucht dezelve Casele voor de neeringe ende oock wat schaede hy den suplliante in het vercreghen octroy zoude moghen doen [...]." Posthumus, *Bronnen*, 4:83 [no. 65, 20 January 1614, in the margins].

48 "[...] enigsints onse neringe dienstich is ende zulcx strecken zal tot vorderinge ende gerijf van dezelve, in welck t'ondervinden, alzoe de suppliant grote costen ende moeyten heeft gehadt van ons bedunckens, achten dat hem enige beneficie behoort te werden waerine wy ons zeer gaerne gedragen willen tot discretie van de Staten-Generael." The letter was directed to the States of Holland. Posthumus, *Bronnen*, 4:83 [no. 67, 14 March 1614].

49 The nine-year privilege from the States General followed 15 May 1614. The privilege had a validity of nine years. When it expired in 1623, it was renewed on the strong insistence of the city of Leiden. NL-HaNA, States General, 1.01.02, inv. no. 12301, fol. 55v [G131, 15 May 1614].

50 For another example where the Leiden governors played a central role in the evaluation of a privilege, see Posthumus, *Bronnen*, 4:114 [no. 101].

For it raises the question to what degree the authorities used invention privileges to bypass existing legislation. Notwithstanding the claim that Kayseel's technique did "not have anything in common" with the technique of Maritgen Duysten, the latter effectively lost her monopoly on textile polishing by the introduction of the new technique. One could therefore wonder to what degree the claim to novelty was not just a rhetorical trick to make Maritgen understand that she could no longer hold on to her exclusive monopoly. This conjecture is supported by the fact that, a couple of years later, the city council gave permission to the new burgomaster of Leiden, Daniel Symonsz van Alphen, for the construction of a second calender mill.[51] It seems that, after the death of her husband, Maritgen van Duysten was simply no longer in a position of power that allowed her to hold on exclusively to her once dearly won privilege.

The idea that privileges played a role in outplaying opponents and thus opened the way for competitors to access the market is not something normally emphasized in the secondary literature on early modern economics. Because privileges were a typical phenomenon of *ancien régime* legislation, they are regarded as a natural contrast to later ideas about free market competition. But privileges opened up new possibilities for those to whom they applied, and so the idea that privileges could serve as the basis of economic liberty is not as contradictory as it seems.[52] On the contrary: invention privileges enticed the very thought of competition.

To flesh out this argument, it is worth considering the story of an economical cooking device. On 12 October 1594, the States General granted a twenty-year privilege to the Italian engineer Nicolaes Romero for the invention of an economical cooking device that was "very much needed in the present time of cities under siege, and in other situations."[53] Nothing indicated at this point

51 Posthumus, *Bronnen*, 4:98 [no. 84, 5 September 1619].

52 See on this issue also the unsurpassed work by Bossenga, *The Politics of Privilege*; Deyon and Guignet, "The Royal Manufactures."

53 "[…] eene sake seer noodich in desen jegenwoordigen tyt van belegerde steden, ende anderssins." NL-HaNA, States General, 1.01.02, inv. no. 12298, fol. 122v [G18, 12 October 1594]. Besieged cities had become a fairly common occurrence during the Dutch Revolt as part of a new set of war strategies that had been developed throughout the sixteenth century. Going by the date of Romero's privilege application, reference is probably made to the besiegement of Coevorden; a crucial stronghold in the province of Drenthe that, after it was conquered by Maurice in 1592, remained under permanent siege by the Spanish army general Verdugo. It was relieved by the armies of Maurice and William Louis in 1594. Besieged cities were at a constant loss for supplies and, as one can imagine, cutting down on the use of peat and other raw materials was therefore a permanent point of concern.

in time that Romero had not yet been able to reduce his invention to practice. Little less than a year later, however, on 14 September 1595, Romero's invention was again discussed in a meeting of the States General. It appeared that Romero's invention had dried up in the project phase and he was therefore put under the obligation to crystallize his ideas within a year on penalty of having his privilege rights annulled.[54] As we have seen elsewhere (p. 120) there was nothing exceptional about such an exploitation clause. What is remarkable, in this case, is that during the same meeting, the States General granted a privilege to a certain Claes Gerritsz and Jan Caussens for an invention that would cut down on the use of peat as well.[55] Like Romero, Claes Gerritsz and Jan Caussens were given one-year respite to 'effectuate' (*effectueren*) their privilege claim as well.

Little less than a month later, yet another inventor obtained a privilege for an invention that promised to cut down on the use of peat. This privilege was made out to Eustace Rogge, a medical doctor from Brabant, and it clarified how the inventor was able to produce "a Brewery" (*een Brouwssel*) with 45 tons of peat, or less, instead of the 90 tons needed according to the brewers of Delft.[56] Since Eustace, moreover, was willing to prove the functionality of his proposal at his own expense, in conformance with the model he had submitted at the registry, he was given three months to put his invention to work. The story continued. Two months later, on 31 January 1596, Esaias Lindt, a former burgomaster of Rotterdam, obtained a ten-year privilege for his invention of a new heating device for brewing-coppers.[57] Lindt too had furnished a "model on paper" (*Patroon, ofte formulier*) that was kept at the registrar's office. Finally, on 9 February 1596, a certain Corstiaen Anthonisz obtained a similar privilege for an economic boiling device. This time, the situation was clearly spelled out: the States issued the privilege to Corstiaen Anthonisz "on the understanding that, in case the invention is not new, herewith previous concessions will not be prejudiced, granted for the same motif to Nicolas Romero, Claes Gerritsz, and Jan Caussens, Eustace Rogge, and Esaias de Lindt."[58]

54 RSG, 8:668.

55 RSG, 8:668. NL-HaNA, 1.01.02, inv. no. 12298, fol.125v [G23, 14 September 1595].

56 NL-HaNA, 1.01.02, inv. no. 12298, fol. 127v [G25, 2 November 1595]. The invention had originally been devised for extracting salt and soap. As the Dutch privilege mentioned, the inventor "had practiced his art in Scotland." Rogge had obtained privilege there in 1588.

57 NL-HaNA, States General, 1.01.02, inv. no. 12298, fol. 134v [G26, 31 January 1596].

58 "Wel verstaende, dat hiermede niet en sullen wordden gepreiudicieert de voorgaende concessien, op gelyck voorgeven geaccordeert ende namentlyck aen Nicolas Romero, Claes Gerritsz, ende Jan Caussens, Eustace Rogge, ende Esaias de Lindt, Ingevalle dat het egeen nieuwe Inventie en soude wesen." Dodt, *Archief*, 5:92; NL-HaNA, States General, 1.01.02, inv. no. 12298, fol. 139v 140 [G27, 9 February 1596].

Now what was going on here? Are we really to believe that such a large number of inventors had suddenly and coincidentally invented a peat-saving machine? That may well have been the case. But this would not have required the inventors to petition simultaneously for a privilege for their new invention. Instead, it seems that the States General was looking for a certain type of technology and then proceeded to grant privileges for the same type of invention. So, they effectively created a kind of contest, using the privileges to incite competition among the different inventors.

When we examine the privilege grants in the Dutch Republic more closely, it appears that the case of the economic boiling device is far from isolated. A similar situation emerged, for example, when the famous sculptor and town architect (*stadsbeeldhouwer*) Hendrick de Keyser (1565–1621) of Amsterdam obtained a privilege for a new type of bridge that "would allow ships to pass with a standing mast."[59] Only several months after Keyser had obtained his privilege, his colleague Henrick Jacobsz Staes (1558–*c.* 1630), the town carpenter (*meestertimmerman*) of Amsterdam, obtained a privilege for the production of a similar type of bridge "on the condition that it will be a new invention, and without prejudice to the privilege we have granted to Henrick de Keyser."[60] Central to this clause are the words "without prejudice". This addition did not mean that the privilege was only valid if it did not conflict with earlier rights, or that the privilege was only valid if a similar invention had not been made earlier (which would presuppose a notion of absolute novelty).[61] Instead, it meant "without any loss or waiver of earlier rights or privileges," i.e. if a privilege conflicted with earlier rights, the holder of these earlier rights still had the

59 The ten-year privilege issued to De Keyser was dated 18 May 1596. NL-HaNA, States General, 1.01.02, inv. no. 12298, fol. 155 [G29, 18 May 1596]. It was furnished with a recommendation by the city of Amsterdam dated 18 April 1596: RSG, 9:353. On the implementation of the invention, see Davids, *The Rise and Decline*, 1:86. For more information about De Keyser as an inventor, see Kossmann, "Hendrick de Keyser."

60 "behoudelyck dat het zy eene nieuwe Inventie, ende sonder preiuditie van het Octroy by ons verleent aen henrick de Keyser, ende dat de voors. henrick Jacobssoon Staes gehouden wordt die steden ende andere die hem sullen employeren te dienen tot redelycken pryse." The privilege had a duration of six years: NL-HaNA, States General, 1.01.02, inv.nr. 12298, fol. 158 [G32, 1 October 1596]. Reccomendation letter by Amsterdam dated 1 October 1596; RSG, 9:353. For Staets, see De Roever, "De kroniek van Staets." In point of fact, Cornelis II Dankertsz (1561–1634) was appointed in the same year as Town mason (*stadsmetselaar*). It is not entirely clear how the three actors cooperated, because the archive of the *stadsfabriekambt* has gone lost. But we do know that Staets had twelve carpenters working for him in 1597, and a hundred men by 1623. Meischke, "Het Amsterdamse fabrieksambt," 104.

61 Cf. Doorman, *Octrooien*, 24.

possibility to file a suit.[62] In some cases, research was carried out to determine whether a new privilege would prejudice the privileges of existing trades. In other cases, earlier inventors were mentioned by name (as in the examples we just considered). In most cases, however, the clause "without prejudice of any general or particular concessions" was included as a general rule to make sure that the rights of third parties were not affected.[63] In a sense, this was nothing but a more explicit variation of the long-standing legal precept that one privilege could not be used against another privilege (*Privilegiatus contra privilegiatum non utatur privilegio*). So, it was very well possible for two privileges to exist side by side, each covering the same type of invention.

Furthermore, one should not forget that a privilege was a gift, not a right, issued by the authorities to achieve a specific goal for the sake of the public good. Privilege applicants sometimes produced evidence that their invention had "nothing in common but was completely" different from earlier inventions, even though the effects were comparable.[64] Yet in other cases there was no ambiguity, either from the side of the authorities or from the side of the inventors, about the fact that the technical basis of the invention was the same (and it is remarkable in this context how well informed certain inventors were of pending privilege applications). It was all about dividing the market share in these cases. For instance, when the merchants Servaes Hellings and Ian de Carpentier obtained a seven-year privilege to produce steel from iron ore, the States General superadded in the privilege act that the applicant had understood that the Count of Horn had already obtained a similar privilege to "solely use the science to produce steel from iron."[65] The inventors, however, had "objected, demanding that [the States General] would not grant the privilege to the Count of Horn, or at least to let the remonstrant in on the same."[66] There

62 Cf. West's Encyclopedia, "without prejudice." See also the entry "prejudicie" in the *Woordenboek der Nederlandsche Taal.*

63 The standard clause was "*sonder preiuditie van eenige voorgaende generale oft particuliere concessien.*"

64 On 25 April 1620, Rooswinckel obtained a privilege for a method to work baleen that was completely new and "nothing in common but was entirely different from the invention by the Englisman Jan Osborn" (*heel verscheyden van de Inventie van Jan Osborn Engelsman*). NL-HaNA, States General, 1.01.02, inv. no. 12302, fol. 171v [G186, 25 April 1620].

65 "verstaen hebbende, dat den Grave van Hoorn aen ons versochte te mogen hebben Octroy om alleen te moghen gebruycken de wetenschap om uuyt Iser Stael te maecken [...]." For the privilege of Hellings and Carpentier, see NL-HaNA, States General, 1.01.02, inv. no. 12301, fol. 253 [G142; 7 November 1615]; for that of Rene de Cercleres (Count of Hornes), see NL-HaNA, States General, 1.01.02, inv. no. 12301, fol. 199 [G137, 1 June 1615].

66 "syluyden haer daertegen hadden geopposeert, Versouckende dat wy den voors Grave tselve Octroy oft niet en souden vergunnen, oft ten minsten hun Remonstranten tot het

are numerous similar examples in the sources.[67] It resulted in continuous competition, not unlike that experienced in a modern patent race, though with the difference that invention privileges were usually granted in a relatively early stage of development.[68]

Competitors and Companions

Privileges had a clear geographical component at all times. The local character of the privilege, however, did not prevent lawmakers from being completely unaware of a possible difference between the categories of novelty and imitation. In several instances, privilege acts mentioned that a particular invention was "never used before, neither in these Lands, nor in other foreign Lands"[69] or

---

selve mede inne laten, dan dat gebeurt was dat wij den voors. Grave hadden verleent Octroy, om van gegoten Iser alleen Stael te maecken [...]." NL-HaNA, States General, 1.01.02, inv. no. 12301, fol. 253v [G142; 7 November 1615].

67 On 7 July 1610, the inventors Johan van Slingelant Boudewijnsz (based in Dordrecht) and Adriaen van Slingelant (based in Gorinchem) recieved a privilege for a water mill "on the same conditions as the Italian" (*op deselve conditien als eerst aen den Italiaen gegunt is*). RSG NR 1:172 (7 July 1610), Dodt, 5:21. The Italian was Piedro Sardy, who had just negotiated a new deal for his invention of a water mill. RSG NR 1:87 (16 April 1610). See also pp. 70–71.

68 Early modern 'privilege races' were not unique to the Dutch Republic. Deborah Harkness noted, in her study of Elizabethan London, that "[c]ompetition was bound to happen in the race for project support. Elizabeth gave both men their patents for different novel engines in the same month, content to let the results speak for themselves." Harkness, *The Jewel House*, 153.

69 When Jan Maertsz Ghisp obtained a privilege for a bolting machine on 17 August 1589, it was specified that he, "by his Industry and subtlety of mind, has invented and reduced to practice a certain instrument that had never been seen or used in neither these, nor in other foreign lands, very convenient for all the inhabitants and the bakers of these lands" (*by zyn Industrie ende subtylhyt van verstande heeft geinventeert ende gepractizeert zekere Instrument twelck noyt in dese noch in andere vrempde landen gesien noch gebruyckt is geweest zeer bequaem gemackelyk en gerieffelyck wesende voor allen ingesetenen ende backers van dese landen*). NL-HaNA, States General, 1.01.02, inv. no. 12298, fol. 6v [G4, 17 August 1589]. Another example was that of Cornelis Cornelisz van Uytgeest, who had invented a mechanism that "has never been seen used or applied before now in any works in other Lands or in these lands, but only in the sawmills originally [sic!] found by the petitioner three or four years ago" (*voor desen tydt noyt in andere Landen, nochte in desen gesien nochte in eenige Wercken gebruyckt of geappliceert, dan alleenlyck in de SaeghMolens by den Suppliant over drie of vier jaren originelyck gevonden*). NL-HaNA, States of Holland, 3.01.04.01, inv. no. 363, fol. 567 [H23, 6 December 1597].

that the invention "had not been found by anyone in the whole of Europe."[70] Such additions hint at the awareness of a difference between an invention that was 'really new' and something that was merely an imitation of an invention that existed elsewhere. All the same, imitation did not translate into a lesser reward, reflected for instance in the duration of the privilege. On the contrary. Particularly on an international level, the authorities went to substantial trouble to lay their hands on existing technologies.

Driven by the limited legal validity of privileges, inventors sometimes travelled from one country to another to amass as much legal protection as possible.[71] Historians have also interpreted this phenomenon as a manifestation of "competition between states," and in the process portrayed inventors as a type of profiteer who "were prepared to commit themselves to foreign service in return for good pay."[72] Inventors, however, rarely ran into trouble when they returned to their home country and the receiving authorities were very open about where they got their knowledge from. One privilege granted by the Dutch authorities, for example, mentioned that the inventor had already successfully put his newly invented water-lifting device to work in the moat of the castle of the King of Denmark.[73] Another inventor referred with pride to his *copie autentycq* of a privilege granted by the King of France.[74] Had all these cases been examples of early modern industrial espionage, it would have been illogical for authorities to speak so openly about where they had obtained their knowledge.[75]

---

70 "[…] by nyemanden in gehele Europa gevonden hadde kunnen wordden." NL-HaNA, States General, 1.01.02, inv.nr. 12300, fol. 124. It concerned the privilege for Willem Kick; see also pp. 72, 201–202.

71 There are many examples of inventors who traveled around Europe working for different patrons. A well-known Dutch case is that of Cornelis Jansz Meijer. See Van Berkel, "'Cornelis Meijer Inventor'"; Berveglieri, "Tecnologia idraulica"; Berveglieri, *Inventori stranieri a Venezia*, 152–66.

72 Belfanti, "Between Mercantilism and Market," 328; De Vries and Woude, *The First Modern Economy*, 348.

73 NL-HaNA, States General, 1.01.02, inv. no. 12298, fol. 257 [G45, 12 December 1598]. It concerned Kronborg castle.

74 NL-HaNA, States General, 1.01.02, inv. no. 12301, fol. 341 [G148, 2 November 1616].

75 On industrial espionage in the early modern Republic, see Davids, "Openness or Secrecy?"; Davids, "Public Knowledge and Common Secrets." To be sure, in some cases new inventions were encompassed by secrecy, particularly when it involved military instruments. In those cases, however, the authorities often chose not to grant a privilege at all, but instead to buy off the invention or to employ the inventor in the service of the State. See pp. 91–92, 96.

Inventions and privileges were more than just means to score points off competing countries; they also belonged to a 'gift economy' where states exchanged favors in the hope of strengthening future relationships.[76] The fact that inventions had been implemented abroad simply added prestige to the project. Moreover, early modern European States occasionally made agreements on the exchange of technological knowledge, even if they weren't always transparent. The Flemish inventor Passchier Lammertijn, for instance, after obtaining a privilege in the Dutch Republic, traveled to England to obtain a privilege from the Prince of Wales for his art of weaving (see Figure 53). At the request of the inventor, the States General agreed to write a recommendation letter on his behalf. What Lammertijn did not know, however, was that the States had simultaneously instructed their agent Caron to ensure that Lammertijn did not obtain an English privilege, as it "would be very harmful to the inhabitants of these lands."[77]

In order to put into clearer perspective the nature of the knowledge exchange as well as the role of privileges within, it is worth mentioning the story of the Dutch knight Theodoor Rodenburg (*c.* 1574–1644). In 1619, Rodenburg, who had earlier been a representative for Dutch merchants in Spain, offered his services to King Christian IV of Denmark.[78] Posing as a sales agent, Rodenburg proposed to establish several new-fangled business undertakings on Danish territory.[79] His proposal was aimed at preventing emigration and the export of raw materials and instead boosting the Danish economy by importing artisanal knowledge. The proposal consisted of several inventories of artisans and manufactures that were supposed to be transported from the Dutch Republic. These inventories, which have been remarkably well preserved, mentioned over a hundred people (artisans, teachers, businessmen, and others) and a large number of objects (mills, ovens, ships, paintings, cabinets, etc.). Just

76 I focus here exclusively on inventions made in the framework of the privilege system; the use of inventions and novelties as part of diplomatic gift giving is a much larger topic. See, for instance, Swan, "Birds of Paradise."

77 "[…] d'ingesetenen deser Landen sou syn seer prejudiciabel …". Dodt, 4:126. For an overview of the literature on Lammertijn, see Kaptein, "Passchier Lammertijn." On Lammertijn's later adventures in Denmark, see Paludan, "Passchier Lammertijn and the Silk Factory in Copenhagen."

78 This case study is predominantly based on Kernkamp, "Memoriën." Rodenburg was a poet, playwright, diplomat, and a merchant. He was a member of a chamber of rethoric called *De Egelantier* and would later settle in England. For an extensive bibliography on Rodenburg, see "Theodoor Rodenburg."

79 "[…] manefacturen in trayn zullen komen […]." Kernkamp, 220.

FIGURE 53 Napkin made by Passchier Lammertijn, 1601

for illustration, Rodenburg offered the services of the schoolmaster Bernard Zunders, since it had come to his attention that the elite of Copenhagen was in need of:

> a good school teacher, for the benefit of the education and instruction of their children, both sons and daughters, taking into consideration that they do not have the opportunity to teach their children arithmetic, Italian book keeping and other necessary sciences for the youth.[80]

80 "[…] een goet schoolmeester, tot educatie en onderwysinghe van hun kinderen, zo zonen als dochteren, ten aensien dat zy hier daertoe gheen ghelegentheyt hebben, om hun kinderen d'arithmetica, italians boeckhouden en andere noodsakelijcke scientiën voor de jeugden te laten onderwysen […]." Kernkamp, 234.

The employment of Zunders would make it unnecessary to send the poor children to Bremen, Hamburg, or even the Netherlands to get a good education.

The question of privileges came up when Rodenburg mentioned the services of Hendrick Schoormans, "a merchant of good credit, name and fame, living in Dordrecht."[81] Schoormans had expressed his willingness to set up the steel industry in Denmark in return for a privilege guaranteeing that no one else was allowed to erect steel mills for a couple of years and that any leftover steel could be exported "toll free to other quarters."[82] The King agreed to this proposal. Schoormans was also in the possession of a new mill that could "pump up 6000 tons of water in 24 hours," for which he requested a twenty-year privilege.[83] To press home his claims, the inventor had offered to provide Rodenburg with a model "to show to your majesty."[84] The King, however, was of the opinion that "one should first know how high the water can be elevated by such a mill."[85] When Rodenburg informed the King after a couple of months that the mill would elevate "more water than two windmills, raising the water six feet high, without wind, horses or human help,"[86] the King decided that "the height [was] not sufficient and the petition therefore denied."[87]

The second time privileges came into play, was when Rodenburg mentioned that he had found a company of fifty persons prepared "to transport all the preparations from Holland [...] in order to start up the brick-making factories within a year."[88] Two people from the company had already been sent over to investigate whether the land around Copenhagen was suitable for cutting stone. But in order to implement the project, it was reasonable:

81 "[...] koopman van goed credyt, naem en faem, wonende tot Dort [...]." Ibid., 232.

82 "[...] dat 't stael. 't welck hier in 't land niet verkost noch vertiert kan werden, tolvry na ander quartieren versonden mach werden [...] Ten anderen, dat voor dezelfde tijt van jaeren niemant in 't koningrijck van Denemarken eenighe staelmolens zal moghen opstellen." Ibid., 232.

83 "[...] heeft een nieuwe inventie om een molen te stellen, die in vier en twintch uyren uyt zal wercken zes duysent tonnen water." Ibid., 233.

84 "[...] zal het model van de voornoemde molen my leveren om u ko.mat. te laten besichtigen." Ibid., 233.

85 Kernkamp, "Memoriën," 232–33. "Bey diesem puncte muste man zuvor weiszen wie viel fusz hoch das waszer durch sothane muhle konne auffgetrieben werden." Ibid.

86 "[...] zes voeten rijst; szonder windt, paerden oft menschenhulp [...] meer water als twee windtmolens [...] versoecken daarover octroy voor twintich jaeren." Kernkamp, 253.

87 "[...] die höhe nicht sufficient [ist], dannenhero abgeschlagen." Kernkamp, 233.

88 "op eyghen kosten van Holand hier laten komen [...] alle preperativen [...] om dit jaer de steenbackeryen in 't werck te stellen." The privilege was needed for the "large expenses that must be done to construct houses, ovens and all that is connected, so it may please your majesty that no-one may erect stone bakeries for a certain amount of time." Kernkamp, 220.

> that your Royal Highness would gracefully be willing to grant, that no-one would be allowed to erect brick-making factories for the time of several years [because of] the large expenses that must be done to construct houses, ovens and everything connected to it.[89]

This remark perfectly illustrates what early modern privileges were all about: they functioned as economic guarantees to attract entrepreneurs who were willing to bring in private investments.

The same idea returns in the third example, where a privilege was discussed in the context of setting up the production of soap in Denmark. The King had already granted Rodenburg an eight-year privilege to erect several oil mills in combination with the injunction that others would no longer be allowed to import soap from abroad.[90] He had also allocated a piece of land in the proximity of Copenhagen where Rodenburg was to erect his mills and he had authorized the company of Rodenburg to (once-only) import white salt from Holland free of taxes. Based on his privilege, Rodenburg had decided to buy three mills, the first of which had recently been sent to Copenhagen whereupon the rest were soon to follow. The whole operation had cost a fair bit of money, Rodenburg pointed out, but "based on the privileges granted by the king *the advancement of the money* had not troubled me."[91] This remark shows again how privileges were a way of attracting investors, who hoped that the proceeds of the privilege would outweigh their investment.

But there are other reasons to dwell for a while on Rodenburg's proposals. First of all, they illustrate the remarkable homogeneity between the different privilege systems throughout Europe. Inventors and brokers apparently had little problem understanding 'national' legal procedures, because the different actors spoke the same legal language. This homogeneity facilitated the process of negotiation and made it possible for projectors to proceed to international

89 "dat u kon.mat. ghenadelijck mocht ghelieven te vergunnen, dat niemant voor de tijt van enighe jaeren zodanighe steenbackeryen zullen moghen opstellen [vanwege de] groote ongelden, die gheschieden moeten tot de bouwing van de huysen, ovens en het aenkleven van dien." Kernkamp, 222. The response of the King decided at this instant is unknown, but a couple of months later Rodenburg requested an extension of the privilege "because it is impossible to bring across 50 persons from Holland, without giving them a little more compensation" (*want my niet moghelijck [is] vijftich personen van Hollandt hierwaerts te doen transporteren, zonder hun eyts meer contentements te gheven*). It concerned the production of "Steinsoffen, allerhande steine, pannen, astrach, figursteine, etc. so bisz dato ausz andern landen gebracht." Ibid., 256.

90 Kernkamp, 195. Dated 22 February 1620.

91 "[…] doch fondament makende in de vergunde octroyen van u ko.ma^{t}., verdriet my niet het voorschot van penninghen." Kernkamp, *Memoriën*, 237.

'patent hopping'. The second reason, to elaborate somewhat on Rodenburg's proposals, is because they put a different complexion on the idea that the institution of privileges was purely driven by competition and mercantilist ideas of competition. Of course, the Danish authorities wished to limit the export of raw materials and increase the production of finished commodities. But the examples in Rodenburg's proposal also mentioned the possibility of exporting steel to other countries and importing salt from Holland in return for human capital. It confirms the impression that the exchange of knowledge among friendly States was, at least in part, driven by concrete economic advantages. This idea returns in a number of other examples as well; for instance, the case of Pieter van Vossele, who asked the States General in 1621 to instruct their ambassadors in England to help him obtain a twenty-one-year privilege from King James I (1566–1625).[92] Van Vossele's project consisted of bringing an iron mine in Ireland into operation for the production of iron castings; in return for the assistance offered by the States, the inventor would export the fabricated artillery exclusively to the Dutch Republic. In short, part of the entire mercantilist enterprise was not confrontation but cooperation.

Notions of Efficiency

Invention privileges offer a unique window onto the mercantilist attitude regarding import and export capabilities. Yet they also offer a valuable insight into how contemporary authorities regarded labor distribution and cost efficiency. One recurrent theme in the privilege acts was that new machinery saved on labor costs. Another recurrent theme was that inventions used less energy or reduced the need for raw materials.[93] Over and above this understanding of utility hovered a changing notion of efficiency shaped by tradesmen

92 RSG NR, 5:20 [no. 105, 16 January 1621]. The request by Pieter van Vossele was denied by the States General.

93 The different topics were usually closely intertwined. Economical ovens, for instance, used less energy and reduced the need for raw materials. Labor-saving could also mean that less ox or horse power was needed, for instance. For illustration, it was adduced that "the same amount of work can now be done with one horse, whereas nowadays two horses must be used" (*dat alwaer nu jegenwoordich twee peerden moeten gebruyckt werden, zoo veel wercx te sullen doen met een peert*) NL-HaNA, States General, 1.01.02, inv. no. 12302, fol. 1v [G149, 2 January 1617]. It is highly questionable if contemporary made any distinction between primary and of secondary energy sources, or between human (labor) power, animal power, and other sources of energy. "Mills, however, unlike humans, horses, and oxen, did not have to be fed." Cambell, "The Uses and Exploitation of Human Power," 189. For a broad overview of interconnected relationships between energy consumption and

and merchants. One of those merchants was the son of a distinguished Flemish émigré family, Abraham van Tongerloo (*c.* 1569–1623).[94]

Van Tongerloo trafficked in practically everything that he could lay hands on. He had been involved with the construction of a huge ship for the Venetian merchant Francesco Morosini; was one of the first investors in the drainage project of the *Zijpepolder* (1597); and had lived and worked in Venice for a while, where in 1603 he had obtained a privilege for land-reclamation from the Venetian Senate.[95] In 1605, Van Tongerloo was back in the United Provinces. On 3 May of that year, he obtained an exclusive privilege for the production of a new type of silk mill. The text of this privilege spoke of an:

> Invention of silk-mills with which all sorts of raw silk can be treated, so that it will not be necessary [...] that Merchants send that [silk] to Italy or other Kingdoms. [There are] for two reasons that. First, because of this invention all the world will be given the possibility to start up a silk company [...] whereas nowadays to the contrary that can only be done by a few, powerful people [...]. And secondly, because by the old way of doing things it will cost the Merchant seventeen or eighteen shilling to convert the raw silk into Tear, Stick, or sew silk; whereas this can be done in Italy for nine or ten shilling. [The current process is more expensive, because] the silk, on the old instruments, or mills, could not be labored, without the aforementioned silk having passed through four [pairs of] hands; namely first, through [the hands of] the "winsters"[96] who bind it on the bobbins; secondly it is put in the mill in order to weave it, thirdly it is doubled by the Winsters, and fourthly the doubled Bobbins are put at last on the mills, where the silk can then [finally] be properly labored.

economic development in Europe, see also Kander, Malanima, and Warde, *Power to the People.*

94 Abraham van Tongerloo came from a distinguished family. His father was a wealthy merchant who had had fled the Southern Netherlands during the Revolt. His older brother Cornelis van Tongerloo (1555–1617) was connected to the Baptist circle surrounding Jaques de Clercq (c1555–1609). About Abraham van Tongerloo, see Schulting and Jackson, *Sant'Agata Morosina, an Argosy*, 54, note 22.

95 For the privilege of the Ventian Senate, see Roberto Berveglieri, *Inventori stranieri a Venezia*, 84–85 [no.84, 6 September 1603]. The name was spelled Abraam Tangherlo (Abraham van Tongerloo). In his Venetian privilege, Tongerloo claimed to have been involved in several drainage projects in the Republic "di quel principe." Reference must have been made here to the reclamation of the *Zijpepolder* (1597). In later life, Van Tongerloo also worked for the Danish King Christian IV.

96 A "winster" was a female worker who was winding silk thread on spools, before it was twisted. See *Woordenboek der Nederlandsche Taal.*

> All of these things can now be done perfectly on the new Invention of the Suppliant, without one [activity] getting in the way of the other, by one person, at once, and by an instrument.[97]

Framing the benefit of new technologies in a framework of labor reduction may not surprise us nowadays. But economic historians generally agree that such a vision was not that common in the early modern period. What was missing, it is argued, was a "faith in the elasticity of demand", i.e. the notion that less labor pushed the price down, which would in turn increase demand.[98]

Yet, when analyzing the Dutch privilege acts in more detail, it appears that Tongerloo's privilege was anything but exceptional. Time and again, privilege acts pointed out that machinery could replace human labor and hence provided an economic advantage in terms of labor costs. In some cases, these advantages were not explicitly mentioned, as in the example of a new type of

---

97 "Inventie van zyde molens [...] mette welcke alle de rouwe zyden, [...] zullen kunnen gereedet wordden, sonder dat deselve (als jegenwoordich dagelycx gebeurt) in Italien ofte in andere Coninckrycken, by de Coopluyden sullen derven wordden versonden, ende dat om twee redenen, als Ierst, dat door dese Inventie alle de werelt commoditeyt gegeven sal wordden om de zyde reederie int cleyn ofte groot te beginnen, daer ter contrarien nu tselve alleene by machtige, ende weynige persoonen kan ter hant genomen wordden, sonder dat het mogelyck soude zyn, op dien voet, in vele jaren, de zyde alreede int Lant zynde, dezelve Reederie behoirlyck te beneficieren ende ten tweeden omdat op de oude maniere van doen de rouwe zyde, in egeen organsyn, Traen, Stick, ofte naeysyde, en kan geconverteert wordden, sonder den Coopman te costen seventhien, of Achthien schellingen tpondt, dat in Italien om negen ofte thien schellingen gedaen wiert [...] Ende dat daerom overmits de zyde, op de oude Instrumenten, ofte moelens niet en konde getwerent worden, sonder dat de voors. zyde vier handen zy gepasseert, Te weten Ierst by den winsters op Babinen te winden, daernaer op ten Moelen te setten, ende te spinnen, Ten derdden door de winsters wederom gedoubleert, ende ten vierden mette gedoubleerde Babinen voor het Laeste op de molens te setten, alwaer de zyde alsdan behoorlyck getwerent wordt, Alletwelcke door des Suppliants nyewe Inventie volcommelyck sonder dat d een d ander verhindert, by een persoon, op een tyt, ende door een Instrument sal kunnen wordden gedaen." Interpunction added. NL-HaNA, States General, 1.01.02, inv. no. 12299, fol. 218v [G90, 3 May 1605].

98 MacLeod, *Inventing the Industrial Revolution*, 212. Of course, there was a rich tradition of *automata*, machines that could perform human opera, but the goal of these machines had not been the reduction of labor costs. Compare also the observations by Margeret C. Jacob, who claimed that in England the steam engine "forced a change in economic thinking about inventions and work as early as the 1740s." Whereas up to then, inventions were aimed (aside from replacing imports) at employing the poor, it was only then that the one realized the "obvious economic benefits to be derived from the labor of one less man." Jacob, *The First Knowledge Economy*, 78.

bridge that could be operated "without the help of human hands."[99] In other cases, the competitive advantages of labor-saving technologies were clearly spelled out. The drainage device invented by the company of the Seigneur Louys de Escuyer, for instance, was considered beneficial, because "one boy can throw out as much water as hitherto [...] six or seven men."[100] The invention of a new pile-driver, to give another example, was considered advantageous because it could:

> easily be operated by six or seven men [...] instead of by 28 or 30 men who were till now needed per pile, [and] who each had to strike 20 or 22 times, and who often needed six or seven times rest before the pile was finally in the ground. [The effect of the new invention was] that the beer and the daily wages of twenty persons can be saved per pile, aside from saving a lot of time and work, so that [the invention] would contribute quite a lot to big works such as sluices, valves, gates, dams, dikes, and pile works.[101]

Clearly, in these cases the authorities did not see any contradiction between labor-saving devices and the general well-being of society.

This is not to say that authorities did not consider the possible loss of jobs because of labor-saving inventions at all. At the Assembly of the States of Holland, for instance, on the morning of Monday 23 May 1624, it was deliberated that:

---

99 "[...] dattet sonder eenich peryckel van den passanten ende *sonder behulp van menschenhanden* [...]." NL-HaNA, States General, 1.01.02, inv. no. 12298, fol. 155 [G29, 18 May 1596]. For the details concerning this invention, see also p. 208.

100 "[...] Inventie, dat eenen Jongen soo veele waters sal connen doen uuytwerpen als de voors. ses, oft seven mannen souden connen doen [...]." NL-HaNA, States General, 1.01.02, inv. no. 12302, fol. 177v [G188, 17 June 1620]. Seigneur Louys Escuyer and Seigneur de Froyennes with their associates. On the use of child labour to reduce labour costs, see Davids, "Beginning Entrepeneurs," 170–72.

101 "heywerck, hetwelcke bequamelyck by sesse ofte seven mannen soo te water, als te lande ganckbaer gemaeckt, [...] in plaetsse daermen tot noch toe XXVIII ofte XXX mannen tot een hey van doen heeft gehadt, die t'allen XX ofte XXII slagen, ende sulcx dickwils sesse, oft seven malen moesten rusten eer die pael in den gront was, soo dat het bier, ende dach-gelden van over twintich persoonen aen elcke hey gespaert, ende daerbeneffens soo veel tyts, ende wercx konste gewonnen wordden, dat het aen groote wercken als sluysen, poorten, dammen, dycken, ende paelwercken niet weynich bedragen en soude." NL-HaNA, States General, 1.01.02, inv. no. 12298, fol. 119 [G21, 1 August 1595]. Invention by the carpenter Lambrecht Gerritsz.

> Jan Barntsz, carpenter of Haarlem, claimed to have found a certain instrument to refine salt with much less effort and costs and in a much shorter time. However, as it is understood that such inventions were very inexpedient for the land, as they weaken other crafts and businesses and take away the bread from the common man, this request has been rejected.[102]

On 23 February 1625, two other inventors came forward with a similar invention to make "a certain small salt." Their request for a 25-year privilege was turned down as well, because "this was an invention that would damage the other salt business considerably; and one simply ought to reject this and similar proposals, because the work would be taken away from sober workmen and as a consequence the bread from their mouth."[103]

When it came to labor-saving technologies, the authorities seem to have made a distinction between economizing on expensive day-laborers and economizing on the job losses among the lower middle classes.[104] So, they may have tried to prevent social unrest. And in fact, it seems that the authorities generally succeeded in making this comparative assessment, since privileges granted for labor-saving machinery did not generally meet with much opposition from the population.[105] The most notorious case where discord *did* arise, was the case of a new type of ribbon-mill invented by the prolific inventor Willem Dirckz Sonnevelt. The details of this case not only offer a unique look at the issues related to labor-saving machinery, but also show how merchants in the Dutch Republic managed to force through their interests by playing the different local authorities against each other.

---

102 "M. Jan Barentsz, timmerman van Haerlem, seyde gevonden te hebben zeeker instrument om met veel minder moeyte ende costen ende binnen veel corteren tijdt het zoudt te raffineren ende versocht daeromme octroy van 'tzelve den tijdt van 45 jaren. Doch also verstaen worde dat sodanighe inventien voor 't landt geheel ondienstich waeren, omdat se andere handwereken ende neeringhen verswackten ende den gemeene man het broodt uytte mondt namen, is dit versouck om dezelve reedenen wille affgeslaeghen." Veenendaal-Barth, Smit, and Vree, *Particuliere notulen*, 1987, 2:385.

103 "[...] dat dit een inventie was daerdoor de ander soudtneeringe merkelijcken soude werden gekrenckt, dat oock daerdoor de sobere arbeytsluyden het werck ende dienvolgens het broot uytten mondt zoude werden ghenomen ende dat men daeromme dit ende gelijcke versoucken simpelijcken behoorden aff te slaen." Veenendaal-Barth, Smit, and Vree, 2:475. This time, it concerned a couple of merchants from Hamburg. It is worth noting that the nobilty gave their fiat, whereas the city of Alkmaar objected.

104 I would like to thank Karel Davids for this suggestion, done in personal conversation. Cf. De Vries, "How Did Pre-Industrial Labour Markets Function?," 46.

105 But compare Van Berkel, "Enige opmerkingen," 137–38. It somewhat unclear to what degree the example of group angry fullers (1637) can be generalized, or what the real reason for the resistance was in this case.

Somewhere in the beginning of 1604, Sonnevelt invented a new type of ribbon mill.[106] On 10 July 1604, he entered into a contract with Gabriel Hanedous and Pieter Pietersz Bodijn for the production of three mills for 550 guilders a piece, upon the stipulation that he would not "make or contract for any other big works in the meanwhile."[107] Furthermore, the parties agreed that, upon finishing the mills, Sonnevelt was under the obligation to make similar mills for the same price, if so desired.[108] In return, Sonnevelt obtained the right to make three additional mills, which he could rent out for the duration of fourteen years at his discretion, for six guilders per week.

The novelty of the mill was that it produced twelve ribbons at a time instead of one. And this is where the trouble started. In July 1604, some ribbon-merchants in Leiden began to complain that the new mill was putting them out of business.[109] They asked the Court of Leiden to prohibit the use of the new mill, claiming that the "many thousands that earned their living [with the honorable profession of ribbon-making], and in particular children from 6 to 14 years old, would be forced into poverty and to live off almsgiving."[110] Hanedous & Co. felt threatened by the unrest the invention had provoked. As a spokesman, Hanedous addressed a letter to the Court of Leiden stating that, although the Court had not forbidden the use of the mill, his company felt forced to leave Leiden, because their opponents might resort to force and violence.[111]

Simultaneously, Hanedous sent a letter to the magistrate of Delft, in which he recounted the problems his company had encountered in Leiden.[112] He claimed that his company had introduced the ribbon industry in Leiden "as first and most special inventors" (*als eerste en bysonderste inventeurs*) around the year 1586. Others had then soon picked up the ribbon industry and taken advantage of their endeavors. A continuing problem was, however, that the boy laborers in the company continued to make many mistakes and so brought

106 Dirckz Sonnevelt originated from Maasland, but was based in the village of Voorshoten, near Leiden. Whether the ribbon mill was really a Dutch invention, or rather an imitation of a technique already in use in Germany, remains in doubt. See Doorman, *Patents*, 75.

107 "[…] zonder dat hy middeler tijt eenige andere groote wercken zal mogen aannemen ofte maeken […]." Posthumus, *Bronnen*, 3:696 [no.350].

108 Posthumus, *Bronnen*, 3:696 [no. 350].

109 Posthumus, *Bronnen*, 3:698 [no. 351]

110 "[…] dat veele duysenden daermede heur broot verdienen ende aldermeest kinderen van 6 off 7 tot 12,13 off 14 jaeren toe […] in armoede ende genootsaect zouden zijn van de aelmissen te leven […]." Posthumus, *Bronnen*, 3:699.

111 Posthumus, *Bronnen*, 3:700 [no. 352].

112 Posthumus, *Bronnen*, 3:701 [no. 353].

"headache and distress."[113] This is why Hanedous and his company had decided to commission the "practitioner" (*practisijn*) Dirckz Sonnevelt to help devise a solution to produce better ribbons. Sonnevelt had passed this assignment with flying colors, and only because of "bitter hatred and jealousy" (*bittere haedt en nijdt*) had others now conspired and filed complaint.[114] When the Court of Leiden summoned the Hanedous company to rebut the accusations that the new invention was harmful, the company had considered it wise to leave Leiden altogether.

Hanedous was looking for a new location to set up his business. His proposal was to start up the ribbon industry in Delft on the condition that Delft would allow his company to use their new ribbon mill. He emphasized in his letter how he feared others would request the States of Holland to prohibit the use of the instrument altogether, and therefore he urged Delft to prevent this in time. On 4 August 1604, the city of Delft and Hanedous & Co. then entered into a contract that stipulated that the magistrate of Delft would offer protection and the exclusive right for ten years to produce ribbons (for the price they would sell at the Leiden market) on the condition that Hanedous would not transport his work elsewhere.

Just a couple of months later, Hanedous wrote another letter to the magistrate of Leiden. He considered it his duty to alert the magistrate to his contract with Delft, indicating that he would rather stay in Leiden because he had been a citizen in that city for nineteen years.[115] All he asked for was a privilege that would be "equal or better" (*gelijk of beter*) than the one that had been granted by the city of Delft. But the matter was urgent, he added, because Sonnevelt had almost finished his second mill: the mill that had already been transported to Delft would have to be picked up as soon as possible in case Leiden was prepared to offer equal conditions, because "all the instruments being together [at the same place], the art to imitate them or [to make] similar [ones] will not be found so easily."[116] On 18 December 1604, the magistrate of Leiden adopted the contract made earlier with Delft, offering the people working for the

113 "[…] sulcx dat (daer)door groote faulte onder de jongens valt van onperfect werck, mitsgaders onseggelicke hooftsweer ende moyte […] tot de meeste perfectheyt ende correcste werck te geraecken […]." Posthumus, *Bronnen*, 3:701 [no. 353, note 3]. The expression "hooftsweer ende quellaige" comes from page 700 [no.352].

114 Posthumus, *Bronnen*, 3:701 [no. 353, note 3].

115 Posthumus, *Bronnen*, 3:704 [no. 354, note 2]. Dated 9 December 1604.

116 "[…] opdat alsoo alle de instrumenten by den anderen sijnde, de const van sulcke of diergelycke nae te maecken soo licht niet gevonden en worde." Ibid.

Hanedous Company safeguard and protection for the duration of twelve years on the condition that the Company donated a sum to the poor in compensation.[117]

The troubles entailed in the introduction of the ribbon mill were not over, though. On 3 October 1605, Sonnevelt obtained a ten-year privilege from the States General for the exploitation of his ribbon mill.[118] The privilege immediately led to problems. Within a couple of days, some merchants in the ribbon industry demanded from the Court of Leiden that "any one would be free to have such instruments made," but their demand was denied.[119] Then, on 15 December 1605, a certain Jacob de Ruysscher de Jonge came into conflict with Hanedous because he had commissioned his own version of the instrument. Hanedous complained that the new instrument was an infringement of his privilege, but the matter was settled amicably. Hanedous would allow the use of the mill provided that De Jonge did not sell ribbons below the normal price; that he did not export the mill outside of the city; and that he donated 150 guilders to the orphans.[120] On 24 January 1606, another conflict emerged. This time, Passchier Segers had infringed upon the privilege. The matter was again settled by the Court on condition that Segers promised not to sell the instrument, or to export it out of the city, and by stipulating that he would pay 24 guilders to the orphans as a form of "recognition."[121] Each time, Hanedous appeared willing to let others use the instrument.[122]

Then, for a few years, no more was heard on the subject of the ribbon mill. Until, in April 1610, new complaints poured in.[123] This time, some merchants in the ribbon industry notified the Court of Leiden that the magistrate had originally allowed for the use of the mill on the express promise that the inventors kept the details of their invention secret, and did not use more than 2 or 3 looms "in order for the poor to keep working too."[124] By then, however, the company of Hanedous was using a total of 31 mills at different locations and, moreover,

117 Posthumus, *Bronnen*, 3:704–706 [no. 354].

118 NL-HaNA, States General, 1.01.02, inv. no. 12299, fol. 268v [G97, 3 October 1605]. For the transcription of the privilege, see Posthumus, *Bronnen*, 3:711–712 [no. 362, note 1].

119 "[...] een yegelijck zoude vrystaen sulcke instrumente te laten maecken [...]." Posthumus, *Bronnen*, 3:707 [no. 355]. The Court asked the applicants to be patient.

120 Posthumus, *Bronnen*, 3:708 [no. 358]. Jacob de Ruysscher de Jonge had employed a carpenter to produce a similar ribbon mill.

121 Posthumus, *Bronnen*, 3:709 [no. 359].

122 A later infringement was also settled amicably. See *Bronnen*, 3:709 [no. 360, 31 January 1606.]

123 Posthumus, *Bronnen*, 3:710 [no. 362].

124 "[...] ten eynde de schamele mede werckende zouden blyven [...]. Posthumus, Bronnen, 3:712 [no. 362].

some of those mills had "not been made by [...] Willem Dirckz, but by other persons, which had been forbidden according to the privilege."[125] The complainants demanded immediate action by the Court, but they did not find a willing ear.[126]

When the privilege expired, a new conflict emerged over the exact width of the ribbons that wholesalers were allowed to produce with the new invention.[127] On 11 August 1623, the States General tried to hush up the situation by issuing a writ to limit the production to only narrow ribbons.[128] Yet the result of the legislation was minimal, amongst other things because Amsterdam continued to "allow and tolerate" (*toestaen ende gedoogen*) the use of the ribbon mill for all purposes.[129] Leiden was unwilling to crack down on offenders of the placard as long as the law was not observed in the other cities.[130] The use of the ribbon mill expanded rapidly and spread to other towns in the Dutch Republic.[131] The introduction of Sonnevelt's invention therefore shows that, despite the necessary bickering, labor-saving machinery found its way to the market after all, partly because the towns in Holland could be played off against one another.[132]

It must be said that, although the authorities recognized the efficacy of labor-saving machinery, they were never interested in upholding a utopian vision of a mechanized world. There were more practical considerations at stake. As one inventor put it, the true point was that "the development of manufactures and merchant companies in these lands is helped and favored by [...] privileges."[133] In that light, any invention that could bring about a reduction in

125 "[...] by den voors. Willem Dircxz. niet en zijn gemaect, maer by andere personen, by 't octroy nochtans verboden, gedaen maecken." Ibid.

126 Posthumus, *Bronnen*, 3:713–714.

127 Posthumus, *Bronnen*, 4:512–521 [no. 382–384, 385, 386, 388]

128 Plakkaet 11 August 1623; *Groot Placaet-boeck*, I, column 1190.

129 Posthumus, *Bronnen*, 4:517 [no. 385, 7 April 1622].

130 The Court of Leiden only punished the biggest offenders to set an example, but let others go ahead. The Court considered that the matter should be settled in The Hague "since it is supposed to be that way" (*daer ende sulx behoort*). Posthumus, *Bronnen*, 4:521, (no. 388, 23 March 1624).

131 For the wide scale introduction of silk ribbon machines on the European continent, see Pfisterer, "Craft Guilds and Technological Change."

132 For similar observations in the English context, see Randall, *Before the Luddites*, 50. Randall argued that "[n]ot all machines provoked resistance" and that the fear of job losses of was not necessarily linked to reality. Moreover, he noted that the treat of violence into postponing the introduction of machinery was sometimes successful, and sometimes not. Randall, 83. On the importance and impact of urban rivalry for commercial devlopment, see Gelderblom, *Cities of Commerce*. A focus on the patent system offers support for Gelderblom's more general thesis.

133 "[...] dat de vorderinge vande manifacturen ende coopmanschappen inde landen wordt geholpen ende vermeerdert [en] begunstigt met [...] octroy." From the privilege

terms of production cost was deemed worthy of an honorary privilege. This was by no means always explicitly stated in the documents. It was much more common to assert that an invention was more "convenient" (*bequamelyck*) or "useful" (*dienstelycker*) than the technology presently in use; in practice, this would mean that less labor, time, and energy was needed to achieve a specific result. In a nutshell, then, "amelioration" and "progress" were defined from the perspective of the merchant-producer.

In effect, the point was to drive up production capacity. Aside from forthright cost reduction, the possibility of intensifying production within a certain amount of time was therefore an important aspect. The inventor Samuel Maralois, to give just one example, obtained the privilege for his invention of a new type of horse mill that could polish all types of melted steel "without the loss of time and with very little cost."[134] The aim was to make more profit, also in those cases where faster handling of products was needed to guarantee the quality of the final product. A mill that worked both on wind- and horsepower, for example, was considered:

> useful for the drapery businesses (*neeringe van drapperye*) and cloth makers in these Lands, because it often happens that the sheets under treatment in the mill by the ordinary and normal way of proceeding, cannot be finished (*affgevolt*) because of a silence and lack of wind, which means that they degrade, and become rotten.[135]

Production capacities were expanded under the cloak of better product quality. In a slightly different way, this was also the case with large-scale reclamation projects, involving powerful investors, where the continuous operation of

---

application of Jacob van Aecken, NL-HaNA, States General, 1576–1796, 1.01.04, inv. no. 7474. The privilege for the wind and water mill for melting iron was granted on 28 July 1601. NL-HaNA, States General, 1.01.02, inv. no. 12298, fol 368 [G60, 28 July 1601].

134 "[…] volgende het model, dat daermede *bequamelyck sonder tyt verlies ende seer weynich costen ende moyeten* voortaen by de viere gesmolten […] zoude kunnen worden." NL-HaNA, States General, 1.01.02, inv. no. 12300, fol. 234 [G107, 8 February 1611].

135 "[…] dienstich voor de neeringe van drapperye ende lakenmakeryen in dese Landen, doordien dat in de ordinaris ende gewoonelycke maniere van volmolens dickwils gebeurt, dat de lakens in de molen, ende onderhanden synde, by stilte ende gebreck van winde nyet cunnende affgevolt werden, by hen selven verslimmen, ende bederven, Twelck by dese nyeuwe molens bequamelyck verhoet ende voorcomen sal cunnen werden, doordien de meesters van de selve molens by faute van winde terstont het paert sullen mogen gebruycken, om haer werck te continueren, ende aff te wercken, oock somtyts ende als 't werck soo vereyscht alle beyde te gelyck […]." NL-HaNA, States General, 1.01.02, inv. no. 12302, fol. 7v [G154, 13 February 1617].

machines was essential for achieving good results. The carpenter Ghysbrecht Pietersz hence described his invention as a mill "that can drain a large quantity of water without the help, agency, or government of humans (indeed, every hour continuously throughout the day, more than other mills, and with fewer costs)."[136] In all these cases, time was irrefutably connected to money.

The uniform basis for evaluating new technologies was effectiveness. The relatively high number of inventions considered useful because they could be operated by *a variety of energy sources* should be interpreted in the same light.[137] One mill, for example, was praised because it worked "without wind or human labor, but with the help of a horse."[138] In another example, however, a mill could be "governed by the wind instead of doing the same by means of a horse."[139] The main point was not to replace horse-mills by windmills, for instance, or vice versa, but rather to build machines that were not dependent on one particular energy source. Thus, the use of manpower to operate a multifunctional mill was justified in one example with the argument that the new instrument "not being subject or subordinate [...] to the necessity of wind, nor to the course and power of the rivers, could be operated by one man or a horse."[140] By supplementing natural energy sources such as wind and water

136 "[...] Molenken, omme daermede water uuyten lande te malen, het welcke *sonder hulpe, toedoen, noch Gouvernement van menschen*, eene seer groote quantiteyt van wateren, jae meer als d'andere molens met minder cost yder ure continuelycken dach, ende nacht uuyt den lande cost malen." NL-HaNA, States General, 1.01.02, inv. no. 12299, fol. 291v [G98, 13 December 1606].

137 The diversification of energy use has, according to my knowledge, so far not yet received any attention in secondary literature. Instead, secondary literature seems to focus on identifying the primary energy source. Richard Unger, for instance, stressed the importance of (imported) coals, thus challenging the idea forwarded by De Zeeuw that peat alone allowed the existence of heating energy intensive industries in Holland. Unger, "Energy Sources." Karel Davids, on the other hand, underlined the importance of the windmill industry. Davids, "Innovations in Windmill Technology."

138 "Instrument, ofte molen, gaende met drye raden, sonder cammen, oft staven daertoe te gebruyeken, waermede d'Supplianten *sonder wint, ofte mensschen arbeyt, doch met behulp van een paerts werck*, soo veel waters can uuytwerpen vier voeten hooch als twee van de beste achtkante watermolens met volle seylen, ende bequamen wint uuyt ende op werpen kunnen [...]." NL-HaNA, States General, 1.01.02, inv. no. 12302, fol. 136 [G181, 16 August 1619].

139 "[...] inventie om zeem te bereyden door eenen muelen die mette winde geregeert ende gedreven soude wordden inde plaetsse dat men tselve met een peert hadde moeten doen, *welcken wintmolen dat commodienselycker ende gerieffelycker zyn ende oyck t zeem ende leer veel beter bereyden ende maken soude als tgundt dat metten peerde bereydt wierde.*" NL-HaNA, States General, 1.01.02, inv. no. 12298, fol. 54 [G14, 26 May 1592].

140 "[...] nyet en syn subiect noch onderworpen (gelyck geseyt is) de noodsaekelicheyt van wint, noch loop ende cracht der rivieren, maer worden alleen bearbeyt deur een man oft peert [...]." NL-HaNA, States General, 1.01.02, inv. no. 12302, fol. 38v [G160, 15 August 1617].

with the more reliable working power of men and horses, manufacturers could achieve an advantage in terms of production.

Unmistakably, the key to this process of production intensification was the strong position of merchants in the Republic. The common good was equaled to their interests. In a privilege claim for a fulling mill, for example, an inventor adduced that:

> the use of the instrument will serve the convenience of the merchant and will lead to less costs for those who use it to polish their textiles. And as it serves the general cause, that the merchant is served in the best possible manner and with the least costs and [...] that it is reasonable that the finder of such an instrument enjoys some benefit.[141]

Merchants had the capacity to invest the considerable amounts of money needed to reduce novel inventions to practice. It comes as no surprise that they were frequently called upon to vouch for the profitability of particular inventions. What really mattered was that inventions were "to the satisfaction of the merchant."[142]

Conclusions

This chapter has scrutinized the economic importance of invention privileges. It appeared that the Dutch were clearly aware of the possibilities of turning invention privileges into a policy measure to achieve mercantilist ideals. Although this conclusion concurs to some degree with observations made earlier by other scholars, one could perhaps say that mercantilist tendencies were even more strongly apparent than hitherto assumed.

Yet it was not all about mercantilism. A deeper analysis of the privilege documents shows that, despite the lack of any contemporary treatises on the topic, the Dutch authorities were clearly aware of the labor-saving possibilities of novel technologies. This observation brings greater variety to the assumption that the economic advantages of labor-saving machinery was a phenomenon

141 "[...] Dat oock 't gebruyck van 'tzelve instrument zal dienen tot groot gerijf van den coopman ende tot minder costen van degenen, die hem mette voors. stoffen generen om te doen calenderen. Ende alzo de gemene zaecke vorderlick es, dat den coopman best ende mette minste costen wert gedient, efter niettemin [...] redelick es, dat den vinder van zodanich instrument eenich beneficie geniete [...]." Posthumus, *Bronnen* 4:80 [no 64].

142 "[...] tot contentement van den coopman [...]." Posthumus, *Bronnen*, 3:702. From the letter by Hanedous & Co. and the City of Delft.

only fully understood by the end of the eighteenth century. Many inventions, of course, did provide work, and many of the contracts between inventors and city governments contained conditions to keep a certain number of people at work at all times; but this did not mean that privileges in the Dutch Republic were granted on condition that the new invention provided work for the inhabitants. Privileges that would have taken work from the lower middle class, perennially under the threat of poverty, were refused at the provincial level. At a central level, however, there were no conditions imposed as to the number of workers an inventor had to retain or instruction in a certain skill that he had to provide for his fellow citizens (stipulations that were in force in other countries at the time). Negotiations on this matter took place at a lower level, namely between the inventors and the cities. The result was that the inventor found himself in a relatively good bargaining position: if one city or province decided not to implement his invention, he could always try his luck in another location. The Dutch state system led not only to fairly good conditions for the inventor; it also created a situation in which it was easier to get approval for those inventions that reduced production costs and could therefore generate more profit for corporate producers.

The emphasis on 'effectiveness' that continuously surfaced in the privilege acts should be interpreted in this context of Dutch mercantile culture as well. The practical considerations when building machines that could be driven by a variation of energy sources seems to have been much more important than the utopian vision of machines taking over the work of humans. Merchants wanted to build new mills that could be operated by men when there were no horses in stock; by horses when the wind died down; or by the wind and water, when these were freely available anyhow. By doing so, production could be forced up, which directly served the interests of the merchant.

Conclusion

FIGURE 54 The emblem of Invention

> A young woman with beautiful clothes, in white, on which it is written "non aliunde", that is, not from elsewhere; having a veil around her head, embellished with various colors, showing art and beauty by curls and hair braiding.[1]

This is how Cesare Ripa (ca 1555–1622) begins the allegorical description of 'invention' in his influential *Iconologia* (1593). Ripa, who lived at the crossroads of a new era in which the privilege system spread with lightning speed throughout Europe, considered invention to be a complex phenomenon comprising

1 "Een jonge Vrouwe met prachtige kleedinge, in 't wit, waer op sal geschreven staen, non aliunde, dat is, niet elders van daen; hebbende 't hoofd rontom de sluyer, met verscheyden verwen geciert, alwaer, door aerdige krullen en drayen, de konst en de schoonheyt vertoont wert. " Ripa, *Iconologia*, 550.

 DOI:10.1163/9789004320420_008

several elements. He saw invention as a philosophical concept that united understanding and practice; as something that was regulated by law; as a field of tension where transparency and secrecy were in conflict; and as a metaphor expressing the mysterious relationship between man and nature:

> That she keeps the statue of Nature in her left hand is to emphasize that Nature is the Finder of all things. Therefore, it brings nothing to find something new and not to divulge it; therefore, the Legislators say that those who hide their intentions or art cannot have any benefit; and therefore, the Philosophers say "Virtus in actione consistit", that is virtue consists in action: one should divulge and execute one's Findings in order to gain praise from everybody. The woman is depicted with an open somewhat exalted hand. For already among the Egyptians, a painted or sculpted hand is evidence of an industrious architect; it is the hand that has served us to do labor, and which makes the ideas in our mind visible to the eyes— the reason for the words "ad operam" on the right arm. The two naked arms with the golden bracelets symbolize the reward that the Ancients gave to those who found something honorable and useful to the advancement of the arts. Thus is recounted by Pierius Valerianus in Book 15.[2]

Ripa was not the only one with an undeniable admiration for invention. The whole of Europe was bubbling with the fulminating idea that the world could be ameliorated, in an abstract as well as material sense.[3] Humanist princes did

2 "Dat zy het beeld van de Natuyre in de slincker hand houd, is om te betoonen, dat de Natuyre de Vindster is van alle dingen. En daerom iets nieuws te vinden, en niet in 't licht te geven, is een saecke die niemant vorderlijck is, en dat nae de spreucke van de Wetgevers, die zijn voornemen of konst verberght, seggen zy, daer van kan geen nut komen, en hierom is 't geen wonder, want de Philosophen seggen, Virtus in actione consistit, dat is, de Deughd bestaet in de daed: Hier over behoortmen zijne Vindingen in 't licht te stellen en uyt te voeren, om by een yder lof te behaelen. En daerom is dese beeldnisse gemaelt datse de opene hand een weinigh verheven uytstrekt. Want by den Egyptenaers was een gemaelde of gesneden hand, een bewijs van een vlijtigh Man totte Bouwkonst, welcke hand ons veelmaels tot het werck heeft gedient, en door wiens weldaed, de konst van alle dingen, wort gevonden, om de beeldnissen, die in de gedachten ontworpen zijn, sichtbaerlijck voor de oogen te vertoonen, en daerom hebben wy het woord ad operam, aen de rechter arm geschreven. De twee bloote armen mette goude armringen, bedieden het loon, dat de Oude plachten te geven, aen die geene die iet loflijx en treflijx, tot vorderinge der konsten, hadden gevonden. En dit verhaelt Pierius Valerianus in 't XL boeck." Ripa, 550.

3 Crowley, *The Invention of Comfort*; Slack, *The Invention of Improvement*; Warde, "The Idea of Improvement." Obviously, this is not a linear (or typically English) history and it did matter from which perspective one approached the matter: not every "improvement" was actually

not only sponsor great intellectuals who went in search of the sources (innovation did not of course mean that history had become irrelevant), they also built great palaces and new cities where people could construct a new and more comfortable sort of life with all manner of new products (books, stockings, and spices, to name but a few). As a result, the cosmic conflict between what God had provided and what people could add to it was brought into sharp focus. A Dutch professor of mathematics later expressed it as follows, in a book on atronomy and geography where he stressed the importance of his area of expertise in the light of long-distance trade:

> Because just as it has pleased God not to bless one city, country, or realm with altogether everything, but to bless and enrich the one province with some gifts and products, the second with some other, [etc.], whilst making profit common to the use of mankind, so do navigation and maritime shipping make us enjoy everything with which the Lord has enriched the earth and the sea. It is true that the almighty God has given to each a piece of land, where inhabitants can survive and live, but meanwhile there is a difference between living and living well. Ships and maritime navigation can best provide for the things that are missing in one Land for a good and healthy life [...].[4]

The discovery of new worlds and new well-being would not have been possible without the accumulation of large sums of capital. The early modern period in that way gave bankers and industrialists the possibility to claim a central role in the state system. It was their interests, in particular, that had to be served and protected by the law, as can be seen with the advent of new

---

an improvement for the actors involved and there was certainly a lot of resistance to novelties aimed at material progress. Particularly relevant in this regard is the excellent monograph by Yamamoto, who has shown how "distrust" of projects led to a transformation of procedures and strategies that eventually culminated in an early capitalist mode of organization. Yamamoto, *Taming Capitalism*.

4 "Want nadermael het Godt belieft heeft een Stadt, Landt, oft Rijck niet te begiftigen met alles, maer d' een provincie met dese, de tweede met andere gaven ende waren te verrijcken ende te zegenen, ende nochtans de selvige Godt tot het gemeene menschelijcke geslachts profijt heeft gemaekt ende geschapen, soo doet de navigatie ende zeevaert ons alles genieten, waer mede Godt de Heere de aerde ende Zee heeft verrijckt. t Is wel waer dat d'selvighe almachtige Godt yeder Lant heeft gegeven, waer mede d' inwoonders sich by het leven konnen onderholpen, maer ondertusschen isser onderscheyt tusschen leven ende wel leven. 't Gene dan tot het wel ende gesont leven 't eene Lant ontbreeckt, kan bestdoor Schepen ende Zeevaert geholpen ende verbetert worden [...]." Metius, *Astronomische ende geographische onderwijsinghe*, A2-A3 (preface).

kinds of institutions: banks and insurance companies. State and capital institutions gradually merged into one entity and changed the nature of knowledge considered useful and important. It is exactly at this point that privilege rights for inventors emerged.

The quintessential characteristic of knowledge production in the early modern period was the extent to which those in power got involved in the development of technology and science. As Klaus Vögel rightly said, one could illustrate this "by pointing at the employment of scientist at European courts; whereas in the fifteenth century only a few courts employed scientist, in the sixteenth and seventeenth century they could be found everywhere."[5] It could even be said that this was the early advent of a technoscientific society, in which industry, technology, and science, were completely interwoven.[6] The system of invention privileges was no more than a logical and integral part of this. It was the law by which the state could give form to the ideology of technological progress.

Against this background it is easy to see why the privileges pit "art" against "nature". The reseller of a privileged waterwheel claimed, for instance, that "the most honorable among all arts are those by which we obtain agreeable aids against several deficiencies of nature" (*gebreckelijckheden der natuer*).[7] "Art" was also contrasted with "labor", as in the case of a dredging device, which was defined as an "instrument, with which one could deepen with little effort (*aerbeyt*) and great art, all harbors, Rivers [etc.]."[8] Art was the ability to outwit nature, or as the scientific ideologist Francis Bacon put it: "it is the duty of art to perfect and exalt nature."[9] The mechanical arts were granted a central role in this venture, partly as well under the influence of the *Mechanica* by Pseudo-Aristoteles, who enjoyed huge popularity in the sixteenth century:

> One marvels at things that happen according to nature, to the extent the cause is unknown, and at things happening contrary to nature, done through art for the advantage of humanity. Nature, so far as our benefit is concerned, often works just the opposite to it. For nature always has the same bent, simple, while use gets complex. So whenever it is necessary to do something counter to nature, it presents perplexity on account of

5 Vögel, "Cosmography," 471.

6 Cf. Klein, "Technoscience Avant La Lettre."

7 Westerdyck, *Bewys*, from the dedication letter to the States General.

8 "Instrument, met het welcke, dat men met cleynen aerbeyt, ende groote const soude connen diepen alle havenen, Rivieren [...]." NL-HaNA, 1.01.02, inv. no. 12299, fol. 75 [G73, 10 July 1602].

9 Bacon, *Of the Proficience and Advancement of Learning*, 189.

> the difficulty, and art [*techne*] is required. We call that part of art solving such perplexity a *mechane*. As the poet Antiphon puts it: *We win through art where we are beaten through nature.*[10]

The marvel at "things happening contrary to nature" was a radical about-turn with regard to the way in which the church fathers had seen the relationship between man and environment. In their view, which would largely dominate the Middle Ages, the will to know, and possibly even to change, the world was considered a sign of hubris that dissuaded men from the simple admiration of God's creation. Novelty was all too readily associated with a strayed sense of curiosity. Saint Augustine was particularly influential in this area, with his remark that *curiositas* was nothing other than "a lust of the eyes" (*concupiscentia oculorum*).[11]

The negative attitude of Augustine and others has led a wide range of scholars to assume that 'theoretical curiosity' was buried under the yoke of Christianity before it was unleashed in the early modern period and ushered in modern science. But although this train of thought has spawned voluminous in-depth books, grand narratives like these have lost much of their appeal nowadays. In a superb study of curiosity in early modern France and Germany, the scholar Neil Kenny has shown how important it is to distinguish between the different contexts in which the word 'curiosity' was used.[12] Thus, whereas 'curiosity' in an ecclesiastical context was seen as an intellectual vice (going back to the story of Adam and Eve in Genesis), curiosity could have quite a different set of (more positive) connotations in other contexts. Similar observations could be made about the concept of "invention".[13] Where it could sometimes signify a "fabrication" (in poetry) and sometimes "the selection of objects" (in fine art), where it sometimes carried the negative connotation of "renewal" (in a religious context) and sometimes a more neutral meaning within the system of rhetoric, there can be no doubt whatsoever that invention within the privilege system had an extremely positive meaning as the discovery of new products and methods that the authorities considered to be the benefit of the common good.

10 As cited in Winter, "The Mechanical Problems," 1.

11 Augustinus *Bekentnisse X* 35, as cited in Niklaus Largier 'Rethorik' der Erfahrung', in Daston and Krüger, *Curiositas*, 141.

12 Kenny, *The Uses of Curiosity*.

13 Nota bene that I am talking about invention, not innovation or ingenuity. For the latter, see Marr et al., *Logodaedalus*. For innovation, see Godin, *Innovation Contested*.

This book has shown that during the emergence of the Dutch Republic, it was mainly the States General that shaped the outlines of privileged knowledge. The system of invention privileges was an integral part of what has been identified as the first modern economy; it allowed the authorities to determine which knowledge was desirable, and which knowledge was of subordinate importance. Privileges prompted the development of innovative knowledge on a local level and acted as a magnet for intellectual capital that was available abroad. Entrepreneurs and merchants who operated at the heart of the State thus turned the patent system into an indispensable part of the knowledge infrastructure that underpinned the Republic. They managed to set up a stable institution that combined a belief in technical progress with financial gain.

Studying invention privileges from this perspective offers a unique insight into state formation processes in the early modern period. First of all, they might help to show the ongoing discussion about the political structure of the Dutch Republic in a different light. The Republic is often regarded as an absolute anomaly in early modern Europe, where rivalry among cities and provinces frustrated every claim for a central authority. Yet, whereas in some areas the federal authorities may not have managed to assert themselves, in other areas, such as the conferment of invention privileges, their authority was clearly present.

The problem with theories that emphasize the independence of the provincial states and the merchant cities seems to be that they confound 'centralization' with a central location, a specific place.[14] If, indeed, a state is defined by centralization guided from a specific location, the Dutch Republic would not have been a state at all. There was no center of power in the Dutch Republic in the sense that Paris or London was the center of France or England. But this did not mean that local authorities could not embody a centralist ideology. As Michael Braddick formulated it:

> The problem here seems to be the identification of the state as the institutions at the 'center' – there is an elision here of 'centralized' and 'centrally located'. Clearly, though, the institutions of a centralized state are not all centrally located. By the same token, the institutions of local

---

14 To be clear, this problem occurs both with theories that emphasize decentralization, as well with those that emphasize centralization. Michael Mann is a good example of someone who stresses the importance of centralization for state formation, claiming that "states can be recognized by the central location of their differentiated institutions." Mann, "The Autonomous Power of the State," 112.

> governement can, in principle at least, be component parts of a centralised state.[15]

This clearly stands out when examining the privilege system in the Dutch Republic. Although there was a strong degree of competition among the cities, local institutions eventually turned into components of the centralized state; which is to say, even if they tried to defend their own interests, the result of their efforts was to the benefit of the Generality. Seen from that perspective, it mattered less whether there was a specific location from where the federal authorities could claim undivided sovereignty. What mattered, instead, was the way in which the ideology of a central state permutated the different localities.

A second observation to be made on the basis of examining the privilege documents is that, although 'warfare' and 'international competition' were important elements in the development of a Dutch mercantile state, they were not the only driving mechanisms for the state to become stronger. Of course, it must be said from the outset that invention privileges say very little about the total expenditure of the state, and they make up but a small part of the overall state apparatus. When taking privilege documents as a gateway to reveal prevailing thought patterns, however, it appears that concrete local concerns were a focal point for the Dutch authorities. This observation brings greater variety to ideas about state formation forwarded especially by Charles Tilly, who famously advanced the thesis that "war made the state, and the state made war."[16] Tilly sketched the image of a purely intrusive state apparatus, neglecting for instance the importance of cooperation between the authorities and the population of which the state was composed.[17] The formation of the state

15 Braddick, *State Formation*, 14.

16 Tilly, "Reflections," 42. Tilly proposed a model of three alternative (and to some degree parallel) modes of state formation. He classed the Republic as a *capital-intensive* mode, where "rulers relied on compacts with capitalist – whose interests they served with care – to rent or purchase military force, and thereby warred without building vast permanent state structures. [...] As the history of the Dutch Republic illustrates, at the extreme this mode produced federations of largely autonomous city-states, and constant negotiation among them over state policy." Tilly, *Coercion*, 30. But within the United Provinces, there never existed anything similar to the independent city-states on the Italian peninsula. On the long-lasting impact of the Tilly thesis, see Kaspersen and Strandsbjerg, *Does War Make States?* Tilly's ideas can also be read against the background of the infamous debates on the Military Revolution; see Rogers, *The Military Revolution Debate*; Teschke, "Revisiting."

17 Tilly does mention the process of negotiation of power, but again defines this bargaining process as a-symmetrical; as an act of conflict (or 'coercion') between an oppressor and the population. Tilly, *Coercion*, 99–103. He seems to neglect that the subjects that were

was, however, not just the outcome of a top-down movement, but also, in part, the response to a particular demand. In the Dutch scenario, this meant finding a response to the continuous threat of water which had to be controlled in some way. It also meant finding methods to make more efficient use of wind and to find solutions to make life in the cities more bearable. State control was not only about the threat of power, but about the multifarious organization of resources.

One of the most important resources that the state apparatus had to manage was 'knowledge' in the broadest sense of the word. Knowledge management was obviously not just about pushing innovation. It also included the establishment of universities and schools, the diffusion of ideas through print, the smooth running of apprenticeships, etc. Invention privileges were just one element of the entire knowledge management process, albeit an important one, because they allowed the authorities to turn technological knowledge into a political weapon. A privilege functioned as a hallmark showing that the States had given their approval for the project. The authorities vouched, as it were, for the quality of the inventions which they had deemed worthy of privilege. For this reason, it was in their interests to carefully examine the proposals put forward by different inventors: the very functionality of the technology fostered a sense of compliance and so ultimately contributed to the political legitimacy of the authorities. Another issue of continuous concern was to maintain the rule of law, so that the privilege would not be "rendered unfruitful [and] illusory," as one inventor put it, and "others would be discouraged from stimulating the business and traffic in these lands."[18]

This brings us to the question, whether there was anything typically Dutch about the things that have been described in this book. The approach taken so far has deliberately remained somewhat 'insular' based on the view that only a symmetrical account (i.e. a comparison whereby each of the parties to be compared is assigned equal attention) can deliver worthwhile results. Despite the fact that a number of good studies have appeared in recent years on specific

---

tabled were often in the interests of both parties who participated in the process of bargaining. Moreover, contrary to what Tilly seems to suggest, bargaining not only took place between the state and groups of individuals (who refused to cooperate with the state) but between members of the State apparatus and particular individuals as well: the privilege system is good evidence thereof.

18 "[…] octroy ten deele infructueus wert gemaeckt […] versoeckende dat wy om te verhoeden dat by dusdanige indirecte middelen t voors. octroy mette ampliatie van dien niet illusoir en mochte werden gemaeckt, ende andere den moet benomen, van met nieuwe inventien, de neringen ende trafique deser Landen te vorderen […]." Doorman, *Octrooien*, 163 [G247, 15 February 1625].

aspects of invention privileges in early modern Europe, I regarded the source material as too extensive and too complex for one person to take a proper comparative perspective. Nevertheless I shall make an attempt, on the basis of the information available in the secondary literature, to compare the system in the Dutch Republic with the situation in Habsburg Spain, England, and Venice. The central focal points in this comparison are the legal and institutional framework, examinations, and the role of merchants in the establishment of privileged knowledge.

Previous studies have shown that "the act of invention took place within a context of multiple interests defined by entrepreneurs, experts, and royal officials" in Habsburg Spain.[19] At first sight, there seem to have been no major difference here, except of course that in the Dutch case there were no royal officials (there is no evidence that this made the procedures any less bureaucratic, by the way). What is striking, though, is that the Spanish policy was in every respect extremely fragmented: while in the Dutch Republic the States General (and later the States of Holland) clearly formulated the policy when it came to conferring privileges, in the Habsburg kingdom there was a system of viceroys and Councils (such as the Council of Aragon, the Council of Castile, etc.), each of which had their own policy concerning the assessment and assurance of patents, which in turn "prevented coordinated and effective legal action by inventors to defend their monopolies and cut their costs in terms of both time and money."[20] In fact, it could even be argued that the Spanish policy was less centralized than the Dutch policy until the awarding of patents was centralized in the Council of Castile (in 1715–1716) and patents became applicable throughout Spain. Moreover, there seemed to be more focus on issues of economic ethics, at least in the Council of Aragon.[21]

In spite of the important differences, there were many similarities between the systems. Privileges were awarded, also in Spain, to both foreigners and indigenous inventors for the discovery of technologies and methods that were new and "tan útiles y necessarios a nuestro servicio y al bien público."[22] As in

19 Barrera-Osorio, *Experiencing Nature*, 67.

20 Royo, "State Policy," 157.

21 Royo, 157. As in England, tax stipulations were an important part of the eventual privilege grants in Spain: the profits made by the inventor had to be shared with the Crown with the idea that the loss of import tax (because of import substitution), or the implicit tax on the retail price (as a result of the monopoly position), had to be compensated for somehow. As we have seen, this was not the case in the Dutch Republic. This is because taxes were only collected at the local level.

22 García Tapia, *Patentes de Invención Españolas*, 255. See also Yun-Casalilla, "Social Networks."

the Republic, the main areas of interest were the "construction of mills, shipping and diving, and public works and building, providing mechanisms for lifting and transport, construction materials and tools."[23] Unfortunately, since there are no exact figures for this, it is still unclear whether there were more or fewer merchants involved in the Spanish patent applications, but in terms of inventions and examining procedures, the system in Habsburg Spain and in the Dutch Republic was largely the same. And thus, also in Habsburg Spain:

> The crown expanded empirical practices already in use among artisans and incorporated these testing practices into the state's regulation of instruments, technology, and inventors. The establishment of testing rules [...] compelled artisans to take their instruments before royal officials, where experts would test their actual performance. Inconclusive tests led to more tests under better conditions. Test results provided royal officials with the information necessary to validate the claims made by the inventor.[24]

The Spanish system, through its implementation in the Burgundian lands, served as the model for the later developments in the Dutch Republic. The main difference was that the *Chambre des Comptes* played a key role in the Habsburg Provinces, whereas there was no central accounting office in the Republic (until 1608). Financial considerations regarding invention privileges were made directly in the Assembly of the States General in which all provinces, and by extension all cities, were represented. And it was the General Assembly that directly supervised the examinations.[25]

Privilege examinations were the rule rather than the exception all throughout Europe, and also for instance in England, where royal councilor Sir William Cecil, 1st Baron Burghley (1520–1598) was notorious for being "extremely careful in granting patents, often going to inordinate trouble to discover the merits of new inventions."[26] The English system of privileges, which fully flourished

23 Royo, "State Policy," 150.

24 Barrera-Osorio, *Experiencing Nature*, 62–63.

25 Furthermore, in the Dutch case, the Stadtholder of Holland was directly involved in the handling of military innovations in the Republic: this was completely different under Philip II (even if he was a renowned patron of technology who was in direct contact with inventors), but also for example in England, where the Royal family rarely entertained inventors or their representatives directly. Harkness, *The Jewel House*, 147; Goodman, "Philip II's Patronage of Science and Engineering."

26 Davies, "The Early History," 106. Also in England, the use of models was a common phenomenon, especially under the reign of James I. See Davies, 268ff. The habit of providing models in England slowly disappeared 1637 to 1670, just as in the Republic.

during the reigns of Elizabeth I (1558–1603) and James I (1603–1625), has been the focus of much attention in the historiography because it ultimately ended in the famous Statute of Monopolies (1624). Following the work of Joan Thirsk it has become common to make a distinction between the Constructive phase (1540–1580) and the Scandalous Phase (1580–1601, 1601–1621) of the monopoly system. Thirsk asserted that privileges served primarily to produce foreign products locally and thus increase well-being at home. Whilst in the constructive phase, privileges were granted to start up projects "deemed most suitable for employing the poor", this changed in the scandalous phase:

> In the 1580s possibly, and by the 1590s certainly, some of the patentees were no longer inventors and skilled craftsmen, but courtiers, merchants, and speculators who planned to hire the services of such craftsmen, while they themselves shouldered the main financial risk. On the face of it, this development was sensible, and perhaps inevitable, but it brought the patent system into disrepute.[27]

It is worth noting that in the Republic merchants liked the privileged monopolies a lot too, but that this did not result in major opposition.[28] Perhaps this was because of the limited duration of privileges in the Republic. It is also possible that there was less opposition against monopolies in the Republic because the privilege holder was not per se connected to the central powers, in contrast to England where the patentees were often confidants of the Court. Another major difference with the English system was that in the Dutch case, from the very beginning, existing laws were respected, and the central authorities seldomly granted monopolies for products that were not new to the country.

It was precisely the mercantile character of the Dutch system that provided a point of difference. This is not to say that authorities in other countries did not look explicitly after the interests of merchants; yet, in contrast to those

27 Thirsk, *Economic Policy and Projects*, 57. Over the years, specific aspects of Joan Thrisk's work have been challenged as well; for a comprehensive and subtle analysis of the English historiography in the field, see Yamamoto, *Taming Capitalism*, 9–19.

28 In contrast to England, there was no elaborate public culture in the Dutch Republic in which the projectors were ridiculed, for example in plays or mock poems. For a broader discussion of the rhetoric concerning the so-called projectors in Britain, see Ratcliff, "Art to Cheat"; Thébaud-Sorger and Yamamoto, "Inventeurs et hommes à projets"; Yamamoto, "Reformation and the Distrust"; Yamamoto, *Taming Capitalism*. The word "projector" did not occur in the Dutch language until the second half of the eighteenth century (*Project-maaker*). Also the word "project", in the sense of a more or less elaborate plan or idea, did not make its appearance until later in the seventeenth century; before that, the word "project" was used in the sense of design for a work of construction or water engineering, i.e. a construction drawing.

other countries, questions of economical ethics and technological unemployment rarely played a prominent role in the Republic. The driving force behind the Statute of Monopolies, the English barrister Sir Edward Coke (1552–1634), for instance, gave as one example of an "inconvenient" patent, a new type of mill that would have replaced workers and thus threatened "to turn so many labouring men to idleness."[29] As a result of the strong bargaining power of inventors, who could promote their inventions in various cities, the creation of employment seemed to be less central in the Dutch Republic. At the level of the cities, those starting a new project were required to keep a certain number of people employed, or the inventor had to divulge his knowledge by taking on apprentices (if the privilege was for a new method). But on an interprovincial level scant attention was paid to the issue of technological umemployment. Industrial innovation revolved entirely around commercial interests, and "useful inventions" were those that best served the merchant.

Incidentally, the local authorities awarded trade monopolies too in the form of exclusive contracts aimed to regulate the local market. In this respect there was a clear difference with the practices later institutionalized by the eminent French statesman Jean-Baptiste Colbert (1619–1683), who was known to strongly support the use of temporary monopolies to boost the local economy. Whereas, in France, the central government awarded invention privileges as well as trade monopolies, in the Republic this occurred on two different levels; the interprovincial authorities awarded the invention privileges that were valid in the whole Republic, while the cities granted trade monopolies, whose validity only extended to the city walls. The system of representation in the Republic made all the difference: this ensured a balance between local interests and federal institutions.

It is in this sense of institutional infrastructure that we may speak of a republican patent system. But it is important to mention straightaway that early modern republics came in many shapes and sizes. During the founding years of the Dutch Republic, the Republic of Florence was an example of how not to do things, and the Republic of Venice was less valued than for example the Swiss Confederation.[30] The eventual composition of the United Provinces was indeed unique; it is best understood as a federal republic, where, due to the internal interconnectedness of city and provincial interests, knowledge could spread quickly. Based on the information so far available, it seems that the diffuse power structure in the Republic enabled inventors to come into contact more quickly with the higher echelons of political power, while in

29 Coke, *The Third Part of the Institutes of the Laws of England*, 184.

30 Haitsma Mulier, *The Myth of Venice*, 64; Van Gelderen, "Het gemenebest".

other countries with a more pronounced court culture several intermediaries got involved before the sovereign power was even aware of a new invention. Perhaps there was less ostentatious posturing in the Republic and more emphasis on the supposed utility of inventions. But the background to this was institutional and it is complete nonsense to suggest that the privileges awarded in the Republic were therefore not a "political measure, but wisely accommodating the emerging needs of the Dutch inventive spirit."[31]

Notwithstanding the difficulties of talking about a 'Republican model', it is worthwhile to briefly compare inventors' practices in the Republic of the United Provinces with practices in the Republic of Venice, which played a pioneering role in the development of the patent system in early modern Europe. Like the Dutch, the Venetians built a powerful state that ruled over the seas and comprised a vast colonial empire (though the colonies were in Greece, not the tropics). The Venetian realm was founded on merchant shipping, and it is not unlikely that it was these mercantile interests that initially steered the legitimacy of monopolies. Alas, the exact details of this are unavailable. In contrast, it is obvious that among the Venetians already it was common practice to test new inventions extensively.[32] It also seems that, as in the Dutch Republic, it was not necessary to go through a great many middlemen to reach the central institutions of the state in Venice.[33] As Giuseppe Ceredi, the court engineer in Parma, recounted around the middle of the sixteenth century:

> I have seen many other models of machines that have never been reduced to practice; and in particular in the secret rooms of the *Ufficio de' Provveditori di commune* in Venice: where everyone who is convinced to have found some ingenious and beautiful thing takes his models in order to obtain a privilege.[34]

---

31 The entire quotation is: "Nicht als wirtschafts-politische Massnahme, sondern den machtig aufstrebenden Bedürfnissen des niederländischen Erfindergeistes verständnisvoll entgegenkommend, erteilten die Generalstaaten wie auch die Provinzialstaaten (von Holland vor allem) eine wachsende Anzahl von Privilegien, die hier ganz eindeutig - wie schon in Venedig - als Erfindungspatente moderner Prägung bezeichnet werden können. " Silberstein, *Erfindungsschutz*, 74–75.

32 See, for instance, Mandich, "Venetian Patents (1450–1550)," 176, 184.

33 A whole series of inventors from all kinds of backgrounds were able to navigate the corridors of Venetian power in order to obtain a privilege; Berveglieri, *Inventori stranieri a Venezia*.

34 "Molte altre machine in modelli, che non sono mai stato poste in opera reale, ho io visto in diversi luoghi; & specialemente nelle camere secrete dell' ufficio de Proveditori di commune in Venetia: ove ognuno, che si persuade havere ritrovato alcuna ingeniosa, & bella cosa porta i modelli, per ottenere qualche privilegio." Ceredi, *Tre Discorsi* (1567), 19.

There is no doubt that the Venetian system served as an example that would play a role across Europe in the visualization of privileges as an effective means of advancing the political economy. And just as Venetian richness had taken on mythical proportions in the sixteenth century, the sudden prosperity of the Dutch Republic after its separation from Habsburg Spain had not escaped the notice of contemporaries.[35] The use of privileges to favor innovative industries seems to have attracted attention too in that context. At least, it figured prominently in the writings of one writer, the dramatist Antoine de Montchrestien (*c.* 1575–1621), who at the beginning of the seventeenth century wrote a major treatise on the French political economy (*Traité de l'économie politique,* 1615). The book was an appeal to the young King Louis XIII (with his Regent and Queen Mother, Marie de Medici, in the background) to invest in the local economy and to reduce the need for imported goods. A special role in this context was also attributed to the artisans, who, in the author's view, formed the basis of industrial progress.[36] In that account, the author upheld the image of the Dutch, the English, and the Germans, as a model for France:

> Your Majesty should only accommodate somewhat better and more willingly than others, those who, for the benefit of our citizenry, can communicate some profitable and advantageous industry to us. To this end, the examples of our neighbors, who make better use of us than we do ourselves, lead and instruct us.[37]

Other nations, so Montchrestien argued, had surpassed France in terms of economic activity, only because many French artisans had left their homeland to seek more favorable conditions elsewhere. If France would only treat

35 Somewhat surprisingly, the impact of the Dutch Republic as ideological paradigm for economic development has received only scattered attention. An important exception is Reinert, "Emulating Success." Reinert does not mention invention privileges.

36 A counterpart to Montchrestien within Portuguese political thought can be found, for instance, in Manuel Severim de Faria, who argued that the solution to the economic crisis, that Portugal went through in the mid-seventeenth century, would be "bringing excellent artisans from other provinces, and giving them convenient salaries and comforts, favouring good skills, prizing inventions and works that show singularity and rarity, and designating rewards for perfection and excellence." (As cited in Almodovar and Cardoso, *A History of Portuguese Economic Thought*, 28.

37 "Vos Majestez y doivent seulement un peu mieux recevoir que les autres et plus volontiers ceux qui, comme pour le paiment de nostre bourgeoisie, nous peuveut communiquer quelque industrie profitable et advantageuse. A cela nous induisent et instruisent les exemples de nos voisins, qui se servent de nousmesme mieux que nous ne faisons nousmesmes." Montchrestien, *Traicté*, 36.

its artisans somewhat better, it would soon outdo all other countries, for it was clear that "Les arts nourrissent bien les hommes ; mais l'honneur nourrit les arts."[38] And this was the point at which privileges came into play: for only when artisans were given the possibility to make a profit, would they apply themselves to the creation and production of useful things.[39]

The Dutch in particular had understood the vital importance of artisanal industries to increase the wealth of a country. One of the things Montchrestien admired most – besides the introduction of child labor and the clever use of a staple market – was that the Dutch had managed to introduce a division of labor.[40] What France needed, then, was specialized manufacturing ateliers protected by privileges, where work was divided according to the capabilities of each and every one.[41] The importance of privileges in this context was clearly acknowledged by Montchrestien. In a long passage he recounted that:

> The people of Holland [...], as I recall having touched upon elsewhere, have proven in various ways especially over the last twenty-five or thirty years, that necessity gives rise to much spirit and inventiveness, [...] and, because it is necessary for the public hand to help the private one in matters of great actions [≈ when it comes to great projects], as soon as one of them, acknowledged as a hard-working man with sound judgment, wants to demonstrate and establish a useful industry, he requests from the States some convenient place to do so, and obtains it almost instantly, without the length [of the process] causing him to fall in languor; at the same time he obtains reasonable privileges to secure for himself the fruits of his invention, and he is never defrauded under the pledge of public faith. [...] Thus, one should not be surprised if industry makes greater progress there than among us; for the most gentle spirits, the men most accommodated to goods and fortune, take pride in finding and profiting from some ingenious and useful artifice, by means of which the most common yet indispensable arts can be practiced easier, and with greater output. Because of this, they have achieved great success, since

38 Montchrestien, *Traicté*, 120. Cf. idem, 35: 'L'honneur nourrit les arts et les arts nourrissent les hommes" (*Honor nourishes the arts and the arts nourish men*).

39 Montchrestien, *Traicté*, 16.

40 Montchrestien, *Traicté*, 37–38. About child labor in the Dutch Republic, see Montchrestien, *Traicté*, 103–105; about the staple market Ibid, 107–108. About the use of child labor to reduce labor costs, see Davids, "Beginning entrepeneurs," 170–72; Posthumus, *Nederlandsche prijsgeschiedenis*, 57; Van Zanden, *Arbeid*, 20–23; Van Nederveen Meerkerk and Schmidt, *Wage Labor and Vocation*.

41 Montchrestien, *Traicté*, 73.

> by [using] machines and tools of mechanical invention, they infinitely relieve the work of men and consequently reduce the cost of the work, which allows them to offer goods at such a low price, rather than [for instance because of] the great abundance or the diligence of the craftsmen. It should be added that, now that they are able to supply us, they do not aim so much at doing [things] well as at doing a lot. Their industry is tailored and deployed towards making products look good, [thus making it possible] to sell them easier, rather than towards making them good [and of high quality].[42]

The text by Montchrestien suggests that invention privileges played an important part in the realization of the Dutch economic miracle: the Dutch had a cutting-edge advantage because they freely managed with privileges that encouraged inventors to search for profitable machinery. As Montchrestien saw it, there was no natural contradiction between savings in labor and the production of wealth: mechanical inventions would bring down the production costs, which would in turn allow for an increase in the volume of production. This mindset had largely determined the wealth of the Dutch Republic.

What can be concluded at the end of this brief survey of the various systems throughout Europe is that "republican patents" were possibly different in terms of economic ethics; yet, that there was nothing unique about the formal procedures that constituted the Dutch privilege system. The European

42 "Les Holandois qui, comme il me soufvient d'avoir desja touché ailleurs, ont prouvé en plusieurs façons, depuis vingt cinq ou trente ans principallement, que la necessité donne beaucoup d'esprit et d'invention, [...] et, à cause qu'en matiere de grandes actions il faut que la main publique aide à la particuliere, si tost que quelqu'un d'entr eux, reconnu pour homme de labeur et de jugement, veut faire la preuve et l'establissement d'une utile industrie, demandant pour cest effet aux Estats quelque place commode, il l'obtient quand et quand, sans que la longueur le face tomber en langueur; quand et quand 'il a les privileges raisonnables pour s'asseurer le fruict de son invention, et n'est jamais fraudé sous le gage de la foy publique. [...] Ainsi ne faut-il point s'estonner si l'industrie fait là de plus grands efforts que parmy nous ; car les plus gentils esprits, les hommes plus accommodés et de bien et de fortune y font gloire de cercher et profit de trouver quelque ingenieux et utille artifice, au moyen duquel les arts plus vulgairement et plus necessairement en usage puissent venir à une practique plus facile, à une plus prompte expedition ; *de là leur reussit un grand fruict, d'autant que par engins et utils d'invention méchanique ils soulagent infiniment le labeur des hommes et par consequent diminuent les frais de la besongne,* ce qui leur permet, plustost que la grande abondance ou que la diligence des artisans, de nous donner les marchandises à si petit prix. A joindre que maintenant qu'ils sont en possession de nous fournir, ils ne visent pas tant à bien faire qu'à beaucoup faire, et que leur industrie s'employe et desploye plustost à farder les ouvrages, pour les mieux vendre, qu'à les rendre bons." Montchrestien, *Traicté*, 118–119. Emphasis added.

privilege system was a truly international one, even if it had some local quirks and specificities. Indeed, it was the homogeneous language of the law that facilitated the process of negotiation and, for instance, made 'patent hopping' possible. If anything, the Dutch system of invention privileges was different because it was run by merchants. These merchants saw the benefit of cutting down on manpower and running production processes more efficiently. They appreciated the power of mechanics and technical knowledge, knowing what this knowledge could do for them while expanding their trade ventures. And eventually, it would be this commercial rationale of the privilege system that served as a paradigm pointing the way ahead for the later interplay of knowledge, patents, and power.

Bibliography

The bibliography mentions both primary sources and secondary literature. In case the name of the author is unknown, sources are alphabetized by title, ignoring articles in the title name. Place names have been given in their modern form.

Achermann, Eric. "Unähnliche Gleichungen. *Aemulatio, imitatio* und die Politik der Nachahmung." In *Aemulatio. Kulturen des Wettstreits in Text und Bild (1450–1620)*, edited by Jan-Dirk Müller, Ulrich Pfisterer, Anna Kathrin Bleuler, and Fabian Jonietz, 35–73. Berlin: De Gruyter, 2012.

Adams, Julia. "The Familial State: Elite Family Practices and State-Making in the Early Modern Netherlands." *Theory and Society* 23, no. 4 (1994): 505–39.

Adams, Julia. *The Familial State: Ruling Families and Merchant Capitalism in Early Modern Europe*. Ithaca, NY: Cornell University Press, 2005.

Aerts, Erik, and others, eds. *De centrale overheidsinstellingen van de Habsburgse Nederlanden (1482–1795)*. 2 vols. Brussel: Algemeen Rijksarchief, 1994.

Agamben, Giorgio. *Homo sacer: Die souveräne Macht und das nackte Leben*, trans. Hubert Thüring, 11th ed. Frankfurt am Main: Suhrkamp, 2002.

Ahsmann, Margreet. *Collegia en colleges: Juridisch onderwijs aan de Leidse Universiteit 1575–1630 in het bijzonder het disputeren*. Groningen: Wolters-Noordhoff/Egbert Forsten, 1990.

Alberti, Leon Batista. *The Architecture of Leon Batista Alberti in Ten Books*. London: Edward Owen, 1755. Web: http://echo.mpiwg-berlin.mpg.de/MPIWG:342WTAWE (last accessed 21 February 2021).

Almodovar, António, and José Luís Cardoso. *A History of Portuguese Economic Thought*. London: Routledge, 1998.

Amir R. Alexander. "Exploration Mathematics: The Rhetoric of Discovery and the Rise of Infinitesimal Methods." *Configurations* 9, no. 1 (2001): 1–36.

Andrewes, William J. H, ed. *The Quest for Longitude: The Proceedings of the Longitude Symposium, Harvard University, Cambridge, Massachusetts, November 4–6, 1993*. Cambridge, MA: Collection of Historical Scientific Instruments, Harvard University, 1996.

Antunes, Caatia, and Jos Gommans, eds. *Exploring the Dutch Empire: Agents, Networks and Institutions, 1600–2000*. London: Bloomsbury, 2015.

Appuhn, Karl. *A Forest on the Sea: Environmental Expertise in Renaissance Venice*. Baltimore: Johns Hopkins University Press, 2010.

Austin, J. L. *How to Do Things with Words: The William James Lectures Delivered at Harvard University in 1955*. London: Oxford University Press, 1962.

Ash, Eric H. *Power, Knowledge, and Expertise in Elizabethan England*. Baltimore: Johns Hopkins University Press, 2004.

Ash, Eric H. "Introduction: Expertise and the Early Modern State." *Osiris* 25, no. 1 (2010): 1–24.

Ash, Eric H. *The Draining of the Fens: Projectors, Popular Politics, and State Building in Early Modern England*. Baltimore: Johns Hopkins University Press, 2017.

Atkinson, Catherine. *Inventing Inventors in Renaissance Europe: Polydore Vergil's* De Inventoribus Rerum. Tübingen: Mohr Siebeck, 2007.

Baarsen, Reinier. "Kistjes van Kick? Hollands lakwerk uit de vroege 17de eeuw." *Bulletin van het Rijksmuseum* 56, no. 1/2 (2008): 12–27.

Bacon, Francis. *The Works of Francis Bacon, Lord Chancellor of England: A New Edition*. Edited by Basil Montagu. 16 vols. London: W. Pickering, 1825–34.

Bacon, Francis. *Of the Proficience and Advancement of Learning*. Edited by Basil Montagu. London: W. Pickering, 1838.

Bailly, Maria Charlotte Le, and Christel Madeleine Odile Verhas. *Hoge Raad van Holland, Zeeland en West-Friesland (1582–1795): De hoofdlijnen van het procederen in civiele zaken voor de Hoge Raad zowel in eerste instantie als in hoger beroep*. Hilversum: Verloren, 2006.

Barrera-Osorio, Antonio. *Experiencing Nature: The Spanish American Empire and the Early Scientific Revolution*. University of Texas Press, 2006.

Baudaert, Willem. *Memoryen ofte Cort Verhael der Gedenck-Weerdichste so kercklicke als werltlicke Gheschiedenissen van Nederland, Vranckrijck, Hooghduytschland, Groot Britannyen, Hispanyen, Italyen, Bohemen, Savoyen, Sevenburghen ende Turkyen: Van den Jaere 1603. tot in het Jaer 1624*. Arnhem: Jan Jansz, 1624.

Baudry, Jérôme. "Examining Inventions, Shaping Property: The Savants and the French Patent System." *History of Science* 57, no. 1 (2019): 62–80.

Bauer, Clemens. "Conrad Peutinger und der Durchbruch des neuen ökonomischen Denkens in der Wende der Neuzeit." In *Augusta 955–1955*, 219–28, 1955.

Baxter, Hugh. "Bringing Foucault into Law and Law into Foucault," *Stanford Law Review* 48, no. 2 (1996): 449–479.

Beauchamp, Christopher. "Patenting Nature: A Problem of History." *Stanford Technology Law Review* 16, no. 2 (2013): 257–312.

Beck, Ulrich. *Risikogesellschaft: auf dem Weg in eine andere Moderne*, Frankfurt am Main: Suhrkamp, 1986.

Becker, Peter, and William Clark, eds. 2000. *Little Tools of Knowledge: Historical Essays on Academic and Bureaucratic Practices*. Ann Arbor, Mich: University of Michigan Press.

Beeckman, Isaac. *Journal tenu par Isaac Beeckman de 1604 à 1634*. 4 vols. The Hague: Martinus Nijhoff, 1939–1953. Edited by Cornelis de Waard. *dbnl.org*. https://www.dbnl.org/tekst/beec002jour00_01/ (accessed 15 April 2021).

Beerninck, Fridericus. *Dissertatio juridica inauguralis. De Privilegiis.* Hardewijk: Johannem Moojen, 1739.

Belfanti, Carlo Marco. "Between Mercantilism and Market: Privileges for Invention in Early Modern Europe." *Journal of Institutional Economics* 2, no. 3 (2006): 319–338.

Belfanti, Carlo Marco. "Guilds, Patents, and the Circulation of Technical Knowledge: Northern Italy During the Early Modern Age." *Technology and Culture* 45, no. 3 (2004): 569–589.

Beltran, Alain, Sophie Chauveau, and Gabriel Galvez-Behar. *Des brevets et des marques. Une histoire de la propriété industrielle.* Paris: Fayard, 2001.

Bennett, James Arthur. *Navigation: A Very Short Introduction.* Oxford: Oxford University Press, 2017.

Bennett, James Arthur. *The Divided Circle: A History of Instruments for Astronomy, Navigation and Surveying.* Oxford: Phaidon - Christie's, 1987.

Benoist, Caspar. *De gehele Conste tmiddel ende Belyet om de sijdewormen te winnen, oock thaeren onderhoude witte moerbeye bomen te sayen, te planten ende op t brengen door der natueren ondersoeck ende veler cloecke geesten ervarenheyt metter tijt vercregen.* N.d. Amsterdam University Library, Ms IC 37.

Bently, Lionel and Brad Sherman, *The Making of Modern Intellectual Property Law: The British Experience* (Cambridge: Cambridge University Press, 1999).

Berger, Adolf. *Encyclopedic Dictionary of Roman Law.* Reprint. Transactions of the American Philosophical Society v. 43, p. 2. Phildelphia: The American Philosophical Society, 1980.

Bergsma, Wiebe. "Willem Lodewijk en het Leeuwarder hofleven." *It Beaken* 60, no. 3/4 (1998): 191–256.

Berlin, Michael. "Guilds in Decline? London Livery Companies and the Rise of a Liberal Economy, 1600–1800." In *Guilds, Innovation, and the European Economy, 1400–1800*, edited by Stephan R. Epstein and Maarten Prak, 316–342, 2008.

Berveglieri, Roberto. *Inventori stranieri a Venezia, (1474–1788): Importazione di tecnologia e circolazione di tecnici artigiani inventori.* Venezia: Istituto veneto di scienze lettere e arti, 1995.

Berveglieri, Roberto. "Tecnologia idraulica olandese in Italia nel secolo XVII: Cornelis Janszoon Meijer a Venezia (gennaio-aprile 1675)." *Studi Veneziani* no. 10 (1998): 81–91.

Beukers, Eelco, and Thimo de Nijs, eds. *Geschiedenis van Holland.* 4 vols. Hilversum: Verloren, 2002–2003.

Bevir, Mark, and R. A. W. Rhodes. *The State as Cultural Practice.* Reprint. Oxford: Oxford University Press, 2015.

Biagioli, Mario. "From Print to Patents: Living on Instruments in Early Modern Europe." *History of Science* 44, no. 2 (2006): 139–186.

Biagioli, Mario. "Patent Republic: Representing Inventions, Constructing Rights and Authors." *Social Research: An International Quarterly of Social Sciences* 73, no. 4 (2006): 1129–1172.

Biagioli, Mario. "Patent Specification and Political Representation: How Patents Became Rights." In *Making and Unmaking Intellectual Property*, edited by Mario Biagioli, Peter Jaszi, and Martha Woodmansee, 25–40. Chicago: University of Chicago Press, 2011.

Biagioli, Mario. *Galileo's Instruments of Credit. Telescopes, Images, Secrecy*. Chicago: University of Chicago Press, 2006.

Biagioli, Mario, and Marius Buning. "Introduction: Technologies of the Law/ Law as a Technology." *History of Science* 57, no. 1 (2019): 3–17.

Bijker, Wiebe E. "How Is Technology Made?—That Is the Question!" *Cambridge Journal of Economics* 34, no. 1 (2010): 63–76.

Bittel, Carla, Christine Von Oertzen, and Elaine Leong, eds. *Working with Paper: Gendered Practices in the History of Knowledge*. Pittsburgh, Pa: University of Pittsburgh Press, 2019.

Blair, Ann. *The Theater of Nature: Jean Bodin and Renaissance Science*. Princeton, NJ: Princeton University Press, 1997.

Blockmans, Wim P. "De representatieve instellingen in het Zuiden 1384–1482." In *Algemene geschiedenis der Nederlanden*, edited by Dirk Peter Blok and others, 4:156–163. Haarlem: Fibula-Van Dishoeck, 1977–1983.

Blockmans, Wim P., André Holenstein, Daniel Schläppi, and Jon Mathieu. *Empowering Interactions: Political Cultures and the Emergence of the State in Europe, 1300–1900*. Burlington, VT: Ashgate, 2009.

Blok, Petrus J., Laurentius Knappert, Friedrich Karl H. Kossmann and Philipp C. Molhuysen, eds. *Nieuw Nederlandsch biografisch woordenboek*. 10 vols. Leiden: Sijthoff, 1911–1937.

Bowker, Geoffrey C., and Susan Leigh Star. *Sorting Things Out: Classification and Its Consequences*. MIT Press, 2000.

Bouman, Jacob. *Bedijking, opkomst en bloei van de Beemster; voorafgegaan door eene beschouwing van den vroegeren toestand van Noord-Holland*. Purmerende: Schuitemaker, 1857.

Bouwsma, William James. "Lawyers in Early Modern Culture." *American Historical Review* 78 (1973): 303–327.

Bossenga, Gail. *The Politics of Privilege: Old Regime and Revolution in Lille*. Cambridge: Cambridge University Press, 1991.

Braddick, M. J, and John Walter. "Grids of Power: Order, Hierarchy and Subordination in Early Modern Society." In *Negotiating Power in Early Modern Society: Order, Hierarchy, and Subordination in Britain and Ireland*, 1–42. Cambridge, UK; New York: Cambridge University Press, 2001.

Braddick, Michael J. *State Formation in Early Modern England, c.1550–1700*. Cambridge: Cambridge University Press, 2000.

Brakensiek, Stefan. "New Perspectives on State-Building and the Implementation of Rulership in Early Modern European Monarchies." In *Structures on the Move*, edited by Antje Flüchter and Susan Richter, 31–41. Berlin: Springer, 2012.

Braudel, Fernand. *Civilization and Capitalism, 15th-18th Century: The Perspective of the World*. Vol. 3. 3 vols. New York: Harper & Row, 1984.

Braunstein, Philippe. "A l'origine des privileges d'invention aux XIVe et XVe siècles." In *Les brevets. Leur utilisation en histoire des techniques et de l'économie*, edited by Francois Caron, 53–60. Paris: Éditions du CNRS.

Brendecke, Arndt, Stefan Ehrenpreis, and Susanne Friedrich, eds. *Transformations of Knowledge in Dutch Expansion. Transformations of Knowledge in Dutch Expansion*. Berlin: De Gruyter, 2015.

Briggs, Martin. "Architectural Models." *Burlington Magazine* 54 (1912): 174–83 and 245–52.

British History Online. "Calendar of the Cecil Papers in Hatfield House: Volume 7, 1597. Ed. R. A. Roberts. London: Her Majesty's Stationery Office, 1899. 500–517." Web: http://www.british-history.ac.uk/cal-cecil-papers/vol7/pp500–517 (last acessed 12 April 2021).

Brodhead, John Romeyn. *Documents Relative to the Colonial History of the State of New York: Procured in Holland, England, and France*. Edited by E. B. O'Callaghan. 15 vols. Albany: Weed, Parsons, 1853.

Broekman, Inge. "Constantijn Huygens, de kunst en het hof." PhD thesis, Universiteit van Amsterdam, 2010.

Brown, Archibald. *A New Law Dictionary and Institute of the Whole Law: For the Use of Students, the Legal Profession, and the Public*. London: Stevens & Hayes, 1880. Web: https://archive.org/details/anewlawdictiona00browgoog (accessed 12 April 2021).

Brown, Harcourt. *Scientific Organizations in Seventeenth Century France (1620–1680)*. Baltimore: Williams & Wilkins, 1934.

Bruinvis, C. W. "Caspar Benoist. Zijdewever te Alkmaar." *Oud-Holland* 12 (1904): 131–135.

Buning, Marius. "Inventing Scientific Method: The Privilege System as a Model for Scientific Knowledge-Production," *Intellectual History Review* 24, no. 1 (2014): 59–70,

Buning, Marius. "Between Imitation and Invention. Inventor Privileges and Technological Progress in the Early Dutch Republic (*c.* 1585–1625)," *Intellectual History Review* 24, no. 3 (2014): 415–27,

Buning, Marius. "Promoting Technical Knowledge. Printing Privileges and Technical Literature in the Early Dutch Republic," in *Le livre et les techniques avant le XXe siècle. À l'échelle du monde*, ed. Liliane Hilaire-Pérez et al. (Paris: CNRS éditions, 2017), 453–62.

Buning, Marius. "Stench and the City. Urban Sensescapes and Technological Innovation in the Early Modern Period," in *Knowledge and the Early Modern City: A History of Entanglements*, ed. Bert de Munck and Antonella Romano (London: Routledge, 2019), 101–25.

Burk, Dan L., and Jessica Reyman. "Patents as Genre: A Prospectus." *Law & Literature* 26, no. 2 (2014): 163–90.

Burton, M. Diane, and Tom Nicholas. "Prizes, Patents and the Search for Longitude." *Explorations in Economic History* 64 (2017): 21–36.

Cambell, Bruce. "The Uses and Exploitation of Human Power from the Thiteenth to the Eighteenth Century." In *Economia e energia secc. XIII-XVIII. Atti della "Trentaquarttresima Settimana di Studi" 15–19 Aprile 2002*, edited by Simonetta Cavaciocchi (Firenze: Le Monnier, 2003), 183–212.

Cambell, James Stuart. "The Alchemical Patronage of Sir William Cecil, Lord Burghley." MA-thesis, Victoria University of Wellington, 2009.

Cau, Cornelis, and others, eds. *Groot Placaet-boeck, inhoudende de Placaten ende Ordonnantien van de Hoogh Mog. Heeren Staten Generael der Vereenighde Nederlanden: ende vande Ed: Groot-Mog. Heeren Staten van Hollandt ende West- Vrieslandt; mitsaders. van de Ed. Mog. Heeren Staten van Zeelandt.* 9 vols. The Hague: Widow of H.J. van Wouw/ Paulus Scheltus, 1658–1796.

Ceredi, Giuseppe. *Tre discorsi sopra il modo d'alzar acque d'Luoghi bassi.* Parma: S. Viotti, 1567.

Clarke, Desmond M. *Descartes. A Biography.* Cambridge: Cambridge University Press, 2006.

Cockshaw, Pierre. "A propos des Pays de par deçà et des Pays de par delà." *Revue belge de Philologie et d'Histoire* 52, no. 2 (1974): 386–88.

Coke, Sir Edward. *The Third Part of the Institutes of the Laws of England: Concerning High Treason, and Other Pleas of the Crown, and Criminal Causes.* London: A. Crooke, 1669.

Coleman, D. C. "Eli Heckscher and the Idea of Mercantilism." *Scandinavian Economic History Review* 5, no. 1 (1957): 3–25.

Con Díaz, Gerardo. *Software Rights: How Patent Law Transformed Software Development in America.* New Haven: Yale UP, 2019.

Constable, Marianne. *Our Word Is Our Bond: How Legal Speech Acts.* Stanford: Stanford University Press, 2014.

Consultatien advysen en advertissementen, gegeven en geschreven bij verscheiden treffelijke rechtsgeleerden in Holland en elders. Vol.3. Rotterdam, Joannes Naeranus, 1662.

Cook, Harold J. "Amsterdam, entrepôt des savoirs au XVIIe siècle." *Revue d'histoire moderne et contemporaine* 55, no. 2 'Sciences et villes-mondes, XVIe-XVIIIe siècles' (2008): 19–42.

Cook, Harold J. *Matters of Exchange: Commerce, Medicine, and Science in the Dutch Golden Age.* New Haven, Conn.: Yale University Press, 2007.

Coombe, Rosemary J. *The Cultural Life of Intellectual Properties: Authorship, Appropriation, and the Law*. Durham: Duke University Press, 1998.

Coster, Abraham. *De grouwelijke ongehoorde blasphemien ende rasernijen van Thomas Leamer Enghelsman, die hem valschelijcknoemt Liefhebber der waerheyt, in syn Clare vertooninge (alsoo hyse noemt)*. Rotterdam: Mathijs Bastiaensen, 1613.

Coren, Jacob Gerritsz. *Observationes Rerum in Senatu Iudicatarum; Item Consilia Quaedam*. The Hague: Aert Meuris, 1633.

Crawford, Catherine. "Legalizing Medicine: Early Modern Legal Systems and the Growth of Medico-Legal Knowledge." In *Legal Medicine in History*, edited by Catherine Crawford and Michael Clark, 89–116. Cambridge: Cambridge University Press, 1994.

Crowley, John E. *The Invention of Comfort: Sensibilities and Design in Early Modern Britain and Early America*. Baltimore, Md.: Johns Hopkins University Press, 2003.

Creutz, Hans-Jürgen. "Die Herausbildung des Erfindungsschutzes in Sachsen im 15. und 16. Jahrhundert." *Jahrbuch für Wirtschaftsgeschichte* 2 (1983): 91–110.

Daston, Lorraine, and Michael Stolleis, eds. *Natural Law and Laws of Nature in Early Modern Europe*. Aldershot: Ashgate, 2006.

Daston, Lorraine, and Klaus Krüger, eds. *Curiositas: Welterfahrung und ästhetische Neugierde in Mittelalter und Früher Neuzeit*. Göttingen: Wallstein, 2002.

Davids, Karel. "Beginning Entrepeneurs and Municipal Governements in Holland at the Time of the Dutch Republic." In *Entrepeneurs and Entrepreneurship in Early Modern Times. Merchants and Industrialists Within the Orbit of the Dutch Staple Market*, edited by Clé Lesger and Leo Noordegraaf, 24:167–183. Den Haag: Stichting Hollandse Historische Reeks, 1995.

Davids, Karel. "Openness or Secrecy? Industrial Espionage in the Dutch Republic." *The Journal of European Economic History* 24, no. 2 (1995): 333–348.

Davids, Karel. "Public Knowledge and Common Secrets. Secrecy and Its Limits in the Early-Modern Netherlands." *Early Science and Medicine* (2005): 411–427.

Davids, Karel. "Shifts of Technological Leadership in Early Modern Europe." In *A Miracle Mirrored: The Dutch Republic in European Perspective*, edited by Karel Davids and Jan Lucassen (Cambridge: Cambridge University Press, 2010), 329–365.

Davids, Karel. "Guilds, Guildsmen and Technological Innovation in Early Modern Europe: The Case of the Dutch Republic." *Economy and Society of the Low Countries Working Papers* 2 (2003): 2003–2.

Davids, Karel. "Innovations in Windmill Technology in Europe, ca. 1500–1800." In *Economia e energia secc. XIII-XVIII. Atti della "Trentaquarttresima Settimana di Studi" 15–19 Aprile 2002*, edited by Simonetta Cavaciocchi (Firenze: Le Monnier, 2003), 271–291.

Davids, Karel. "Patents and Patentees in the Dutch Republic Between *c.* 1580 and 1720." *History and Technology* 16, no. 3 (2000): 263–284.

Davids, Karel. *The Rise and Decline of Dutch Technological Leadership: Technology, Economy and Culture in the Netherlands, 1350–1800*. Leiden: Brill, 2008.

Davids, Karel. *Zeewezen en wetenschap: De wetenschap en de ontwikkeling van de navigatietechniek in Nederland tussen 1585 en 1815*. Amsterdam: De Bataafsche Leeuw, 1986.

Davids, Karel, and Jan Lucassen. "Conclusie." In *Een wonder weerspiegeld. De Nederlandse Republiek in Europees perspectief*, edited by Karel Davids and Jan Lucassen, 409–431. Trans. Amsterdam: Aksant, 2005.

Davies, D. Seaborne. "The Early History of the Patent Specification." *Law Quarterly Review* 50 (1934): 260–274.

Davies, D. Seaborne. "The Early History of the Patent Specification." *Law Quarterly Review* 50 (1934): 86–109.

Davis, John A. "The European Economies in the Eighteenth Century." In *Economic History of Europe*, edited by Antonio Di Vittorio, 92–129. London: Routledge, 2006.

Dear, Peter. "Mysteries of State, Mysteries of Nature: Authority, Knowledge and Expertise in the Seventeenth Century." In *States of Knowledge: The Co-Production of Science and the Social Order*, edited by Sheila Jasanoff, 206–24. London: Routledge, 2004.

De Albuquerque, Louis. "Portuguese Navigation: Its Historical Development." In *Circa 1492: Art in the Age of Exploration*, edited by Jay A. Levenson, 35–39. New Haven: Yale University Press, 1991.

De Bruin, G. "De soevereiniteit in de Republiek. Een machtsprobleem." *BMGN* 94 (1979): 27–40.

De Bom, Erik, Randall Lesaffer, and Werner Thomas, eds. *Early Modern Sovereignties: Theory and Practice of a Burgeoning Concept in the Netherlands*. Leiden: Brill, 2020.

De Damhouder, Joost. *Practycke in civile saecken, seer nut, profijtelijck ende nodigh allen schouten, borghemeesteren, magistraten ende andere rechteren*. 1626. Antwerp: Kluwer, 1999.

De Damhouder, Joost. *Praxis Rerum Criminalium Iconibus Illustrata*, 1562.

De Groot, Hugo. *Inleydinge tot de Hollandsche regts-geleertheyt*. Delft: Arnold Bon, 1664.

De Jong, Michiel. *"Staat van oorlog": Wapenbedrijf en militaire hervorming in de Republiek der Verenigde Nederlanden, 1585–1621*. Hilversum: Verloren, 2005.

De Jonge, C. "Franciscus Junius (1545–1602) en de Engelse separatisten te Amsterdam." *Nederlands Archief voor Kerkgeschiedenis* 59, no. 2 (1979): 133–159.

De Molina, Luis. *R.P. Lodovici Molinae ... de justitia et jure opera omnia*. Vol. 2. 5 vols. Coloniae Allobrogum: Bousquet, 1733.

De Munck, Bert, and Karel Davids, "Innovation and Creativity in Late Medieval and Early Modern European Cities: An Introduction," in *Innovation and Creativity in Late Medieval and Early Modern European Cities*, ed. Karel Davids and Bert de Munck (Burlington, VT: Ashgate, 23), 1–34.

De Munck, Bert, Piet Lourens, and Jan Lucassen. "The Establishment and Distribution of Craft Guilds in the Low Countries, 1000–1800." In *Craft Guilds in the Early Modern Low Countries: Work, Power, and Representation*, 32–73. Aldershot: Ashgate, 2006.

De Mullenheim, Marie. "Les privilèges pour invention à Florence à la fin XVIe et au début XVIIe Siècle." In *Les archives de l'invention. Écrits, objets et images de l'activité inventive*, edited by Marie-Sophie Corcy, Christinane Demeulenaere-Douyère, and Liliane Hilaire-Pérez (Toulouse: CNRS, 2003), 333–339.

De Renzi, Silvia. "Medical Expertise, Bodies, and the Law in Early Modern Courts." *Isis* 98, no. 2 (2007): 315–22.

De Roever, N. "De kroniek van Staets. Een bladzijde uit de geschiedenis van het fabriekambt der stad Amsterdam. 1594–1628." 1886. *Jaarverslag Koninklijke Oudheidkundig Genootschap* 28 (1980).

De Roover, Raymond. "The Concept of the Just Price: Theory and Economic Policy." *The Journal of Economic History* 18, no. 4 (1958): 418–434.

De Roover, Raymond. "Scholastic Economics: Survival and Lasting Influence from the Sixteenth Century to Adam Smith." *The Quarterly Journal of Economics* 69, no. 2 (1955): 161–190.

De Ruysscher, Dave. "'Naer het Romeinsch recht alsmede den stiel mercantiel'. Handel en recht in de Antwerpse rechtbank (16de-17de eeuw)." PhD thesis, University of Brussels, 2009; Kortrijk: UGA, 2009.

De Schepper, Hugo. "Beleid en bestuur inzake de hoge overheidsfinanciën in het Bourgondische-Habsburgse Nederland." In *Van tresorier tot thesaurier-generaal: Zes eeuwen financieel beleid in handen van een hoge Nederlandse ambtsdrager*, 17–52. Hilversum: Verloren, 1996.

De Vos, Annemie. "Popaganda voor staat en huis. Maurits en de bouwkunst." In *Maurits Prins van Oranje: Dienaar met vorstelijke allure*, edited by K. Zandvliet, 123–136. Zwolle: Waanders, 2000.

De Vries, Jan. *Economy of Europe in an Age of Crisis, 1600–1750*. Cambridge: Cambridge University Press, 1976.

De Vries, Jan. "How Did Pre-Industrial Labour Markets Function?" In *Labour Market Evolution. The Economic History of Market Intergration, Wage Flexibility and the Emplyment Relation*, edited by George Grantham and Mary MacKinnon, 39–62. London and New York: Routledge, 1994.

De Vries, Jan. "Luxury in the Dutch Golden Age in Theory and Practice." In *Luxury in the Eighteenth Century: Debates, Desires and Delectable Goods*, edited by Maxine Berg and Elizabeth Eger, 41–57. London: Palgrave Macmillan, 2003.

De Vries, Jan, and Ad van der Woude. *The First Modern Economy: Success, Failure, and Perseverance of the Dutch Economy, 1500–1815*. Cambridge: Cambridge University Press, 1997.

De Vries, Simon. *Historische oeffeningen ontrent allerley aenmercklijcke, gewichtige, nauwkeurige en gemoed-vervrolijckende stoffen: doorvloghten, met een goed deel bysonderheden van natuer en konst.* Amsterdam: Aert Dircksz Ooszaen, 1686.

Dekker, E. *Rijker dan Midas: Vrijheid, genade en predestinatie in de theologie van Jacobus Arminius (1559–1609).* Zoetermeer: Boekencentrum, 1993.

Den Tex, Jan. *Oldenbarnevelt.* 5 vols. Haarlem: Tjeenk Willink, 1960–72.

Der Weduwen, Arthur. *Dutch and Flemish Newspapers of the Seventeenth Century, 1618–1700.* Leiden: Brill, 2017.

Deyon, P., and P. Guignet. "The Royal Manufactures and Economic and Technological Progress in France before the Industrial Revolution." *Journal of European Economic History* 9, no. 3 (1980): 611–632.

Devreese, J. T, and Guido Vanden Berghe. *"Magic Is No Magic": The Wonderful World of Simon Stevin.* Southampton, UK: WIT, 2008.

Dodt van Flensburg, Jens J. *Archief voor kerkelijke en wereldsche geschiedenissen, inzonderheid van Utrecht.* 7 vols. Utrecht: N. van der Monde, 1838–1848.

Dölemeyer, Barbara, and Heinz Mohnhaupt, eds. *Das Privileg im europäischen Vergleich.* 2 vols. Frankfurt am Main: Klostermann, 1997–1999.

Doorman, Gerard. *Octrooien voor uitvindingen in de Nederlanden uit de 16e-18e eeuw: Met bespreking van enkele onderwerpen uit de geschiedenis der techniek.* The Hague: Nijhoff, 1940.

Doorman, Gerard. *Octrooien voor uitvindingen in de Nederlanden uit de 16e-18e eeuw: Met bespreking van enkele onderwerpen uit de geschiedenis der techniek. Eerste aanvullingen.* The Hague: Nijhoff, 1942.

Doorman, Gerard. *Patents for Inventions in the Netherlands during the 16th, 17th and 18th Centuries: With Notes on the Historical Development of Technics.* Nijhoff, 1942.

Doorman, Gerard. *Techniek en octrooiwezen in hun aanvang: Geschiedkundige aanvullingen bij I. Octrooien voor uitvindingen in de Nederlanden uit de 16e-18e eeuw; II. Het Nederlandsch octrooiwezen en de techniek der 19e eeuw.* The Hague: Nijhoff, 1953.

Drahos, Peter. *A Philosophy of Intellectual Property.* Aldershot: Dartmouth, 1996.

Duffy, J. F. "Rethinking the Prospect Theory of Patents." *U. Chi. L. Rev.* 71 (2004): 439–510.

Dumolyn, Jan. *Staatsvorming en vorstelijke ambtenaren in het graafschap Vlaanderen (1419–1477).* Antwerp: Garant, 2003.

Duncan, Gregor Duthie. "Monopolies Under Elizabeth I, 1558–1585." PhD thesis, Cambridge University, 1977.

Dutton, Harold. I. *The Patent System and Inventive Activity During the Industrial Revolution, 1750–1852.* Manchester: Manchester University Press, 1984.

Duve, Thomas. *Sonderrecht in der Frühen Neuzeit: Studien zum ius singulare und den privilegia miserabilium personarum, senum und indorum in Alter und Neuer Welt.* Frankfurt am Main: Klostermann, 2008.

Duyvendak, J. J. L. "Simon Stevin's 'Sailing-Chariot.'" *T'oung Pao* 36, no. 3/5 (1942): 401–7.

Eggen, J. L. M. *De invloed door Zuid-Nederland op Noord-Nederland uitgeoefend op het einde der XVIe en het begin der XVIIe eeuw*. Gent: Siffer, 1908.

Elias, Johan E. *De vroedschap van Amsterdam, 1578–1795*. 2 vols. Amsterdam: Israel, 1963.

Epstein, Stephan R, and Maarten Prak. "Introduction." In *Guilds, Innovation, and the European Economy, 1400–1800*. Cambridge: Cambridge University Press, 2008.

Epstein, Stephan R. *Freedom and Growth: The Rise of States and Markets in Europe, 1300–1750*. London: Routledge, 2000.

Evans, Michael. "The Geometry of the Mind." *Architectural Association Quarterly* 12 (1980): 32–55.

Evenhuis, Rudolf B. *Ook dat was Amsterdam, II: De kerk der hervorming in de Gouden Eeuw*. Amsterdam: Ten Have, 1967.

Federico, P. J. "Origin and Early History of Patents." *JPOS* (1929): 292–305.

Feingold, Mordechai. "Projectors and Learned Projects in Early Modern England," The Seventeenth Century 32, no. 1 (2017): 63–79.

Feingold, Mordechai, Joseph S Freedman, and Wolfgang Rother, eds. *The Influence of Petrus Ramus: Studies in Sixteenth and Seventeenth Century Philosophy and Sciences*. Basel: Schwabe, 2001.

Felten, Sebastian, and Christine von Oertzen. "Bureaucracy as Knowledge." *Journal for the History of Knowledge* 1, no. 1 (2020): 1–16.

Ferguson, E. S. *Engineering and the Mind's Eye*. Cambridge, Mass: MIT Press, 1992.

Ferguson, Stephen. "System and Schema: Tabulae of the Fifteenth to the Eighteenth Centuries." *Princeton University Library Chronicle* 49 (1987): 9–30.

Finlay, Robert. *The Pilgrim Art: Cultures of Porcelain in World History*. University of California Press, 2010.

Flechsig, Alexander J. *Frühneuzeitlicher Erfindungsschutz: eine Untersuchung unter besonderer Berücksichtigung der Reichsstadt Augsburg*. Münster: Lit Verlag, 2013.

Fleischer, Alette. "The Beemster Polder: Conservative Invention and Holland's Great Pleasure Garden." In *The Mindful Hand: Inquiry and Invention from the Late Renaissance to Early Industrialisation*, edited by Lissa Roberts, Simon Schaffer, and Peter Dear, 145–66. Amsterdam: KNAW, 2007.

Fleming, James Dougal, ed. *The Invention of Discovery, 1500–1700*. Burlington, VT: Ashgate, 2011.

Fockema Andreae, Sybrandus Johannes. *De Nederlandse staat onder de Republiek*. 7th ed. Amsterdam: Noord-Hollandsche Uitgevers Maatschappij, 1975.

Fortunati, Maura. "La lex mercatoria nella tradizione e nella recente ricostruzione storico-giuridica." *Sociologia del diritto* 32, no. 2 (2005): 29–41.

Foucault, Michel. "The Incorporation of the Hospital in Modern Technology." In *Space, Knowledge and Power: Foucault and Geography*, edited by Stuart Elden and J.W. Cramton, translated by Edgar Knowlton Jr., New ed., 141–52. Burlington, VT: Routledge, 2007.

Franceschelli, Remo. "Lo origini e lo svolgimento del diritto industriale nei primi secoli dell'arte della stampa." *Riv. Dir. Ind.* 1 (1952): 151–195.

Franklin, James. *The Science of Conjecture: Evidence and Probability Before Pascal*. Baltimore: Johns Hopkins University Press, 2001.

Friedrich, Susanne. "The Importance of Being a Good Employee: Georg Everhard Rumphius, the Dutch East India Company, and Knowledge in the Late Seventeenth Century." *Early Modern Low Countries* 3, no. 2 (2019): 183–207.

Fruin, Robert Jacobus. *Geschiedenis der staatsinstellingen in Nederland tot den val der Republiek*. 2nd revised edition by Herman Theodoor Colenbrander, 1922. The Hague: Nijhoff, 1980.

Frumkin, M. "Early History of Patents for Invention." In *Proceedings of the Newcomen Society*, 26:47–56. The Newcomen Society, 1947.

Frumkin, M. "Les anciens brevets d'invention: Les pays du continent européen au XVIIe siècle." *Archives internationales d'histoire des sciences* 7 (1954): 315–323.

Führer, Jochen A. *Die Kirchen- und die antireformatorische Religionspolitik Kaiser Karls V. in den siebzehn Provinzen der Niederlande, 1515–1555*. Leiden: Brill, 2004.

Gaastra, F. S. *De geschiedenis van de VOC*. Leiden: Walburg Pers, 1991.

Gabbey, Alan. "Between Ars and Philosophia Naturalis: Reflections on the Historiography of Early Modern Mechanics." In *Renaissance and Revolution: Humanists, Scholars, Craftsmen and Natural Philosophers in Early Modern Europe*, edited by Judith Veronica Field and Frank A.J.L. James, 133–146. Cambridge: Cambridge University Press, 1993.

Gabbey, Alan. "The Mechanical Philosophy and Its Problems: Mechanical Explanations, Impenetrability, and Perpetual Motion." In *Change and Progress in Modern Science*, edited by Joseph C. Pitt, 9–84. Dordrecht: Springer, 1985.

García Tapia, Nicolás. *Patentes de invención españolas en el siglo de oro*. Madrid: Ministro de Industria y Energia centro de Publicaciones, 1990.

García Tapia, Nicolás. *Tecnica y poder en Castilla durante los siglos XVI y XVII*. Estudios de Historia de la Ciencia y de La Técnica, No. 3. Salamanca: Junta de Castilla y León, Consejería de Cultura y Bienestar Social, 1989.

Garnier, Edouard. *Histoire de la verrerie et de l'emaillerie*. Tours: A. Mame, 1886.

Gelderblom, Oscar. *Cities of Commerce: The Institutional Foundations of International Trade in the Low Countries, 1250–1650*. Princeton: Princeton Univers. Press, 2013.

Gerbino, Anthony. *François Blondel: Architecture, Erudition, and the Scientific Revolution*. London: Routledge, 2010.

Giddens, Anthony. *Modernity and Self-Identity: Self and Society in the Late Modern Age*. 1991 first edition. Cambridge: Polity Press, 2010.

Gilbert, Neal W. *Renaissance Concepts of Method*. New York: Columbia University Press, 1963.

Girard, Albert. *Invention nouvelle en l'algèbre*. Edited by Bierens de Haan. Amsterdam: Guillaume Iansson Blaeuw, 1629. Leiden: Muré frères, 1884.

Gitelman, Lisa. *Paper Knowledge: Toward a Media History of Documents*. Durham: Duke University Press, 2014.

Gleitsmann, Rolf-Jürgen. "'Wir wissen aber, Gott lob, was wir thuen': Erfinder-privilegien und technologischer Wandel im 16. Jahrhundert." *Zeitschriftf für Unternehmensgeschichte* 30 (1985): 690–95.

Godin, Benoît. *Innovation Contested: The Idea of Innovation Over the Centuries* (New York, NY: Routledge, 2015).

Godin, Benoît. *Models of Innovation: The History of an Idea*. Cambridge MA: MIT Press, 2017.

Golan, Tal. *Laws of Men and Laws of Nature: The History of Scientific Expert Testimony in England and America*. Cambridge MA: Harvard University Press, 2004.

Golder, Ben and Peter Fitzpatrick, *Foucault's Law* (Abingdon: Routledge, 2009).

Goldthwaite, Richard A. *The Building of Renaissance Florence: An Economic and Social History*. Baltimore: Johns Hopkins University Press, 1982.

Gomme, Allan. *Patents of Invention: Origin and Growth of the Patent System in Britain*. London: Longmans, 1946.

Goodman, David C. *Power and Penury: Government, Technology, and Science in Philip II's Spain*. Cambridge: Cambridge University Press, 1988.

Goodman, David C. "Philip II's Patronage of Science and Engineering." *The British Journal for the History of Science* 16, no. 1 (1983): 49–66.

Gordon, Barry. *Economic Analysis Before Adam Smith: Hesiod to Lessius*. London: Macmillan, 1975.

Goris, J. A. *Étude sur les colonies marchandes méridionales (Portugais, Espagnols, Italiens) à Anvers de 1488 à 1567. Contribution à l>histoire des débuts du Capitalisme moderne*. Louvain: Louvain Librairie Universitaire, 1925.

Grafton, Anthony. *New Worlds, Ancient Texts: The Power of Tradition and the Shock of Discovery*. Cambridge MA: Harvard University Press, 1995.

Greenberg, Abraham S. "The Ancient Lineage of Trade-Marks." *Journal of the Patent Office Society* 33 (1951): 876–887.

Greene, Thomas M. *The Light in Troy: Imitation and Discovery in Renaissance Poetry*. New Haven, Conn.: Yale Univiversity Press, 1986.

Groenveld, S., and J.B. den Hartog. "Twee musici, twee stromingen. Een boekoctrooi voor Anthoni van Noordt en een advies van Constantijn Huygens, 1659." In *Veelzijdigheid als levensvorm: Facetten van Constantijn Huygens' leven en werk*, edited by A. Th. van Deursen, E.K. Grootes, and P.E.L. Verkuyl, 109–27. Deventer: Sub Rosa, 1987.

Haak, Sikko P. *Paullus Merula, 1558–1607*. Zutphen: Thieme, 1901.

Harding, Alan. "The Origins of the Concept of the State." *History of Political Thought* 15, no. 1 (1994): 57–72.

Haitsma Mulier, Eco O. G.. *The Myth of Venice and Dutch Republican Thought in the Seventeenth Century*. Van Gorcum, 1980.

Hall, Bert S. "The Didactic and the Elegant: Some Thoughts on Scientific and Technological Illustrations in the Middle Ages and Renaissance." In *Historical and Philosophical Problems Concerning the Use of Art in Science*, edited by B. S. Baigrie, 3–39. Toronto: Toronto University Press, 1996.

Handvesten ende privilegien der Graven van Hollandt, mitsgaders autentycke stucken, behelsende 't recht, dat beyde de Hoven van Justitie, alsmede de particuliere steden ende collegien in Holland ende West-vrieslant, is competerende, in cas van judicature. Soo in kerckelycke als politycke saecken. Waer van veele noyt in druck geweest. The Hague: Johannes Tongerloo, 1663.

't Hart, Marjolein C. "Cities and Statemaking in the Dutch Republic, 1580–1680." *Theory and Society* 18, no. 5 (1989): 663–687.

't Hart, Marjolein C. *The Making of a Bourgeois State: War, Politics, and Finance during the Dutch Revolt.* Manchester: Manchester University Press, 1993.

't Hart, Marjolein C. *The Dutch Wars of Independence: Warfare and Commerce in the Netherlands 1570–1680* (London: Routledge, 2014).

Harkness, Deborah E. *The Jewel House: Elizabethan London and the Scientific Revolution.* New Haven, Conn.: Yale University Press, 2007.

Hartig, E. "Zur Statistik der Erfindungsprivilegien im Königreiche Sachsen." In *Der Civilingenieur*, columns (not pages!) 51–66, 365–396, 563–576. Leipzig, 1878.

Heal, Felicity, and Clive Holmes. "The Economic Patronage of William Cecil." In *Patronage, Culture and Power. The Early Cecils*, edited by Pauline Croft, 199–229. New Haven, Conn.: Yale University Press, 2002.

Heckscher, Eli F. *Mercantilism.* 2 vols. First edition in English. London: George Allen & Unwin, 1935.

Heller, Henry. *Labour, Science and Technology in France, 1500–1620.* Cambridge: Cambridge University Press, 2002.

Hilaire-Pérez, Liliane. *L'invention technique au siècle des Lumières.* Paris: Albin Michel, 2000.

Hill, Thomas A. "Origin and Development of Letters Patent for Invention." *Journal of the Patent Office Society* 6 (1924): 405–422.

Höffner, Joseph. *Wirtschaftsethik und Monopole im fünfzehnten und sechzehnten Jahrhundert.* Jena: Fischer, 1941.

Howell, Martha C. *Commerce before Capitalism in Europe, 1300–1600.* Cambridge: Cambridge University Press, 2010.

Huff, Toby E. *The Rise of Early Modern Science: Islam, China and the West.* Cambridge: Cambridge University Press, 2003.

Hulme, E.W. "The History of the Patent System Under the Pregorative and the Common Law." *Law Quarterly Review* 12, no. 141 (1896): 141–154.

Hulme, E.W. "The History of the Patent System Under the Pregorative and the Common Law: A Sequal." *Law Quarterly Review* 16, no. 44 (1900): 44–56.

Hunt, Alan and Gary Wickham, *Foucault and Law: Towards a Sociology of Law as Governance* (London: Pluto Press, 1994).

Hurx, Merlijn. "Architecten en gildendwang. Vernieuwingen in het ontwerpberoep in de 15de en 16de eeuw." *Bulletin van de Koninklijke Nederlandse Oudheidkundige Bond* 108 (2008): 1–18.

Hurx, Merlijn. "The Most Expert in Europe: Patents and Innovation in the Building Trades in the Early Dutch Republic (1580–1650)," *Architectural Histories* 7, no. 1 (2019). Web: http://doi.org/10.5334/ah.337.

Iliffe, Rob. "'In the Warehouse': Privacy, Property and Priority in the Early Royal Society." *History of Science* 30 (1992): 29–67.

Instructiën vanden Hove van Hollandt, Zeelandt, ende Vrieslandt. The Hague: Jacobus Scheltus, ca. 1699.

Isoré, Jaques. "De L'existence des brevets d'invention en droit français avant 1791." *Revue historique de droit français et étranger* 4, no. 16 (1937): 94–130.

Israel, Jonathan Irvine. *Dutch Primacy in World Trade, 1585–1740*. Oxford: Oxford University Press, 1989.

Israel, Jonathan Irvine. *The Dutch Republic. Its Rise, Greatness, and Fall, 1477–1806.* Oxford: Clarendon Press, 1995.

Jacob, Margaret C. *The First Knowledge Economy: Human Capital and the European Economy, 1750–1850*, 2014.

Jasanoff, Sheila. *Science at the Bar: Law, Science, and Technology in America*. Cambridge, Mass.: Harvard Univ. Press, 1995.

Janssen, Geert H. *Het stokje van Oldenbarnevelt*. Hilversum: Verloren, 2001.

Janssen, Geert H. *Princely Power in the Dutch Republic: Patronage and William Frederick of Nassau (1613–64)*. Manchester: Manchester University Press, 2008.

Janssen, Noud. "Maria van Oosterwijck." Web: https://mariavanoosterwijck.nl/sites/default/files/stambomen/vLinschoten%20stamboom.pdf (last accessed 20 April 2021)

Jonkers, A. R. T. "Parallel Meridians: Diffusion and Change in Early-Modern Oceanic Reckoning." In *Noord-Zuid in Oostindisch perspectief*, edited by J. Parmentier, 17–42. Zutphen: Walburg Pers, 2005.

Kaptein, Herman. "Een unieke economie? De Zaanse nijverheid op windkracht 1600–1800." In *Geschiedenis van de Zaanstreek*, edited by Eelco Beukers, C. van Sijl, and Piet de Rooy, Vol. 1. Zwolle: Wbooks, 2012.

Khan, B. Zorina. "Inventing Prizes: A Historical Perspective on Innovation Awards and Technology Policy." *Business History Review* 89, no. 4 (2015): 631–60.

Khan, B. Zorina. *The Democratization of Invention: Patents and Copyrights in American Economic Development, 1790–1920*. Cambridge: Cambridge University Press, 2005.

Kander, Astrid, Paolo Malanima, and Paul Warde. *Power to the People: Energy in Europe over the Last Five Centuries*. Princeton, N.J: Princeton UP, 2014.

Kaptein, Herman. "Passchier Lammertijn. Uitvinder of bedrieger?" In *Artistiek en ambachtelijk. Architectuur, kunsten en nijverheid in Alkmaar, 14e-20e eeuw*, edited by Leo Noordegraaf, 57–77. Hilversum: Verloren, 1999.

Kaptein, Herman, and P. Schotsman. "Alkmaar als bakermat van de Hollandse indrustriemolens omstreeks 1600." In *Alkmaar, stad en regio: Alkmaar en omgeving in de Late Middeleeuwen en Vroegmoderne Tijd*, edited by Leo Noordegraaf, 183–226. Hilversum: Verloren, 2004.

Kaspersen, Lars Bo, and Jeppe Strandsbjerg, eds. *Does War Make States? Investigations of Charles Tilly's Historical Sociology*. Cambridge: Cambridge University Press, 2017.

Keller, A. "A Renaissance Humanist Looks at 'New' Inventions: The Article 'Horologium' in Giovanni Tortelli's 'De Orthographia'." *Technology and Culture* 11, no. 3 (1970): 345–365.

Keller, Vera. "Mining Tacitus: Secrets of Empire, Nature and Art in the Reason of State." *The British Journal for the History of Science* 45 (2012): 189–212.

Keller, Vera. *Knowledge and the Public Interest, 1575–1725*. Cambridge: Cambridge University Press, 2015.

Keller, Vera, and Ted McCormick. "Towards a History of Projects." *Early Science and Medicine* 21, no. 5 (2016): 423–44.

Kelly, Tobias. "Legal Anthropology." *Oxford Bibliographies*. Web: http://www.oxfordbibliographies.com/view/document/obo-9780199766567/obo-9780199766567-0049.xml (last accessed 6 April 2021).

Kevles, Daniel J. "Inventions, Yes; Nature, No: The Products-of-Nature Doctrine from the American Colonies to the U.S. Courts." *Perspectives on Science* 23, no. 1 (2014): 13–34.

Kevles, Daniel J. "Patents, Protections, and Privileges: The Establishment of Intellectual Property in Animals and Plants." *Isis* 98, no. 2 (2007): 323–31.

Kenny, Neil. *The Uses of Curiosity in Early Modern France and Germany*. Oxford: Oxford University Press, 2004.

Kernkamp, G. W. "Memoriën van ridder Theodorus Rodenburg betreffende het verplaatsen van verschillende industriën uit Nederland naar Denemarken met daarop genomen resolutiën van Koning Christiaan IV (1621)." *Bijdragen en Mededelingen van het Historisch Genootschap* 23 (1902): 189–257.

Keuning, Johannes. *Petrus Plancius: Theoloog en geograaf 1552–1622*. Amsterdam: Van Kampen, 1946.

Khan, B. Zorina. *The Democratization of Invention: Patents and Copyrights in American Economic Development, 1790–1920*. Cambridge: Cambridge University Press, 2009.

Kitch, Edmund W. "Elementary and Persistent Errors in the Economic Analysis of Intellectual Property." *53 Vand. L. Rev.* 1727 (2000).

Kitch, Edmund W. "The Nature and Function of the Patent System." *Journal of Law and Economics* 20, no. 2 (1977): 265–290.

Klein, P. W. "De Nederlandse handelspolitiek in de tijd van het mercantilisme: Een nieuwe kijk op een oude kwestie." *Tijdschrift voor de geschiedenis* 102 (1989): 189–212.

Klein, P. W. *De Trippen in de 17e eeuw: Een studie over het ondernemersgedrag op de Hollandse stapelmarkt*. London: Routledge, 1999.

Klein, Ursula. "Introduction: Technoscientific Productivity." *Perspectives on Science* 13, no. 2 (2005): 139–141.

Klein, Ursula. "Technoscience Avant La Lettre." *Perspectives on Science* 13, no. 2 (2005): 226–266.

Klein, Ursula. "Hybrid Experts." In *The Structures of Practical Knowledge*, edited by Matteo Valleriani, 287–306. Cham: Springer International, 2017.

Knevel, Paul. *Het Haagse bureau: Zeventiende-eeuwse ambtenaren tussen staatsbelang en eigenbelang*. Amsterdam: Prometheus/Bakker, 2001.

Kocher, Paul H. "Francis Bacon on the Science of Jurisprudence." *Journal of the History of Ideas* 18, no. 1 (1957): 3–26.

Koenigsberger, Helmut G. *Estates and Revolutions: Essays in Early Modern European History*. Cornell University Press, 1971.

Koenigsberger, Helmut G. *Monarchies, States Generals and Parliaments: The Netherlands in the Fifteenth and Sixteenth Centuries*. New York: Cambridge University Press, 2001.

Koenigsberger, Helmut G. "The Power of the Deputies in Sixteenth-Century Assemblies." In *Estates and Revolutions: Essays in Early Modern European History*, 176–210. Studies Presented to the International Commission for the History of Representative and Parliamentary Institutions, vol. 40. Ithaca: Cornell University Press, 1971.

Koldeweij, E. F. "Goudleer in de Republiek der Zeven Verenigde Provinciën: Nationale ontwikkelingen en de Europese context." PhD thesis, Leiden University, 1998.

Kossmann, E.F. "Hendrick de Keyser als uitvinder." *Oud Holland* 46 (1929): 284–288.

Kossmann, E.F. "Volkssouvereiniteit aan het begin van het Nederlandse ancien régime." BMGN 95, no. 1 (1980): 1–34.

Kossmann, Ernst H. "Dutch republicanism." In *Età dei lumi : studi storici sul Settecento europeo in onore di Franco Venturi*, 455–86. Naples, 1985.

Krajewski, Markus. *Paper Machines: About Cards & Catalogs, 1548–1929*. Translated by Peter Krapp. Cambridge, Mass: The MIT Press, 2011.

Kubbinga, H. H. "Stevin en Maurits." *De zeventiende eeuw* 10, no. 1 (1994): 93–102.

Leamer, Thomas. *Een klaer vertoninge, hoe men door het Vyrwerck van Elohim, ofte den groten Eyghenaer, Namelijck: Door Son, Maen en Sterren in alle plaetsen der Werelt, te Water oft te Lande zijn essen Tijt, ende so sijn Meridiaens lengde te weten comen can. Mitsgaders een klaer bevrijdinghe der selven, teghen de Adversarien, wiens namen op de ander zijde des bladts uyt gedruckt staen*. Kampen: Albert Lieffertsz., 1612.

Lakatos, Imre. "History of Science and its Rational Reconstructions." *PSA: Proceedings of the Biennial Meeting of the Philosophy of Science Association 1970* (1970): 91–136.

Lamberini, Daniela. "Inventori di macchine e privilegi cinque-seicentechi dall'Archivio Fiorentino delle Riformagioni." *Journal de la Renaissance* 3 (2005): 177–192.

Landes, David S. *Revolution in Time: Clocks and the Making of the Modern World.* Cambridge, Mass.: Belknap, 1983.

Langholm, Odd. "Monopoly and Market Irregularities in Medieval Economic Thought: Traditions and Texts to A.D. 1500." *Journal of the History of Economic Thought* 28, no. 4 (2006): 395–411.

Latour, Bruno. *An Inquiry into Modes of Existence: An Anthropology of the Moderns.* Cambridge, Massachusetts: Harvard University Press, 2013.

Latour, Bruno. *Reassembling the Social: An Introduction to Actor-Network-Theory.* Oxford: Oxford University Press, 2008.

Latour, Bruno. *The Making of Law: An Ethnography of the Conseil d'Etat.* Cambridge: Polity, 2010.

Lawne, Christopher, *The Prophane Schisme of the Brovvnists or Separatists: With the Impietie, Dissensions, Levvd, and Abhominable Vices of That Impure Sect. Discouered by Christopher Lavvne, Iohn Fovvler, Clement Sanders, Robert Bulvvard. Lately Returned from the Companie of M. Iohnson, That Wicked Brother, into the Bosome of the Church of England, Their True Mother*. London: Printed by William Stansby for Walter Burre, 1612.

Lemire, B., and G. Riello. "East & West: Textiles and Fashion in Early Modern Europe." *Journal of Social History* 41, no. 4 (2008): 887–916.

Leng, Rainer. "Social Character, Pictorial Style, and the Grammar of Technical Illustration in Craftsmen's Manuscripts in the Late Middle Ages." In *Picturing Machines, 1400–1700*, edited by Wolfgang Lefèvre, 85–114. Cambridge, Mass: MIT Press, 2004.

Lesger, Clemens M. *The Rise of the Amsterdam Market and Information Exchange; Merchants, Commercial Expansion and Change in the Spatial Economy of the Low Countries, c.1550–1630*. Aldershot, 2006.

Lefèvre, Wolfgang. "Architectural Knowledge." In *The Structures of Practical Knowledge, edited by Matteo Valleriani*. Cham: Springer International, 2017.

Lincoln, Evelyn. "The Jew and the Worms: Portraits and Patronage in a 16th Century How-to Manual." *Word & Image* 19, no. 1&2 (2003): 86–99.

Loevinger, Lee. "Law and Science as Rival Systems." *Jurimetrics Journal* 8 (1966): 63–82.

Long, George. "Accessio." In *A Dictionary of Greek and Roman Antiquities*, edited by William Smith. London: John Murray, 1875. Web: https://penelope.uchicago.edu/Thayer/E/Roman/Texts/secondary/SMIGRA*/Accessio.html (last accessed 18 April 2021).

Long, Pamela O. "Invention, Authorship, 'Intellectual Property,' and the Origin of Patents: Notes Toward a Conceptual History." *Technology and Culture* 32, no. 4 (1991): 846–884.

Long, Pamela O. *Openness, Secrecy, Authorship: Technical Arts and the Culture of Knowledge from Antiquity to the Renaissance*. Baltimore: Johns Hopkins University Press, 2004.

Long, Pamela O. *Artisan/Practitioners and the Rise of the New Sciences, 1400–1600*. Corvallis, OR: Oregon State UP, 2011.

Long, Pamela O. "Trading Zones: Arenas of Exchange during the Late-Medieval/Early Modern Transitions to the New Empirical Sciences." In *History of Technology*, edited by Ian Inkster, 31:5–25. London: Bloomsbury, 2012.

Long, Pamela O. "Multi-Tasking 'Pre-Professional' Architect/Engineers and Other Bricolagic Practitioners as Key Figures in the Elision of Boundaries Between Practice and Learning in Sixteenth-Century Europe: Some Roman Examples." In *The Structures of Practical Knowledge*, edited by Matteo Valleriani, 223–46. Cham: Springer International, 2017.

Lucassen, Jan, and Piet Lourens. "Ambachtsgilden in Nederland: Een eerste inventarisatie." *NEHA* 57 (1994): 34–62.

Lucassen, Jan, and Piet Lourens. "De oprichting en ontwikkeling van ambachtsgilden in Nederland (13de-19de eeuw)." In *Werelden van verschil. Ambachtsgilden in de Lage Landen*, edited by C. Lis and Hugo Soly: 43–77. Brussels: VUB Press, 1994.

Machielsen, Jan. "Bodin in the Netherlands." In *The Reception of Bodin*, edited by Howell A Lloyd, 157–92. Leiden: Brill, 2013.

Magreta, Todd Jerome. "The Development of Orange-Nassau Princely Artistic Activity, 1618–1632." PhD thesis, City University of New York, 2008.

MacLeod, Christine. *Inventing the Industrial Revolution: The English Patent System, 1660–1800*. Cambridge: Cambridge University Press, 2002.

Mandich, Guilio. "Le privative industriali veneziane (1450–1550)." *Rivista del diritto commerciale del diritto generale delle obbligazioni* 34 (1936): 511–47.

Mandich, Guilio. "Primi riconoscimenti veneziani di un diritto di privativa agli inventori." *Rivista di diritto industriale* 7 (1958): 101–55.

Mandich, Guilio. "Venetian Origins of Inventors' Rights." Translated by F. D. Prager. *Journal of the Patent and Trademark Office Society* 42 (1960): 378–382.

Mandich, Guilio. "Venetian Patents (1450–1550)." Translated by F. D. Prager. *Journal of the Patent and Trademark Office Society* 30 (1948): 166–224.

Mann, Michael. "The Autonomous Power of the State: Its Origins, Mechanisms, and Results." In *States in History*, edited by John A. Hall, 109–136. Oxford: Blackwell, 1986.

May, Christopher. "The Venetian Moment: New Technologies, Legal Innovation and the Institutional Origins of Intellectual Property." *Prometheus* 20, no. 2 (2002): 159–179.

Margócsy, Dániel. *Commercial Visions: Science, Trade, and Visual Culture in the Dutch Golden Age* (Chicago, IL: University of Chicago Press, 2014).

Marr, Alexander, Raphaele Garrod, Jose Ramon Marcaida, and Richard J. Oosterhoff, eds. *Logodaedalus: Word Histories of Ingenuity in Early Modern Europe*. Pittsburgh, Pennsylvania: University of Pittsburgh Press, 2019.

Marr, Alexander, and Vera Keller. "Introduction: The Nature of Invention." *Intellectual History Review* 24 (2014): 283–286.

McGee, David. "The Origins of Early Modern Machine Design." In *Picturing Machines, 1400–1700*, edited by Wolfgang Lefèvre, 53–84. Cambridge MA: MIT Press, 2004.

Meischke, R. "Het Amsterdamse fabrieksambt van 1595–1625." *Bulletin van de Koninklijke Nederlandse Oudheidkundige Bond* 93, no. 3 (1994): 100–122.

Meischke, R. "Het architectonisch ontwerp in de Nederlanden gedurende de late middeleeuwen de zestiende eeuw." *Bulletin van de Koninklijke Nederlandse Oudheidskundige Bond* 5, no. 1 (1952): colum161–230.

Meldau, Robert. "Hauptwurzeln des deutschen Patentrechts." *Technikgeschichte* 26 (1937): 107–114.

Merrill, Elizabeth. "The *Professione di Architetto* in Renaissance Italy." *Journal of the Society of Architectural Historians* 76, no. 1 (2017): 13–35.

Merula, Paullus. *Manier van procederen, in de provintien van Holland, Zeeland ende West-Vriesland, belangende civile zaaken*. 1592. Delft: Adriaan Beman, 1705.

Mertens, Bernd. *Im Kampf gegen die Monopole: Reichstagsverhandlungen und Monopolprozesse im frühen 16. Jahrhundert*. Mohr Siebeck, 1996.

Metius, Adriaan. *Astronomische ende geographische onderwijsinghe in de welcke door 't gebruyck der aertschen globi, ofte gebulte caerten, midtsgaders t'astrolabium catholicum*, ende *platte pas-caerten, de const der zeevaart verlicht ende de schippers ende stuerluyden duydelijck ende cortelijck in haere voyagie onderricht worde*. Amsterdam: Hendrick Lauwerentz, 1632.

Miedema, Hessel. "Over de waardering van architekt en beeldende kunstenaar in de zestiende eeuw." *Oud Holland* 94 (1980): 71–84.

Moerbeeck, Jan Andries. *Redenen, vvaeromme de VVest-Indische compagnie dient te trachten het landt van Brasilia den coninck van Spangien te ontmachtigen, en dat ten eersten. Wesende een ghedeelte der propositie ghedaen door Ian Andries Moerbeeck, aen Zijn Vorstelijcke Ghenade Mauritio prince van Orange, etc. ende eenighe andere heeren ghecommitteerden van de hooghe ende grootmoghende heeren de Staten generael der Vereenichde Nederlanden, in 'sGraven Haghe den 4. 5. ende 6. April, anno 1623*. Knuttel: 3541. Amsterdam, 1624.

Moes, Ernst Wilhelm, and Combertus Pieter Burger. *De Amsterdamsche boekdrukkers en uitgevers in de zestiende eeuw*. Issued in parts, 1896–1915. 2 vols. Utrecht: HES Publishers, 1988.

Mohnhaupt, Heinz. "Unendlichkeit des Privilegienbegriffs. Zur Einführung in das Tagungsthema." In *Das Privileg im europäischen Vergleich*, edited by Heinz Mohnhaupt and Barbara Dölemeyer, 1:1–13. Frankfurt am Main: Klostermann, 1997.

Mokyr, Joel. *The Gifts of Athena: Historical Origins of the Knowledge Economy*. Princeton University Press, 2002.

Mokyr, Joel. "The Intellectual Origins of Modern Economic Growth." *The Journal of Economic History* 65, no. 2 (2005): 285–351.

Mokyr, Joel. "Intellectual Property Rights, the Industrial Revolution, and the Beginnings of Modern Economic Growth," The American Economic Review 99, no. 2 (2009): 349–55.

Molà, Luca. "Il mercato delle innovazioni nell'Italia del Rinascimento." In *Le technicien dans la cité en Europe occidentale, 1250–1650*, edited by Mattieu Arnoux and Pierre Monnet, 215–250. Rome: École Française de Rome, 2004.

Molà, Luca. "States and Crafts: Relocating Technical Skills in Renaissance Italy." In *The Material Renaissance*, 133–153, 2008.

Molà, Luca. *The Silk Industry of Renaissance Venice*. Baltimore: Johns Hopkins University Press, 2000.

Molhuysen, Philip Christiaan, ed. *Bronnen tot de geschiedenis der Leidsche Universiteit. 1574-[1811]*. 7 vols. Rijks Geschiedkundige Publicatiën 20. The Hague: Nijhoff, 1913–24.

Monroe, Arthur Eli. *Early Economic Thought: Selections from Economic Literature Prior to Adam Smith*. 1924c. ed. Cambridge: Harvard University Press, 1945.

Montchrestien, Antoine de. *Traicté de l'œconomie politique*. Edited by Théophile Funck-Brentano. 1615. Paris: E. Plon, Nourrit, 1889.

Monté Verloren, Johan Philip de. *Hoofdlijnen uit de ontwikkeling der rechterlijke organisatie in de Noordelijke Nederlanden tot de Bataafse omwenteling*. Edited by Johannes Emil Spruit. Deventer: Kluwer, 2000.

Montias Database of 17th Century Dutch Art Inventories. Web: https://research.frick.org/montias (last accessed 14 April 2021).

Mout, Nicolette. "Ideales Muster oder erfundene Eigenart. Republikanische Theorien während des niederländischen Aufstands." In *Republiken und Republikanismus im Europa der Frühen Neuzeit*, edited by Elisabeth Müller-Luckner and H.G. Koenigsberger, 196–194. Munich: Oldenbourg, 1988.

Moser, Petra. "Innovation without Patents: Evidence from World's Fairs." *The Journal of Law & Economics* 55, no. 1 (2012): 43–74.

Muller, S. Fz. "John Osborn." *Oud Holland* 33 (1915): 199–207.

Munro, John H. "Industrial Energy from Water Energy." In *Economia e energia secc. XIII-XVIII. Atti della "Trentaquarttresima Settimana di Studi" 15–19 Aprile 2002*, edited by Simonetta Cavaciocchi. Firenze: Le Monnier, 2003.

Nehlsen- von Stryk, Karin. "Die Monopolgutachten des rechtsgelehrten Humanisten Conrad Peutinger aus dem frühen 16. Jahrhundert: ein Beitrag zum frühneuzeitlichen Wirtschaftsrecht." *Zeitschrift für neuere Rechtsgeschichte*, 1988, 1–18.

Nègre, Valérie. "Les figures de l'architecte « savant » (XVe-XVIIIe siècle)." In *L'Europe des sciences et des techniques : Un dialogue des savoirs, XVe-XVIIIe siècle*, edited by Liliane Hilaire-Pérez, Fabien Simon, and Marie Thébaud-Sorger, 133–39. Rennes: Presses universitaires de Rennes, 2018.

Nijenhuis, I. J. A., and others. *Resolutiën Staten-Generaal 1626–1630*. Web: http://resources.huygens.knaw.nl/besluitenstatengeneraal1576-1630/BesluitenStaten-generaal1626-1651 (last accessed 15 April 2021).

North, Douglass C. *Institutions, Institutional Change and Economic Performance*. Cambridge: Cambridge University Press, 1990.

Obreen, Frederik D. O., and Johannes H. Scheffer. *Rotterdamsche Historiebladen. Afd. Genealogische aanteekeningen en levensbeschrijvingen*. Vol. 3. Rotterdam: Nijgh & Van Ditmar, 1880.

Obreen, Frederik D. O., and Johannes H. Scheffer. *Archief voor Nederlandsche kunstgeschiedenis. Verzameling van meerendeels onuitgegeven berichten en mededeelingen betreffende nederlandsche schilders, plaatsnijders, beeldhouwers, bouwmeesters, juweliers, goud- en zilverdrijvers, smeden, stempelsnijders, tapijtwevers, borduurwerkers, plateflbakkers, ivoorsnijders, glasschilders, ingenieurs, landmeters, kaartmakers, verlichters, lettersnijders, schoonschrijvers, boekbinders, enz.* Rotterdam: Van Hengel & Eeltjes, 1877–90.

Octroy by de Hooge Mog: Heeren Staten Generael / verleent aende West-Indische Compagnie / in date den derden Junij 1621. Mette Ampliatie van dien. The Hague: Widow and Heirs H. Jacobsz van Wouw, 1642. Knuttel 4880.

Octroy Vande Ed. Ho, Mo. Heeren Staten Generael der Vereenigde Nederlanden/ Verleent aen Samuel van Pietsen, Heere vander Straten, Royenhuysen, Meulenacker, &c. Opten 20 Decembris/ Anno 1635. Knuttel 133a. Amsterdam: Joost Broersz, 1636.

Ogilvie, Sheilagh. *Institutions and European Trade: Merchant Guilds, 1000–1800* Cambridge; New York: Cambridge University Press, 2011.

Ong, Walter J. *Ramus, Method, and the Decay of Dialogue: From the Art of Discourse to the Art of Reason*. University of Chicago Press, 2004.

Oosterhoff, F. G. *Leicester and the Netherlands, 1586–1587*. Utrecht: HES, 1988.

Ormrod, David. *The Rise of Commercial Empires: England and the Netherlands in the Age of Mercantilism, 1650–1770*. Cambridge: Cambridge University Press, 2003.

Ostkamp, Sebastiaan, and Sjek Venhuis. "'Tot soulagemente van de schamele gemeente' Het Werra-aardewerk uit de werkplaats van Dierck Claesz Spiegel in Enkhuizen opnieuw bekeken (1602–1613)." In *Werra keramiek uit Enkhuizen opnieuw bekeken*, edited by Clevis van Gangelen H., H. Zwolle: SPA, 2009.

Palliser, D.M. *The Age of Elizabeth: England Under the Later Tudors, 1547–1603*. London: Longman, 1983.

Paludan, Charlotte. "Passchier Lammertijn and the Silk Factory in Copenhagen." In *Leinendamaste: Produktionszentren und Sammlungen*, edited by Regula Schorta, 239–49. Riggisberg: Abegg-Stiftung, 1999.

Park, Katharine, and Lorraine Daston. "Introduction: The Age of the New." In *The Cambridge History of Science*, edited by David C. Lindberg, Ronald L. Numbers, Katharine Park, and Lorraine J. Daston. Cambridge: Cambridge University Press, 2006.

Parker, Noel Geoffrey. *The Army of Flanders and the Spanish Road, 1567–1659: The Logistics of Spanish Victory and Defeat in the Low Countries' Wars*. Cambridge: Cambridge University Press, 2004.

Parker, Noel Geoffrey. *The Military Revolution: Military Innovation and the Rise of the West, 1500–1800*. Cambridge: Cambridge University Press, 1988.

Pastore, Alessandro. *Il medico in tribunale: La perizia medica nella procedura penale di antico regime (secoli XVI–XVIII)*. Bellinzona: Casagrande, 1998.

Pastorino, Cesare. "Weighing Experience: Francis Bacon, the Inventions of the Mechanical Arts, and the Emergence of Modern Experiment" (Ph.D., Indiana University, 2011);

Pastorino, Cesare. "The Philosopher and the Craftsman: Francis Bacon's Notion of Experiment and Its Debt to Early Stuart Inventors," *Isis* 108, no. 4 (2017): 749–68.

Petram, Lodewijk. *The World's First Stock Exchange*. New York: Columbia University Press, 2015.

Pérez, Pablo. "Los inventos llevados de España a las Indias en la segunda mitad del Siglo XVI." *Cuadernos de Investigación Histórica* 7 (1983): 35–54.

Pevsner, Nikolaus. "The Term 'Architect' in the Middle Ages." *Speculum* 17 (1942): 549–62.

Pfisterer, Ulrich. "Craft Guilds and Technological Change: The Engine Loom in the European Silk Ribbon Industry in the Seventeenth and Eighteenth Centuries." In *Guilds, Innovation, and the European Economy, 1400–1800*, edited by Maarten Prak and Stephan R. Epstein, 172–198. Cambridge: Cambridge University Press, 2008.

Phillips, Jeremy. "The English Patent as a Reward for Invention: The Importation of an Idea." *The Journal of Legal History* 3, no. 1 (1982): 71–79.

Pieters, Harm, and others. *Uitvinders in Nederland. Vier eeuwen octrooien*. Diemen: G+J Publishers, 2009.

Pigman III, G. W. "Versions of Imitation in the Renaissance." *Renaissance Quarterly* 33, no. 1 (1980): 1–32.

Plasseraud, Yves, and François Savignon. *L'État et l'invention: histoire des brevets*. Paris: Documentation française, 1986.

Pohlmann, H. "Neue Materialien zur Frühentwicklung des deutschen Erfinderschutzes im 16. Jahrhundert." *GRUR* 62 (1960): 272–281.

Pohlmann, H. "The Inventor's Right in Early German Law." *Journal of the Patent and Trademark Office Society* 43 (1961): 121–139.

Popplow, Marcus. "Why Draw Pictures of Machines? The Social Contexts of Early Modern Machine Drawings." In *Picturing Machines, 1400–1700*, edited by Wolfgang Lefèvre, 17–50. Cambridge MA: MIT Press, 2004.

Popplow, Marcus. *Neu, nützlich und erfindungsreich. Die Idealisierung von Technik in der frühen Neuzeit.* Münster: Waxmann Verlag, 1998.

Popplow, Marcus, and Torsten Meyer. "'To Employ Each of Nature's Products in the Most Favorable Way Possible' — Nature as a Commodity in Eighteenth-Century German Economic Discourse." *Historical Social Research / Historische Sozialforschung* 29, no. 4 (2004): 4–40.

Posthumus, Nicolaas Wilhelmus. *De geschiedenis van de Leidsche lakenindustrie*. 3 vols. The Hague: Nijhoff, 1908–1939.

Posthumus, Nicolaas Wilhelmus. *De neringen in de Republiek*. Amsterdam: Noord-Hollandsche Uitgevers-Maatschappij, 1937.

Posthumus, Nicolaas Wilhelmus. *Nederlandsche prijsgeschiedenis*. Leiden: Brill, 1943.

Posthumus, Nicolaas Wilhelmus. *Bronnen tot de geschiedenis van de Leidsche textielnijverheid.* 6 vols. The Hague: M. Nijhoff, 1910–1922.

Pottage, Alain, and Brad Sherman. *Figures of Invention: A History of Modern Patent Law.* Oxford University Press, USA, 2010.

Pottage, Alain, Jacob Copeman, and Alice Street. "Law after Anthropology: Object and Technique in Roman Law." *Theory, Culture & Society* 31, no. 2–3 (2014): 147–66.

Pottage, Alain. "Law Machines: Scale Models, Forensic Materiality and the Making of Modern Patent Law," Social Studies of Science 41, no. 5 (2011): 621–43.

Pottage, Alain. "Foucault's Law by Ben Golder and Peter Fitzpatrick," *The Modern Law Review* 74, no. 1 (2011): 159–68.

Poulsen, Bjorn. *Dutch Herring: An Environmental History, c. 1600–1860*. Amsterdam: Aksant, 2008.

Prager, F. D. "Brunelleschi's Patent." *Journal of the Patent Office Society* 28 (1946): 109–135.

Prager, F. D. "Examination of Inventions from the Middle Ages to 1836." *Journal of the Patent and Trademark Office Society* 46 (1964): 268–291.

Prager, F. D. "The Early Growth and Influence of Intellectual Property." *Journal of the Patent and Trademark Office Society* 34 (1952): 106–140.

Prak, Maarten R. *The Dutch Republic in the Seventeenth Century: The Golden Age*. Translated by Diane L. Webb. Cambridge: Cambridge University Press, 2005.

Prak, Maarten R. "Urbanization." In *The Cambridge Companion to the Dutch Golden Age*, edited by Helmer J. Helmers and Geert H. Janssen, 15–31. Cambridge: Cambridge University Press, 2018.

Prak, Maarten R. "Early Modern Capitalism: An Introduction." In *Early Modern Capitalism: Economic and Social Change in Europe 1400–1800*, edited by Maarten Prak, 1–22. London: Routledge, 2001.

Price, John Leslie. *Holland and the Dutch Republic in the Seventeenth Century: The Politics of Particularism*. Oxford: Clarendon Press, 1994.

Puype, J.P., and A.A. Wiekart, eds. *Van Maurits naar Munster: Tactiek en triomf van het Staatse leger*. Delft: Legermuseum, 1998.

Quaritsch, Helmut. *Staat und Souveränität*. Frankfurt am Main: Athenäum, 1970.

Rabier, Christelle. "Introduction: Expertise in Historical Perespectives." In *Fields of Expertise: A Comparative History of Expert Procedures in Paris and London, 1600 to Present*, edited by Christelle Rabier, 1–14. Cambridge: Cambridge University Press, 2007.

Randall, Adrian. *Before the Luddites: Custom, Community and Machinery in the English Woollen Industry, 1776–1809*. Cambridge: Cambridge University Press, 2002.

Rankin, Bill. "The 'Person Skilled in the Art' Is Really Quite Conventional: U.S. Patent Drawings and the Persona of the Inventor, 1870–2005." In *Making and Unmaking Intellectual Property: Creative Production in Legal and Cultural Perspective*, edited by Mario Biagioli, Martha Woodmansee, and Peter Jaszi, 55–78. Chicago, IL: Chicago University Press, 2011.

Ratcliff, Jessica. "Art to Cheat the Common-Weale: Inventors, Projectors, and Patentees in English Satire, ca. 1630–70." *Technology and Culture* 53, no. 2 (2012): 337–65.

Reinert, Erik S. "Emulating Success: Contemporary Views of the Dutch Economy before 1800." In *The Political Economy of the Dutch Republic*, edited by Oscar Gelderblom, 19–39. Farnham: Ashgate, 2009.

Repertorium van ambtsdragers en ambtenaren 1428–1861. Web: http://resources.huygens.knaw.nl/repertoriumambtsdragersambtenaren1428-1861 (last accessed 15 April 2021).

Ripa, Cesare. *Iconologia, of uytbeeldingen des verstands*. Translated by Dirk Pietersz Pers. Amsterdam: Dirck Pietersz Pers, 1644.

Robbertsz Le Canu, Robbert. *'t Verscheyden antwoordt uyt vele Steden in Hollant, op de vraghe van Numeratio, het eerste A. B. C. der Talkonst / Wt ghegheven door Robbert Robbertsz. Le Canu, Met een corte verantwoordinghe teghen de Leughenen, Lasteringhen, ende Blasphemien, van Thomas Leamer Engelsman*. Hoorn: Willem Andriesz., 1612.

Robberts, Lisa, ed. *Centres and Cycles of Accumulation in and Around the Netherlands during the Early Modern Period*. Zürich: Lit, 2011.

Roberts, Lissa, Simon Schaffer, and Peter Dear, eds. *The Mindful Hand: Inquiry and Invention from the Late Renaissance to Early Industrialisation*. Amsterdam: KNAW, 2007.

Roelevink, J. "Navigating New Waters. the Project Resolutions of the States General 1576–1670." In *Unlocking Government Archives in the Early Modern Period*, edited by Johanna Roelevink and A.J. Veenendaal, 67–83. The Hague: Institute of Netherlands History, 1995.

Rogers, Clifford J., ed. *The Military Revolution Debate: Readings On The Military Transformation Of Early Modern Europe*. Boulder: Westview, 1995.

Rogier, L. J. "Over karakter en omvang van de Nederlandse emigratie in de zestiende eeuw." *Historisch tijdschrift* 16 (1937): 325–367.

Rogier, L. J. "Over karakter en omvang van de Nederlandse emigratie in de zestiende eeuw." *Historisch tijdschrift* 17 (1938): 5–27.

Romano, Antonella. *La Contre-Réforme mathématique. Constitution et diffusion d'un culture mathématique jésuite à la Renaissance (1540–1640)*. Rome: Bibliothèque des Ecoles françaises d'Athènes et Rome, 1999.

Rose, Paul. *The Italian Renaissance of Mathematics: Studies on Humanists and Mathematicians from Petrarch to Galileo*. Genève: Droz, 1975.

Rossi, Paolo. *Philosophy, Technology, and the Arts in the Early Modern Era*. Edited by Benjamin N. Nelson. Translated by Salvatore Attanasio. New York: Harper & Row, 1970.

Rothbard, Murray N. *Economic Thought Before Adam Smith. An Austrian Perspective on the History of Economic Thought*. Vol. 1. 1995. Auburn, AL: Ludwig von Mises Institute, 2006.

Royo, Jose Antonio Mateos. "State Policy, Institutional Framework and Technical Monopoly *in* Early Modern Spain: Invention Patents in the Crown of Aragon during the Seventeenth Century." *History and Technology* 25, no. 2 (2009): 147–62.

Rothenberg, Panther E. "Maurice of Nassau, Gustavus Adolphus, Raimondo Montecuccoli, and the 'Military Revolution' of the Seventeenth Century." In *Makers of Modern Strategy from Machiavelli to the Nuclear Age*, edited by Peter Paret, Gordon A. Craig, and Felix Gilbert, 32–63. Oxford: Oxford University Press, 1986.

Russell, Andrew L., and Lee Vinsel. "After Innovation, Turn to Maintenance." *Technology and Culture* 59, no. 1 (2018): 1–25.

Sanders, George. *Het present van Staat: De gouden ketens, kettingen en medailles verleend door de Staten-Generaal, 1588–1795*. Hilversum: Verloren, 2013.

Scattola, Merio. "Das Privileg des Gesetzes - Francico Suàrez und die alte Lehre des Vorrechts (DL VIII)." In *"Auctoritas omnium legum": Francisco Suàrez' De legibus zwischen Theologie, Philosophie und Jurisprudenz : Francisco Suàrez' De legibus between theology, philosophy and jurisprudence*, edited by Oliver Bach, 333–68. Stuttgart-Bad Cannstatt: Frommann-Holzboog, 2013.

Schmidt, Benjamin. *Innocence Abroad: The Dutch Imagination and the New World, 1570–1670*. Cambrdige, UK: Cambridge University Press, 2001.

Schmitt, Carl. *On the Three Types of Juristic Thought*. Translated by Joseph W. Bendersky. Westport, Conn: Praeger, 2004.

Schulting, To, and Beverley Jackson. *Sant'Agata Morosina, an Argosy: An Episode in the Commercial, Diplomatic and Artistic Relations between Venice, Amsterdam and London, 1595–1609*. Florence: Istituto Universitario Olandese di Storia dell'Arte, 2005.

Schumpeter, Joseph Alois. *The Theory of Economic Development: An Inquiry Into Profits, Capital, Credit, Interest, and the Business Cycle*. New Brunswick: Transaction, 1983.

Schotte, Margaret E. *Sailing School: Navigating Science and Skill, 1550–1800*. John Hopkins University Press, 2019.

Sententie Van de Edele Eerentsteste Heeren Gedelegeerde Rchteren by de Hoge ende Mogende Heeren Staten Generael, gecommitteert over het Silvere ende Goude valsche Draetwerck, dat gemaeckt is by Joan de Carpentier ende Servaes Hellincx, onder pretext van seecker Octroy by de selve van hare Hog: Mog: geimpetreet op den 15. Julij 1619. gepronunchieert in 'sGraven-Hage den 15. Augusti 1624. S.l. : Knuttel 3543.

Shapin, Steven. *A Social History of Truth: Civility and Science in Seventeenth-century England*. Science and its Conceptual Foundations. Chicago: University of Chicago Press, 1994.

Shapin, Steven. *The Scientific Revolution*. Chicago, IL: University of Chicago Press, 1996.

Shapin, Steven, and Simon Schaffer. *Leviathan and the Air-pump: Hobbes, Boyle, and the Experimental Life*. Princeton: Princeton University Press, 1985.

Shapiro, Barbara J. *A Culture of Fact: England, 1550–1720*. Ithaca, NY: Cornell University Press, 1999.

Shapiro, Barbara J. *Probability and Certainty in Seventeenth-century England: A Study of the Relationships between Natural Science, Religion, History, Law, and Literature*. Princeton, NJ: Princeton University Press, 1983.

Sherman, Brad. "Intangible Machines: Patent Protection for Software in the United States." *History of Science* 57, no. 1 (2019): 18–37.

Silberstein, M. *Erfindungsschutz und Merkantilistische Gewerbeprivilegien*. Winterhur: Keller, 1961.

Skinner, Quentin. "A Genealogy of the Modern State." *British Academy*, 2009, 46.

Skinner, Quentin. *The Foundations of Modern Political Thought*. 2 vols. Cambridge: Cambridge University Press, 1979.

Slack, Paul. *The Invention of Improvement: Information and Material Progress in Seventeenth-Century England*. Oxford: Oxford UP, 2014.

Smith, Pamela H. *The Body of the Artisan: Art and Experience in the Scientific Revolution*. Chicago, IL: University of Chicago Press, 2004.

Soll, Jacob. *The Information Master: Jean-Baptiste Colbert's Secret State Intelligence System*. Cultures of Knowledge in the Early Modern World. Ann Arbor: University of Michigan, 2009.

Sprunger, Keith L. *Dutch Puritanism: A History of English and Scottish Churches of the Netherlands in the Sixteenth and Seventeenth Centuries*. Leiden: Brill, 1982.

Sprunger, Keith L. *Trumpets from the Tower: English Puritan Printing in the Netherlands, 1600–1640*. Leiden: Brill, 1993.

Stapel, F. W. "Joris de Bye." *Tijdschrift voor Nederlandse Taal- en Letterkunde* 35 (1916): 28–39.

Star Busmann, Eduard. *Octrooien van uitvinding*. Groningen: Wolters, 1867.

Statute of Monopolies, Westminster (1624). *Primary Sources on Copyright (1450–1900)*, eds L. Bently & M. Kretschmer. Web: http://www.copyrighthistory.org/cam/index.php (last accessed 19 April 2021).

Stevin, Simon. *Principal Works of Simon Stevin. Engineering, Music Civic Life*. Edited by R. J Forbes, A. D Fokker, and Annie Romein-Verschoor. Vol. 5. 5 vols. Amsterdam: Swets & Zeitlinger, 1966.

Stoffele, Bram. "Christiaan Huygens – a Family Affair. Fashioning a Family in Early-Modern Court-Culture." MA-thesis, Universiteit Utrecht, 2006.

Strathern, Marilyn. "Cutting the Network." *The Journal of the Royal Anthropological Institute* 2, no. 3 (1996): 517–35.

Stratmann, Wilhelm. "De militaire hervormingen van de Oranjes: Oorsprong, invoering en navolging." In *Onder den Oranje boom: Nederlandse kunst en cultuur aan Duitse vorstenhoven in de zeventiende en achttiende eeuw*, edited by Markus Schacht, Diete Oudesluijs, and Horst Lademacher, 77–106. München: Hirmer, 1999.

Strieder, Jakob. *Studien zur Geschichte kapitalistischer Organisationsformen; Monopole, Kartelle und Aktiengesellschaften im Mittelalter und zu Beginn der Neuzeit*. New York: B. Franklin, 1971.

Swan, Claudia. "Birds of Paradise for the Sultan. Early Seventeenth-Century Dutch-Turkish Encounters and the Uses of Wonder." *De zeventiende eeuw. Cultuur in de Nederlanden in interdisciplinair perspectief* 29, no. 1 (2013): 49–63.

Taverne, Eduard R. M. *In 't land van belofte: In de nieue stadt - Ideaal en werkelijkheid van de stadsuitleg in de Republiek, 1580–1680*. Maarssen: Schwartz, 1978.

Teschke, Benno. "Revisiting the 'War-Makes-States' Thesis: War, Taxation and Social Property Relations in Early Modern Europe." In *War, the State and International Law in Seventeenth Century Europe*, edited by O. Asbach and P. Schroeder, 35–62. Surrey, England: Ashgate, 2010.

"Theodoor Rodenburg." *dbnl.org*. Web: http://www.dbnl.org/auteurs/auteur.php?id=rode001 (last accessed 14 April 2021).

Thébaud-Sorger, Marie, and Koji Yamamoto. "Inventeurs et hommes à projets." In *L'Europe des sciences et des techniques : Un dialogue des savoirs, XVe-XVIIIe siècle*, edited by Liliane Hilaire-Pérez, Marie Thébaud-Sorger, and Fabien Simon, 125–32. Rennes: Presses universitaires de Rennes, 2018.

Thirsk, Joan. *Economic Policy and Projects: The Development of a Consumer Society in Early Modern England*. New york: Oxford University Press, 1978.

Tieskens, R. W. "Omdat teykeningen niet soo tot ons spreecken als een model." In *Het kleine bouwen: Vier eeuwen maquettes in Nederland*, edited by R.W. Tieskens, 1–24. Utrecht: Central Museum Utrecht, 1983.

Tilly, Charles. *Coercion, Capital, and European States, AD 990–1990*. Cambridge, Mass.: Blackwell, 1990.

Tilly, Charles. "Reflections on the History of European State-Making." In *Formation of National States in Western Europe*, edited by Charles Tilly, 3–83. Princeton, NJ: Princeton University Press, 1975.

Topper, David. "Towards an Epistemology of Scientific Illustration." In *Picturing Knowledge: Historical and Philosophical Problems Concerning the Use of Art in Science*, edited by Brain Scott Baigrie, 215–249. Toronto: University of Toronto Press, 1996.

Tracy, James D. *The Founding of the Dutch Republic: War, Finance, and Politics in Holland, 1572–1588*. Oxford: Oxford University Press, 2008.

Unger, Richard W. "Energy Sources for the Dutch Golden Age: Peat, Wind and Coal." *Research in Economic History* no. 9 (1984): 221–253.

United States Patent and Trademark Office, "MPEP." Web: http://www.uspto.gov/web/offices/pac/mpep/s2138.html (last accessed 12 March 2021).

United States Patent and Trademark Office, "Glossary." Web: https://www.uspto.gov/learning-and-resources/glossary#sec-P (last accessed 19 April 2021).

Usselincx, Willem. *Vertoogh, hoe nootwendich nut ende profijtelick het sy voor de Vereenighde Nederlanden te behouden de vryheyt van te handelen op West-Indien, inden vrede metten Coninck van Spaignen*. Knuttel 1442, 1608.

Van Aitzema, Lieuwe. *Historie of Verhael van Saken van staet en Oorlogh, in, ende ontrent de Vereenigde Nederlanden*. Vol. 7. The Hague: Iohan Veely, 1662.

Van Aller, Karel. *Generale regulen en definitien van beschreve rechten, met verscheyde explicatien en quæstien onder yder behoorende, naer ordre van't corpvs jvris*. Rotterdam: Bastiaen Wagens, 1655.

Van Bavel, B.J.P., and J.L. van Zanden. "The Jump-start of the Holland Economy During the Late-medieval Crisis, C.1350–c.1500." *The Economic History Review* 57, no. 3 (2004): 503–532.

Van Berkel, Klaas. "'Cornelis Meijer Inventor et Fecit'. On the Representation of Science in Late Seventeenth-Century Rome." In *Merchants and Marvels*, edited by Pamela Smith and Paula Findlen, 277–297. New York: Routledge, 2002.

Van Berkel, Klaas. "De geschriften van Rudolf Snellius. Een bijdrage tot de geschiedenis van het Ramisme in Nederland." *Tsch.Gesch.Gnk.Natuurw.Wisk.Techn.* 6, no. 4 (1983): 185–194.

Van Berkel, Klaas. "Enige opmerkingen over de aard van de technische innovatie in de Republiek rond 1600." *Tsch.Gesch.Gnk.Natuurw.Wisk.Techn.* 3 (1980): 123–144.

Van Berkel, Klaas. *Isaac Beeckman (1588–1637) en de mechanisering van het wereldbeeld*. Amsterdam: Rodopi, 1983.

Van Cruyningen, Piet. "Dealing with Drainage: State Regulation of Drainage Projects in the Dutch Republic, France, and England During the Sixteenth and Seventeenth Centuries." *The Economic History Review* 68, no. 2 (2015): 420–40.

Van den Heuvel, Charles. *"De Huysbou": A Reconstruction of an Unfinished Treatise on Architecture, Town Planning and Civil Engineering by Simon Stevin*. Amsterdam: KNAW-Edita, 2005.

Van den Heuvel, Charles."'Tot meerder bewijs'. Een kaart en een model van Daniel Note van 1620 ter demonstratie van een nieuwe uitvinding om de haven van Middelburg zandvrij te schuren," in *Bronnen van kennis. Wetenschap, kunst en cultuur in de collecties van de Leidse Universiteitsbibliotheek*, ed. P Hoftijzer, K van Ommen, and G Warnar (Leiden: Primavera, 2005), 100–108.

Van den Heuvel, Charles. "Mapping Knowledge Exchange in Early Modern Europe: Intellectual and Technological Geographies and Network Representations." *International Journal of Humanities and Arts Computing* 9 (2015): 95–114.

Van der Merwe, Cornelius. "Nova Species." *Roman Legal Tradition* 2 (2004): 96–114.

Van der Schelling, Pieter, and Eduard van Zurck. *Codex Batavus, waar in het algemeen Kerk-, Publyk, en Burgerlyk Recht van Holland, Zeeland, en het Ressort der generaliteit kortelyk is begrepen: Getrocken uit alle de Edicten, Ordonnantien, Placcaten, en Resolutien, voor, en na de Oprechting der Republyk, en wel byzonder der Staatsbestiering van Holland, en West-Vriesland, eerst tot het jaar 1710, en zedert tot 1724. Respective ingesloten, gemaakt, nader verklaard, uitgebreid, en behandeld, volgens de Roomse, en hedendaegse Rechten.* 4th ed. Rotterdam: Jan Daniel Beman, 1758.

Van Deursen, Arie Theodorus. "Tussen eenheid en zelfstandigheid. De toepassing van de Unie als fundamentele wet." In *De Unie van Utrecht: wording en werking van een verbond en een verbondsacte*, edited by S Groenveld and H. L. Ph Leeuwenberg, 136–54. The Hague: Nijhoff, 1979.

Van Deursen, Arie Theodorus. "De Raad van State en de Generaliteit, 1590–1606." *Bijdragen voor de Geschiedenis der Nederlanden* 19 (1964): 1–48.

Van Deursen, Arie Theodorus. *Maurits van Nassau, 1567–1625: De winnaar die faalde.* Amsterdam: Bakker, 2000.

Van Dillen, Johannes G. *Bronnen tot de geschiedenis van het bedrijfsleven en het gildewezen van Amsterdam.* 3 vols. The Hague: Nijhoff, 1929–1974.

Van Eeghen, Isabella H. "Van karthuizerklooster tot karthuizerhof," *Jaarboek Amstelodamum* 81 (1989): 29–62.

Van Gelder, H. E. *De Nederlandse munten.* 5th ed. Utrecht: Spectrum, 1972.

Van Gelderen, Martin. "Freedom Fighters. The Act of Abjuration, Hugo Grotius and Dutch Debates on Liberty." In *The Act of Abjuration. Inspired and Inspirational*, edited by Paul Brood and Raymond Kubben, 155–171. Oosterwijk: Wolf, 2011.

Van Gelderen, Martin. *The Dutch Revolt.* Cambridge Texts in the History of Political Thought. Cambridge: Cambridge University Press, 1993.

Van Gelderen, Martin. *The Political Thought of the Dutch Revolt 1555–1590*. Cambridge University Press, 2002.

Van Gelderen, Martin. "Het gemenebest. Staat en Republiek in de politieke theorie van het Duitse Rijk en de Noordelijke Nederlanden." In *Politieke Representatie*, edited by Henk de Smaele and Jo Tollebeek, 35–50. Leuven: Leuven Univeristy Press, 2002.

Van Helden, Albert. *The Invention of the Telescope*. Philadelphia, Penn.: The American Philosophical Society, 1977.

Van Helden, Albert, Rob van Gent, and Huib Zuidervaart, eds. *The Origins of the Telescope. Proceedings of the Conference 'The Invention of the Dutch Telescope. Its Origin and Impact on Science, Culture and Society, 1550-1650, Held in Middelburg in September 2008*. Amsterdam: Amsterdam University Press, 2010.

Van Leeuwen, Simon. *Het Rooms-Hollands-regt, waar in de Roomse Wetten met het huydendaagse neerlands regt, in alles dat tot de dagelijkse onderhouding kan dienen, met een bysondre kortheit, so wel in de vaste regts-stoffen, als in de manier van regts-vordering over een gebragt werden*. 1664. 11th ed. Amsterdam: Dirk Boom, 1744.

Van Nederveen Meerkerk, Elise, and Ariadne Schmidt. "Between Wage Labor and Vocation: Child Labor in Dutch Urban Industry, 1600–1800." *Journal of Social History* 41, no. 3 (2008): 717–36.

Van Netten, Djoeke. *Koopman in kennis: De uitgever Willem Jansz Blaeu in de geleerde wereld (1571–1631)*. Zutphen: Walburg Pers, 2014.

Van Nierop, Dirk Rembrandtsz. *Des aertrycks beweging, en de sonne stilstant, bewijsende dat dit geensins met de Christelijke religie is strijdende waer in dat alle redenen en argumenten, die tot noch toe hier tegen ingebracht zijn, wederleyt ende beantwoort worden. Met noch verscheyden Aenmerckingen, soo van de vindingh der lenghte van Oost en West, en anders*. Amsterdam: Gerrit van Goedesbergh, 1661.

Van Riemsdijk, Theodorus H. F. *De griffie van Hare Hoog Mogenden: Bijdrage tot de kennis van het archief van de Staten-Generaal der Vereenigde Nederlanden*. The Hague: Nijhoff, 1885.

Van Schelven, A. A. "Engelsch independentisme en Hollandsch Anabaptisme." *Nederlands Archief voor Kerkgeschiedenis* 17, no. 1 (1924): 108–126.

Van Schelven, A. A. *Omvang en invloed der Zuid-Nederlandsche immigratie van het laatste kwart der 16e eeuw*. The Hague: Martinus Nijhoff, 1919.

Van Winter, Pieter Jan. *Hoger beroepsonderwijs avant-la-lettre: Bemoeiingen met de vorming van landmeters en ingenieurs bij de nederlandse universiteiten van de 17e en 18e eeuw*. Amsterdam: Noord-Hollandsche Uitgevers Maatschappij, 1988.

Van Zanden, Jan L. *Arbeid tijdens het handelskapitalisme. Een nieuwe interpretatie van de opkomst en de achteruitgang van de economie van Holland, 1350–1850*. Bergen: Octavo, 1991.

Van Zutphen, Bernhard. *Practycke der Nederlandsche rechten vande daghelijcksche soo civile als criminele questien*. Leeuwarden: Gysbert Sijbes, 1655.

Veenendaal Sr, A. J., and S. P. Haak, eds. *Johan van Oldenbarnevelt. Bescheiden betreffende zijn staatkundig beleid en zijn familie 1570–1620*. 3 vols. Rijks Geschiedkundige Publicatiën 108. The Hague: Nijhoff, 1934–1967.

Veluwenkamp, J. W. "Buitenlandse octrooien en hun betekenis voor de ontwikkeling van het Nederlandse handelsstelsel in de zeventiende eeuw." In *Ondernemers en bestuurders. Economie en politiek in de noordelijke Nederlanden in de late middeleeuwen en de vroegmoderne tijd*, edited by Leo Noordegraaf and Clé Lesger, 547–555. Amsterdam: Aksant, 2000.

Veenendaal-Barth, J. W., A. A. Smit, and V. L Vree, eds. *Particuliere notulen van de vergaderingen der Staten van Holland, 1620–1640 door Nicolaes Stellingwerff en Schot*. 7 vols. The Hague: Nijhoff, 1987–2006.

Venetian Statute on Industrial Brevets, Venice (1474). Primary Sources on Copyright (1450–1900), eds L. Bently & M. Kretschmer. Web: http://www.copyrighthistory.org/cam/index.php (last accessed 19 April 2021).

Vermeir, Koen, and Dániel Margócsy. "States of Secrecy: An Introduction." *The British Journal for the History of Science* 45, no. Special Issue 2 (2012): 153–164.

Vermeir, Koen. "Openness versus Secrecy? Historical and Historiographical Remarks." *The British Journal for the History of Science* 45, no. 2 (2012): 165–88.

Vermij, Rienk H. "The Telescope at the Court of the Stadtholder Maurits." In *The Origins of the Telescope*, edited by Albert van Helden, Sven Dupré, Rob van Gent, and Huib Zuidervaart, 73–92. Amsterdam: KNAW, 2010.

Verqualje, Jan. *Uytvindingh der langhte van Oost en West, Waer van heel klaere onderrechtinge wordt gedaen, om die te konnen bezeylen, begrepen in twee Boeckjes, Handelende (insonderheydt het eerste) van eenighe andere secreten meer, voor desen noyt bekent geweest*. Amsterdam: Weduwe van Joost Broersz., 1661.

Vervolg op de Hollandsche Consultatien en Advysen, door voorname zoo Hollandsche als Zeeuwse en andere rechtsgeleerden. Vol. 2. Amsterdam: Hendrik Gartman, 1782.

Vögel, Klaus A. "Cosmography." In *The Cambridge History of Science*, edited by Lorraine Daston, Katharine Park, David C Lindberg, and Ronald L. Numbers, 3:469–496. Cambridge: Cambridge University Press, 2006.

Voorthuysen, W. D. *De republiek der Verenigde Nederlanden en het mercantilisme*. The Hague: Nijhoff, 1965.

Vranck, François, *Short Exposition (1587)*, reprinted and translated in Martin van Gelderen (ed.), *The Dutch Revolt* (Cambridge: Cambridge University press, 1993), 227–238.

Vukmir, Mladen. "The Roots of Anglo-American Intellectual Property Law in Roman Law." *IDEA: The Journal of Law and Technology* 32 (1991): 123.

Walterscheid, Edward C. *The Nature of the Intellectual Property Clause: A Study in Historical Perspective*. Buffalo, NY: W.S. Hein, 2002.

Warde, Paul. "Law, the 'commune', and the Distribution of Resources in Early Modern German State Formation." *Continuity and Change* 17, no. 2 (2002): 183–211.

Warde, Paul. "The Idea of Improvement, c.1520–1700." In *Custom, Improvement and the Landscape in Early Modern Britain*, edited by Richard W. Hoyle, 127–48. London: Routledge, 2011.

Waterbolk, Edzo H. "Met Willem Lodewijk aan tafel." In *Verspreide Opstellen*, 296–315. Amsterdam: Bakker, 1991.

Weld, Charles Richard. *A History of the Royal Society: With Memoirs of the Presidents*. Vol. 1. 2 vols. London: Parker, 1848.

West's Encyclopedia of American Law, edition 2. S.v. "without prejudice." Web: https://legal-dictionary.thefreedictionary.com/without+prejudice (last accessed 20 April 2021)

Westerdyck, Jan Barentsz, and Willem Wheler. *Bewys van de hoedanigheyt en wercking der geoctroyeerde waterschep-raden geinventeert by Willem Wheler: Vergeleken tegen de tegenwoordigh gebruyckelijcke heefraden*. Amsterdam: J. Blaeu, 1645.

Westra, Frans. "Nederlandse ingenieurs en de fortificatiewerken in het eerste tijdperk van de Tachtigjarige Oorlog, 1573–1604." PhD thesis, Groningen University, 1992; Alphen aan de Rijn: Canaletto, 1992.

Weststeijn, Arthur. *Commercial Republicanism in the Dutch Golden Age. The Political Thought of Johan & Pieter de La Court*. Leiden: Brill, 2012.

Whatley, Stephen. *A General Collection of Treatys*. Vol. 2. 4 vols. London: J. J. and P. Knapton and others, 1732.

Wieacker, Franz. *A History of Private Law in Europe with Particular Reference to Germany*. Oxford: Clarendon, 1995.

Williams, James D. *An Introduction to Classical Rhetoric: Essential Readings*. Malden, Mass.: Wiley-Blackwell, 2009.

Winter, Thomas Nelson. "The Mechanical Problems in the Corpus of Aristotle." *Faculty Publications, Classics and ReligiousStudies Department*, 2007, 25.

Witt, H.A.M. "De Republiek der Verenigde Nederlanden, 1588–1795. Formele positie en bevoegdheden van de Tresaurier-Generaal." In *Van tresorier tot thesaurier-Generaal. zes eeuwen financieel beleid in handen van een hoge Nederlandse ambtsdrager*, edited by Schmidt, 111–36. Hilversum: Verloren, 1996.

Woltjer, J.J. "Dutch Privileges, Real and Imaginary." In *Britain and the Netherlands. Vol. V, Some Political Mythologies : Papers Delivered to the Fifth Anglo-Dutch Historical Conference*, edited by J.S. Bromley and E.H. Kossmann. The Hague: Martinus Nijhoff, 1975.

Woordenboek der Nederlandsche Taal. Web: http://gtb.ivdnt.org (last accessed 28 March 2019).

World Intellectual Property Organization. *Introduction to Intellectual Property: Theory and Practice*. London: Kluwer, 1997.

Wrangel, Ewert, and Tine Beets-Damsté. *De betrekkingen tusschen Zweden en de Nederlanden op het gebied van letteren en wetenschap, voornamelijk gedurende de zeventiende eeuw*. Leiden: Brill, 1901.

Yale, Elizabeth. *Sociable Knowledge: Natural History and the Nation in Early Modern Britain*. Philadelphia: University of Pennsylvania Press, 2016.

Yamamoto, Koji. "Reformation and Distrust of the Projector in the Hartlib Circle." *The Historical Journal* 55, no. 2 (2012): 375–97.

Yamamoto, Koji. *Taming Capitalism Before Its Triumph: Public Service, Distrust, and "Projecting" in Early Modern England*. Oxford: Oxford University Press, 2018.

Yun-Casalilla, Bartolomé. "Social Networks and the Circulation of Technology and Knowledge in the Global Spanish Empire." In *Global History and New Polycentric Approaches: Europe, Asia and the Americas in a World Network System*, edited by Manuel Perez Garcia and Lucio De Sousa, 275–91. Singapore: Springer, 2018.

Zandvliet, Kees. "Het hof van een dienaar met vorstelijke allure." In *Maurits. Prins van Oranje*, edited by K Zandvliet, 37–63. Zwolle: Rijksmuseum, 2000.

Zandvliet, Kees. *Mapping for Money: Maps, Plans, and Topographic Paintings and their Role in Dutch Overseas Expansion during the 16th and 17th Centuries*. Amsterdam: Batavian Lion International, 1998.

Zanetti, Cristiano. *Janello Torriani and the Spanish Empire: A Vitruvian Artisan at the Dawn of the Scientific Revolution*. Leiden: Brill, 2017.

Zeischka, Siger. *Minerva in de polder: Waterstaat en techniek in het Hoogheemraadschap van Rijnland, 1500–1856*. Hilversum: Verloren, 2008.

Zijlstra, Samme. *Het geleerde Friesland - een mythe? Universiteit en maatschappij in Friesland en stad en lande ca. 1380–1650*. Leeuwarden: Fryske Akademy, 1996.

Index

www.ingramcontent.com/pod-product-compliance
Lightning Source LLC
Chambersburg PA
CBHW071325230926
51849CB00002B/316
* 9 7 8 9 0 0 4 3 2 0 3 9 0 *